TALES OF
IMPOSSIBILITY

DAVID S. RICHESON

TALES OF
IMPOSSIBILITY

The 2000-Year Quest
to Solve the Mathematical
Problems of Antiquity

Princeton University Press
Princeton and Oxford

Copyright © 2019 by Princeton University Press

Published by Princeton University Press
41 William Street, Princeton, New Jersey 08540
6 Oxford Street, Woodstock, Oxfordshire OX20 1TR

Requests for permission to reproduce material from this work
should be sent to permissions@press.princeton.edu

press.princeton.edu

All Rights Reserved

Library of Congress Control Number: 2019941488
First paperback printing, 2021
Paper ISBN 978-0-691-21872-4
Cloth ISBN 978-0-691-19296-3
British Library Cataloging-in-Publication Data is available

Editorial: Vickie Kearn, Susannah Shoemaker, and Lauren Bucca
Production Editorial: Karen Carter
Text and Jacket/Cover Design: C. Alvarez-Gaffin
Production: Jacquie Poirier
Publicity: Katie Lewis and Sara Henning-Stout
Copyeditor: Alison Durham
Jacket/Cover art courtesy of Shutterstock

This book has been composed in Palatino

Whenever things sound easy, ...it turns out there's one part you didn't hear.

—Donald Westlake[1]

Contents

Preface

I HAVE BEEN fascinated with the problems of antiquity—squaring the circle, trisecting the angle, doubling the cube, and constructing regular polygons—since I learned about them in my abstract algebra course in my sophomore year of college. One of the goals of the course was to prove the impossibility of these famous problems. I am not alone in my love for them. For over 2000 years, mathematicians and math enthusiasts—including some of history's greatest minds—have been infatuated with these easy-to-state but impossible-to-solve problems.

Although much had been written about these problems, I discovered that no one had written the book I thought they deserved: one that covers the history of each problem—from Greek origins to the eventual algebraic proofs of impossibility; the necessary advancements in geometry, algebra, and the nature of numbers; the people behind the proofs; the playful side of the problems; the generalizations and alternate approaches; and the wild stories of mathematical cranks who thought they'd solved the problems. So I decided to write that book.

The story spans several millennia, and at times it felt like the writing of the book took millennia as well. I did not realize at the start of the project that, because these problems were so beloved and so famous, there would be so much to research and write about. One of the most challenging parts of the writing process was deciding what *not* to include. I could easily have written a multi-volume book on this fascinating topic.

I wrote *Tales of Impossibility* for a general audience. Any reader who has a good grasp of high-school mathematics should be able to understand the material in the book. For most of the book, an understanding of high-school geometry and algebra and a familiarity with the basic types of numbers—integers, rational and irrational numbers, real numbers, and complex numbers—will suffice. Trigonometric and exponential functions make a few appearances. A reader who has not seen or doesn't remember calculus can easily skip the few instances where it arises. A reader can safely skip over or skim the technical arguments and be able to follow the story. Readers who are already familiar with the problems of antiquity might expect the book to end with a discussion of abstract algebra, field theory, and Galois theory; this is, after all, the context in which I first encountered the problems. But I opted to end the book earlier—after the problems were proven impossible. I do not discuss these more advanced and abstract ideas.

Although much of the mathematics is elementary, it *is* a mathematics book, and I do not shy away from discussing and using mathematics. For some people, this would be a reason not to read the book, but I trust that the intended audience will find the mathematics as interesting, deep, and elegant as I do. I aimed to share enough mathematical details to give understanding, but not so much that it becomes technical and dry. I also drew more than 150 figures to make the mathematics more visual.

The book contains extensive endnotes. In these I give references for quotations and certain factual details; I provide more information if I was too brief in the main text (showing the algebraic details behind a calculation, for instance); I mention topics that are too advanced for the intended audience; and I present tangential side notes that I found interesting, but which would be too large a digression in the main body. I read many books and articles while researching this book, and I included many of the essential resources in the bibliography.

I am grateful to Jim Wiseman, Chris Francese, Travis Ramsey, Dan Lawson, Tom Edgar, Rob Bradley, Robert Palais, Bill Dunham, Cotten Seiler, Claire Seiler, Heather Flaherty, Brett Pearson, Gail Richeson, Frank Richeson, Mark Richeson, Ángela Richeson, and the anonymous readers for their help, feedback, encouragement, and support—mathematical and otherwise—while writing this book. I appreciate the many people in the mathematics and history of mathematics communities whose offhand comments after hearing one of my presentations or after reading one of my articles helped me to see my own work from a different perspective. I thank my wonderful editor Vickie Kearn and the staff of Princeton University Press for making this book a reality. I am grateful to Dickinson College and the faculty and students in the Department of Mathematics and Computer Science for their support; I am fortunate to work in such a wonderful environment. And of course, I thank my family, Becky, Ben, and Nora, for their patience while I worked on the book. I can finally answer yes to their oft-asked question, "Are you done with your book yet?"

—David Richeson, August 2018

TALES OF
IMPOSSIBILITY

Introduction

Alice laughed. "There's no use trying," she said:
"one *can't* believe impossible things."
"I daresay you haven't had much practice," said
the Queen. "When I was your age, I always did it
for half-an-hour a day. Why, sometimes I believed as many as
six impossible things before breakfast."
—Lewis Carroll, *Through the Looking-Glass*[1]

"NOTHING IS IMPOSSIBLE." This platitude is used as inspiration by parents, athletic coaches, motivational speakers, and politicians. The hyperbolic news media is constantly alerting us to individuals who have achieved the impossible. It is one of the tenets of the American dream.

In his valedictory high-school commencement speech of June 24, 1904, Robert Goddard, who would later invent the liquid-fueled rocket, said,[2]

> Just as in the sciences we have learned that we are too ignorant safely to pronounce anything impossible, so for the individual, since we cannot know just what are his limitations, we can hardly say with certainty that anything is necessarily within or beyond his grasp.... It has often proved true that the dream of yesterday is the hope of today and the reality of tomorrow.

However, some things *are* impossible, and mathematics can prove that they are. Some tasks cannot be accomplished, regardless of one's intellect, one's perseverance, or the time available. This book tells the story of four impossible problems, the so-called "problems of antiquity": trisecting an angle, doubling the cube, constructing every regular polygon, and squaring the circle. They are arguably the most famous problems in the history of mathematics.

In a geometry course, students are introduced to the Euclidean tools: a compass to draw circles and a straightedge to draw lines

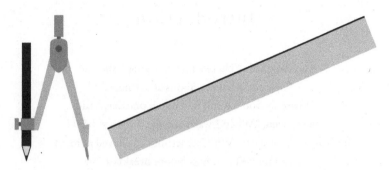

FIGURE I.1. A compass and straightedge.

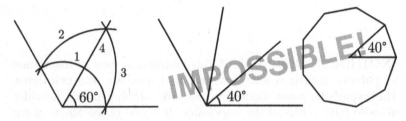

FIGURE I.2. It is possible to bisect a 120° angle using a compass and straightedge, but not to trisect it. Hence, it is impossible to construct a regular nonagon.

(see figure I.1). They learn a variety of basic constructions, such as how to bisect an angle, construct an equilateral triangle, and draw a perpendicular bisector. The problems of antiquity seem—at a glance—just as elementary as those exercises. But they are not. It takes three quick swipes of a compass and a trace along a straightedge to bisect an angle; figure I.2 shows a 120° angle split into two 60° angles. But it is impossible to use these same tools to draw the two rays that trisect a 120° angle; no matter how clever the geometer, it is impossible to construct a 40° angle. Thus, (1) it is impossible to trisect every angle. Moreover, the angle between the center of a regular 9-sided polygon, called a *nonagon*, and two adjacent vertices is 360°/9 = 40°, so it is impossible to construct the polygon. Hence, (2) it is impossible to construct every regular polygon.

Likewise, (3) given a line segment AB, it is impossible to construct a line segment CD so that a cube with side length CD is twice the volume of a cube with side length AB; that is, it is impossible to double the cube. And finally, (4) it is impossible to square the circle: if we begin with any circle, it is impossible to construct a square with the same area.

It is important to point out that these four problems are not *practical* problems. The world is not waiting for a method of constructing a 40° angle or a regular nonagon. The same geometry students could use the protractors in their backpacks to draw a 40° angle. There are other tools that will allow a draftsman or a mathematician to solve each of these problems exactly, and there are numerous techniques that clever craftsmen have devised to get approximations as accurate as desired.

In fact, not only are these problems not practical problems, they are not physical problems at all. They are theoretical problems. More important than the constructions themselves are the *proofs* that they accomplish what they say they accomplish. How do we know that the angle bisection technique truly bisects the angle? For this we need theoretical mathematics. The primary text on geometry in the Greek era and for many centuries afterward is Euclid's 300 BCE masterwork, *Elements*. Euclid began *Elements* with five postulates, and from these he built up all of geometry. The first three postulates are the compass-and-straightedge postulates. The first postulate states that we can draw a line segment between any two points, and the second says that we can extend this line segment beyond its endpoints. The third says that we can draw a circle with a given center and a point on the circle. Euclid wrote,[3]

Let the following be postulated:

1. To draw a straight line from any point to any point.
2. To produce a finite straight line continuously in a straight line.
3. To describe a circle with any center and any distance.

Thus his geometry was built from lines and circles, and to carry out the geometric techniques, one would use a compass and straightedge.

The problems of antiquity were known to be extremely challenging to the ancient Greeks. They were the subject of intense research by the leading mathematicians of the day. The historian of mathematics

Sir Thomas Heath called these problems "rallying points for [Greek] mathematicians during three centuries at least."[4]

The ancient Greeks knew that it was possible to solve the problems if they were allowed to change the rules. What if we have a compass, a straightedge, and a parabola? Or a hyperbola? Or a new mechanical drawing tool? And so on. For instance, Archimedes (ca. 287–212 BCE) proved that if the straightedge had two marks on it, he could trisect any angle. We will present many ingenious ways to solve these problems with an extended toolkit.

The problems were irresistible to mathematicians. For 2000 years, many of the major mathematical developments were directly or indirectly related to these problems. And the list of mathematicians who made contributions to the understanding of these problems is a who's who of the field. In 1913 Ernest Hobson wrote the following about the problem of squaring the circle, although the same could be said about all four problems:[5]

> When we look back, in the light of the completed history of the problem, we are able to appreciate the difficulties which in each age restricted the progress which could be made within limits which could not be surpassed by the means then available; we see how, when new weapons became available, a new race of thinkers turned to the further consideration of the problem with a new outlook.

Although these problems were actively studied for centuries, they were not proved impossible until the nineteenth century. There are several reasons it took over 2000 years for the proofs. First, mathematicians had to realize that the problems were impossible and not just very difficult. Second, they had to realize that it was possible to prove that a problem is impossible. This task is somewhat surprising—that we can use mathematics to prove that something is mathematically impossible. And lastly, the mathematicians had to invent the mathematical tools required to prove the impossibility. All four problems are geometric. The proofs of impossibility did not come from geometry, but from algebra and a deep understanding of the properties of numbers—not just the integers, but rational, irrational, algebraic, transcendental, and complex numbers. Algebra and a sufficient understanding of the real and complex numbers came long after the Greek period ended.

As algebra developed, mathematicians applied it to these problems. François Viète (1540–1603), René Descartes (1596–1650), and Carl Friedrich Gauss (1777–1855) made some headway toward understanding them. But the proofs of impossibility for three of the four—trisecting an angle, doubling the cube, and constructing regular polygons—were due to one man, a man who died young and, sadly, is not nearly as well known as these others: the French mathematician Pierre Wantzel (1814–1848). In his seven-page 1837 article he proved some preliminary results, applied them to the problems, and, in what may be the greatest single page in mathematics, proved all three problems impossible.

The fourth problem—squaring the circle—is somewhat unique among these problems. It is the most famous one, and it was the last to be proved impossible. While geometry and algebra are sufficient to prove that the other three are impossible, squaring the circle has one other wrinkle—it requires an understanding of the nature of π. If a circle has radius 1 cm, for instance, its area is $\pi \cdot 1^2 = \pi$ cm^2. Then to square the circle the geometer must construct a square with area π. Thus, much of our tale involves the history of this famous, enigmatic number. The problem of squaring the circle was eventually proved impossible by Ferdinand von Lindemann (1852–1939) in 1882; he used Wantzel's results and the fact—proved using calculus and complex analysis—that π is a transcendental number.

Because these four problems have been so famous for so long, a full treatment of their history would require multiple volumes. Moreover, even after the problems were proved impossible, their study continued, and they became subsumed into advanced fields such as abstract algebra and Galois theory. We chose to downplay this generalization—both because the mathematics required to understand it is too advanced for our intended audience and to keep the length of the book reasonable.

The book is organized as follows. The chapters follow the history of these fascinating problems—from the introduction of the problems by the Greeks through to the eventual solutions two millennia later. The chapters are roughly chronological. We give the colorful history of the problems, the history of other methods of solving them, and the histories of the fields that were created to finally resolve them.

There are also many interesting and entertaining side stories connected to these problems. So we inserted mini-chapters called Tangents in between each of the proper chapters. For instance, we discuss a bill passed by the Indiana legislature setting an incorrect value for π; we present a variety of unique methods of solving the problems, such as using origami, a drawing tool called a tomahawk, toothpicks, and a clock; we discuss Leonardo da Vinci's elegant contributions; and we present the two sides of the τ vs. π debate.

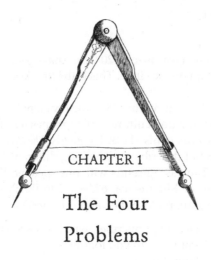

CHAPTER 1

The Four
Problems

Multi pertransibunt et augebitur scientia. (Many will
pass through and knowledge will be increased.)[1]

ALTHOUGH THESE FOUR problems date back to the ancient
Greeks, we know very little about their origins. All we can say with
certainty is that the problems of squaring the circle and doubling the
cube were being studied in the second half of the fifth century BCE.

The difficulty historians face is that, although some Greek sources
come to us fully intact, many are lost or involve partial information of
questionable quality. Much of what we know about mathematics during this time has been cobbled together from these fragments and from
secondary sources often written hundreds of years later. Historians of
mathematics have had to fill gaps and draw conclusions from this lim-
ited information. Indeed, trying to determine who proved what, and
when, is a puzzle that modern historians still chip away at a little at
a time.

Squaring the Circle

The philosopher Anaxagoras (ca. 500–ca. 428 BCE) was a friend and
teacher of Pericles (494–429 BCE). He was imprisoned for claiming that
the sun was a mass of red-hot stone, and according to the biographer

and essayist Plutarch (ca. 46–ca. 120 CE), Anaxagoras spent his time in prison trying to square the circle. This is the first known appearance of the problem.

By the end of the fifth century BCE the problem had become so well known that it entered pop culture. Aristophanes (ca. 446–ca. 386 BCE) made a play on words in his 411 BCE comedy *The Birds*, alluding to the famous problem but describing something different (dividing a circle into four parts). The passage ends with one speaker likening the other speaker to Thales of Miletus (ca. 624–ca. 546 BCE), an early ancient Greek mathematician, philosopher, and astronomer:[2]

METON. By positioning this ruler, which is curved, over its top, inserting a compass—do you follow?

PEISETAERUS. I don't follow.

METON. —and laying a straight ruler alongside it I'll take a measure, so that you will get a circle squared, with a marketplace in the middle, and so there will be straight streets running into it and meeting at the very center, so that just as from a star, itself being round, rays will beam out straight in every direction.

PEISETAERUS. The man's a Thales.

However, in a sense, the problem is much older than this. Ever since humans first drew a circle in the dirt using a rope and two sticks, we have wondered about the nature of this perfect geometric object. What is the area of a circle, and how long is its circumference? What are they *exactly*? Most cultures—the Babylonians, the Egyptians, the Indians, the Chinese, and so on—had a way of approximating the area and circumference of a circle, but they were unable to say exactly how they related to ordinary units.

The problem of squaring the circle is so famous that it has moved out of the realm of mathematics and into the general lexicon. Today the phrase "squaring the circle" is used synonymously for accomplishing the impossible. The statement of the problem, and of its partner problem of rectifying the circle, are simple.

Squaring the circle. Use a compass and straightedge to construct a square with the same area as a given circle.

Rectifying the circle. Use a compass and straightedge to construct a line segment the same length as the circumference of a given circle.

Every schoolchild learns basic techniques for compass-and-straightedge constructions. So this seemingly accessible problem sucked in countless amateur mathematicians. Even Abraham Lincoln (1809–1865) threw his hat into the ring. It is well known that Lincoln was largely self-taught. Less known is that in the 1850s he worked through Euclid's *Elements* while traveling as a circuit court lawyer. Lincoln's law partner Herndon described finding Lincoln in the office early one day:[3]

> He was sitting at the table and spread out before him lay a quantity of blank paper, large heavy sheets, a compass, a rule, numerous pencils, several bottles of ink of various colors, and a profusion of stationery and writing appliances generally. He had evidently been struggling with a calculation of some magnitude, for scattered about were sheet after sheet of paper covered with an unusual array of figures. He was so deeply absorbed in study he scarcely looked up when I entered. I confess I wondered what he was doing.... When he arose from his chair ... he enlightened me by announcing that he was trying to solve the difficult problem of squaring the circle.... For the better part of the succeeding two days he continued to sit there engrossed in that difficult if not undemonstrable proposition and labored, as I thought, almost to the point of exhaustion.

The problem of squaring the circle—also called the quadrature of the circle—was part of a larger class of geometric problems. To *square* a shape means to construct a square with the same area as the shape; this is the Euclidean equivalent of finding the area of the object.

To give a flavor of what's involved, we give an elementary quadrature problem. The triangle in figure 1.1 has base 2, height 1, and thus area 1. To square this triangle we must construct a 1×1 square. This task is straightforward using a straightedge and compass. Figure 1.2 gives the construction: Steps (1) through (5) produce a line perpendicular to AB through B and yield the point D. Steps (6) and (7) produce the point C. Then A, B, C, and D are vertices of the square.

The procedure we described works for only a small selection of triangles—those with a base-to-height ratio of 2-to-1. Later we will look at techniques for squaring more general shapes: rectangles, triangles, polygons with any number of sides, and some objects with curved boundaries. Although the Greeks and future

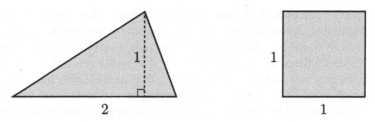

FIGURE 1.1. Squaring a triangle: A 1×1 square has the same area as a triangle with base 2 and height 1.

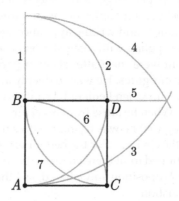

FIGURE 1.2. Steps for constructing a square with side AB.

mathematicians were able to square some very complicated objects, the task of squaring one of the simplest shapes in geometry—the circle—remained elusive.

The problems of squaring and rectifying the circle are intimately related to the constant π, which is typically defined as the ratio of the circumference of a circle to its diameter, $\pi = C/d$. But it also relates its area and radius, $A = \pi r^2$.

If we begin with a circle of radius r, then a square with the same area has side length $r\sqrt{\pi}$ (see figure 1.3). So to square the circle we must be able to construct a line segment of length $r\sqrt{\pi}$. Likewise, to rectify the circle we must be able to construct a line segment of length $2\pi r$. As we shall see, these two problems are equally difficult. If we are able to accomplish one, we are able to accomplish the other. In fact, if we begin with a circle of radius 1, then we can solve these problems if and only if we can construct a line segment of length π.

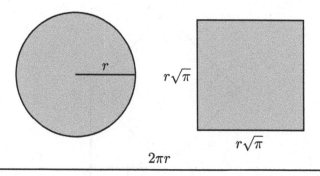

FIGURE 1.3. To square a circle with radius r we must construct a square with side length $r\sqrt{\pi}$, and to rectify the circle we must construct a line segment of length $2\pi r$.

Thus, to understand and resolve the problem of squaring the circle we must examine the history and nature of π.

Doubling the Cube

The name "doubling the cube" may be somewhat surprising. We are discussing planar geometry. Where does the three-dimensional cube come from? Despite the confusing name, the problem of doubling the cube is a problem in planar geometry. To motivate the problem, we describe a simpler version of the problem: doubling the square.[4]

Given a segment AB of length l, is it possible to construct a segment CD such that a square with side length CD has twice the *area* of the square with side length AB? It is possible. Construct a square with side AB and area l^2 (see figure 1.4). Our task is to construct a segment that is the side of a square of area $2l^2$. It is easy to find such a line segment CD: the diagonal of the square! If the diagonal has length d, then by the Pythagorean theorem $l^2 + l^2 = d^2$. So $d = \sqrt{2}l$. Thus, the diagonal would be the side of a square with area $(\sqrt{2}l)^2 = 2l^2$.

In the problem of doubling the cube we are not constructing a cube, but the *side* of a cube (see figure 1.5).

Doubling the cube. Given a segment AB, use a compass and straightedge to construct a segment CD so that a cube with side length CD has twice the volume of one with side length AB.

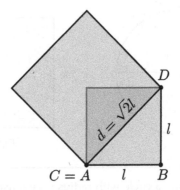

FIGURE 1.4. The square with side CD has twice the area of the square with side AB.

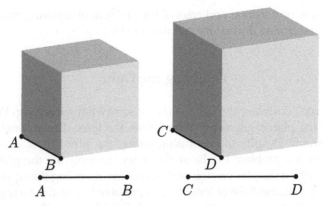

FIGURE 1.5. Given a segment AB, is it possible to construct CD so that $CD^3 = 2AB^3$?

The librarian of Alexandria, Eratosthenes of Cyrene (ca. 276–ca. 195/194 BCE) was a polymath known for his literary, philosophical, and scientific achievements. The "sieve of Eratosthenes"—a method of finding prime numbers—is taught to all schoolchildren. A letter he wrote to King Ptolemy III around 240 BCE contains two legends on the origin of the problem:[5]

It is said that one of the ancient tragic poets brought [the semimythical Cretan king] Minos on the scene, who had a tomb built for [his son] Glaucus. When he heard that the tomb was

100 feet long in every direction, he said: "You have made the royal residence too small, it should be twice as great. Quickly double each side of the tomb, without spoiling the shape."

Clearly, this mathematically challenged poet was mistaken: doubling each side of a cube produces one with eight times the original volume, not twice the volume. Heath wrote that the verses "are the work of some obscure poet, and the ignorance of mathematics shown by him is the only reason why they became notorious and so survived."[6]

In another story, Athens was being ravaged by a plague. Desperate for relief, the Athenians sent a delegation to consult the oracle of Apollo at Delos. The oracle demanded that the cubical altar of Apollo be doubled in size. Later, Theon of Smyrna (fl. 100 CE) wrote,[7]

Their craftsmen fell into great perplexity in their efforts to discover how a solid could be made double of a [similar] solid; they therefore went to ask Plato about it, and he replied that the oracle meant, not that the god wanted an altar of double the size, but that he wished, in setting them the task, to shame the Greeks for their neglect of mathematics and their contempt for geometry.

Because of this anecdote, the problem is often called the *Delian problem*.

One author likened Plato's (427–348/347 BCE) response to that of the Cold War "oracles" who used Russia's launch of the Sputnik satellite as a means of motivating the United States to ramp up their science education, which kick-started the space race.[8] Plato was, of course, a famous and influential philosopher. He was the founder of the Academy in Athens, the first institution of higher learning in the West, which boasts many illustrious alumni. He was not a mathematician, but he was a strong promoter of mathematics. He believed that every philosopher-ruler should be mathematically trained. Above the entrance to the Academy were the famous words "Let no one ignorant of geometry enter here." It is oft-repeated that Plato was not a maker of mathematics, but a maker of mathematicians.

A close look at the time line debunks this legend. Hippocrates Chios (ca. 470–ca. 400 BCE) worked on this problem half a century before Plato could have been involved. One argument is that this story was concocted from within the Academy to push their agenda.[9] It may also have been a dramatic invention of Eratosthenes.[10]

Nevertheless, Abraham Seidenberg argued that this legend (whether it is true, partly true, or completely fabricated) is important because it sheds light on the origin of geometry. It provides evidence that ritual was the ancient source of geometry. "Any suggestion, then, that the story of the Delian oracle is merely anecdotal would be not simply wrong: it would be just the opposite of the likely order of events. That is, the problem of the duplication of the cube starts as a problem in altar construction, with definite underlying theological motives, and ends, when the interest becomes purely mathematical, by dropping the 'anecdote.'"[11] We will encounter this notion again when we discuss the history of π: the Indians were interested in geometric constructions so that they could create altars—combining two square altars into a single altar of the same area, creating a circular altar with the same area as a square altar, and so on.

Just as the problem of squaring the circle could be reduced to the construction of a single line segment—a line segment of length π—so the problem of doubling the cube can be reduced to a single segment. We begin with a segment AB and we must construct a segment CD so that $CD^3 = 2AB^3$. If we arbitrarily assume that our initial segment has unit length, then CD is $\sqrt[3]{2}$ units long. Thus, we will be able to solve the problem of doubling the cube if and only if we can construct a segment of length $\sqrt[3]{2}$ from a segment of length 1.

Regular Polygons

There are many nonmathematical reasons we may wish to construct regular polygons, or equivalently, to divide a circle into an equal number of parts (see figure 1.6). Craftsmen and artists needed such techniques to make clocks, astrolabes, spoked wheels, compasses (the kind that tell the cardinal directions), decorative tiles, flowers, gears, and so on.

The Greeks knew how to construct quite a few regular polygons using only a compass and straightedge. The very first proposition in Euclid's *Elements* gives a technique for constructing an equilateral triangle. Earlier we showed a technique for constructing a square. Constructing a regular pentagon is significantly trickier than these, but is still accessible to a beginning geometer. A hexagon is easy to construct. A natural question arises: Is it possible to construct *every* regular polygon with a compass and straightedge?

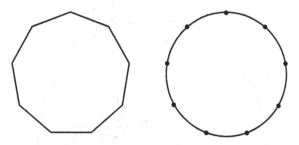

FIGURE 1.6. Constructing a regular n-gon is equivalent to dividing a circle into n equal parts.

Euclid's construction of an equilateral triangle and our construction of a square began with a side of the polygon, and the construction produced the rest of the polygon. We will state the problem, which is also known historically as *cyclotomy* or the *circle division problem*, in a slightly different way. We start with a circle, and we must inscribe the regular polygon in the circle. It turns out that if we can construct the regular polygon, then we can inscribe it in any given circle, so this problem is equivalent to constructing a regular polygon of any size.

Construction of regular polygons. For any integer $n \geq 3$, use a compass and straightedge to inscribe a regular n-gon in a given circle.

We can construct the regular 3-, 4-, 5-, and 6-sided polygons, but not the regular 7-sided heptagon. The octagon? Yes. The nonagon? No. 10, 12, 15, 16? Yes. 11, 13, 14? No. And so on.

Today we have well-defined, checkable criteria to determine whether a regular n-gon is constructible. However, we must evaluate each n separately, and as n gets larger, it gets increasingly more difficult to check. As of 2018, we know the status of all n-gons up to $n = 2^{2^{33}} + 1$ (a number with over 2.5 billion digits). The question of whether this particular n-gon is constructible hinges on whether n is prime; if it is, then the $(2^{2^{33}} + 1)$-gon is constructible, otherwise, it is not. Thus, interestingly, this is the only problem of antiquity that is still open. We do not have a comprehensive list of constructible regular polygons.

Just as the problems of squaring the circle and doubling the cube can be boiled down to the construction of a line segment of a particular length, so can the problem of constructing regular n-gons. If

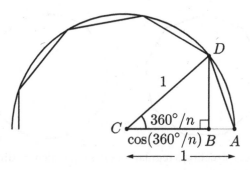

FIGURE 1.7. Constructing a regular n-gon is equivalent to constructing a segment of length $\cos(360°/n)$.

we are given the unit circle (a circle of radius 1), we can inscribe a regular n-gon if and only if we can construct a line segment of length $\cos(360°/n)$.

Let's see why. Let C be the center of the circle and A be a point on the circle (see figure 1.7). Suppose we can find a point B on AC such that BC is $\cos(360°/n)$ units long. Construct a line perpendicular to AC through B. It intersects the circle at a point D. Because CD is the radius of the circle, it is 1 unit long. By elementary trigonometry, $\angle ACD = 360°/n$. Thus the arc from A to D is $(1/n)$th of the circumference, and hence AD is one side of the regular n-gon. Once we have one side, we can construct all the sides. Moreover, these steps are reversible. So if we are able to construct a regular n-gon, then we are able to construct a segment of length $\cos(360°/n)$ simply by dropping the perpendicular from D to AC, as depicted in figure 1.7.

Angle Trisection

The angle trisection problem is easy to state, but also easy to misunderstand. Many amateur mathematicians, hobbyists, and cranks are fooled into thinking that it can be solved—and that they have solved it.

Trisection of an angle. Given an angle $\angle ABC$, use a compass and straightedge to construct a point D so that $\angle ABD = \frac{1}{3}\angle ABC$.

The point of confusion is that the objective is to be able to trisect *every* angle, but some angles *can* be trisected. We constructed a right

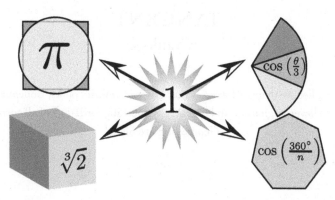

FIGURE 1.8. To solve the problems of antiquity we must start with a segment of length 1 and construct segments of these lengths using only a compass and straightedge.

angle on page 10, so a 270° angle can be trisected. We can construct an equilateral triangle, which has 60° angles, so we can trisect a 180° angle. Likewise, we can trisect 90°, 45°, and in fact infinitely many more angles. But in each of these cases, the procedure for constructing the smaller angle is different. There is no general procedure for angle trisection that applies to a generic angle, as there is for angle bisection.

Using an analysis similar to that for regular polygons, given a segment of length 1, we are able to trisect an angle θ if and only if we can construct a segment of length $\cos(\theta/3)$.

At first glance it was baffling that someone could prove the impossibility of these problems. But now that we have turned a mathematical eye toward them, we see the problems coming into focus. The constructibility of certain figures can be reduced to the constructibility of line segments of specific lengths. Later in the book we will introduce the set of *constructible numbers* and we will see that the four numbers in figure 1.8, π, $\sqrt[3]{2}$, $\cos(360°/n)$ (for certain n), and $\cos(\theta/3)$ (for certain θ) are not in the set.

TANGENT
Cranks

He replied, with great force, ... that I moved amidst cranks, Bohemians,
unbelievers, agitators, and—generally speaking—riff-raff of all sorts.
—George Bernard Shaw[1]

A SERIOUS RISK of writing this book is that I will be inundated
with solutions to the problems of squaring the circle and trisecting the
angle. The problems are catnip for mathematical cranks. Every mathematician who has email has received letters from crackpots claiming
to have solved these problems. They are so elementary to state that
nonmathematicians are unable to resist. Unfortunately, some think
they have succeeded—and refuse to listen to arguments that they are
wrong.

Mathematics is not unique in drawing out charlatans and kooks,
of course. Physicists have their perpetual motion inventors, historians their holocaust deniers, physicians their homeopathic medicine
proponents, public health officials their anti-vaccinators, and so on.
We have had hundreds of years of alchemists, flat earthers, seekers
of the elixir of life, proponents of ESP, and conspiracy theorists who
have doubted the moon landing and questioned the assassination of
John F. Kennedy.

Circle squarers and angle trisectors have been around for as long
as the problems themselves. The ancient Greeks used the word
"$\tau\varepsilon\tau\rho\alpha\gamma\omega\nu\iota\zeta\varepsilon\iota\nu$" ("tetragonidzein"), which translates "to occupy oneself with the quadrature," to describe those trying to solve the circle-
squaring problem.

Augustus De Morgan (1806–1871) wrote numerous columns for the
journal *Athenæum*, which were published posthumously by his wife
as *Budget of Paradoxes*. The subject of these columns? Paradoxers.
De Morgan explained, "I use the word [paradox] in the old sense: a
paradox was something which was apart from general opinion, either
in subject-matter, method, or conclusion."[2] For De Morgan, paradox is

not necessarily a pejorative term, and in fact, he labeled Galileo and Copernicus paradoxers. However, he was particularly interested in when the paradoxers were wrong—when they were cranks.

Given his mathematical background, De Morgan wrote extensively about mathematical cranks. He coined a term for this so-called illness that affects these misdirected enthusiasts: "morbus cyclometricus" ("the circle squaring disease"). He wrote,[3]

> The feeling which tempts persons to this problem is that which, in romance, made it impossible for a knight to pass a castle which belonged to a giant or an enchanter. . . . When once the virus gets into the brain, the victim goes round the flame like a moth; first one way and then the other, beginning where he ended, and ending where he begun.

The first use of the word "crank" in the scientific sense may be in a 1906 book review in *Nature*. The reviewer played off the notion of a crank that you would turn:[4]

> A *crank* is defined as a man who cannot be turned. These men [flat earthers, circle squarers, and trisectors] are all cranks; at all events, we have never succeeded in convincing one of them that he was wrong. The usually accepted axioms, definitions, and technical terms are not for them. Whether they use a term, some-times evidently in two different senses in the same syllogism, it is impossible to find exactly what they mean by it. (Brackets in original)

The term was well established by the middle of the twentieth century. Nobel Laureate John Nash, Jr. (1928–2015), the subject of the popular book and film *A Beautiful Mind*, used the term in a January 1955 letter to the National Security Agency (NSA). Nash was following up on a letter he'd written in which he proposed an encryption–decryption device. The first letter had gone unanswered. His reply, displayed in figure T.1, assured them, "I hope my handwriting, etc. do not give the impression I am just a crank or circle-squarer."[5]

In 1931 the Reverend Jeremiah J. Callahan (1878–1969) became the fifth president of Duquesne University in Pennsylvania, and he almost immediately caused a stir by claiming that he could trisect an angle using only a compass and straightedge. (He also published a controversial book called *Euclid or Einstein: A Proof of the Parallel Theory*

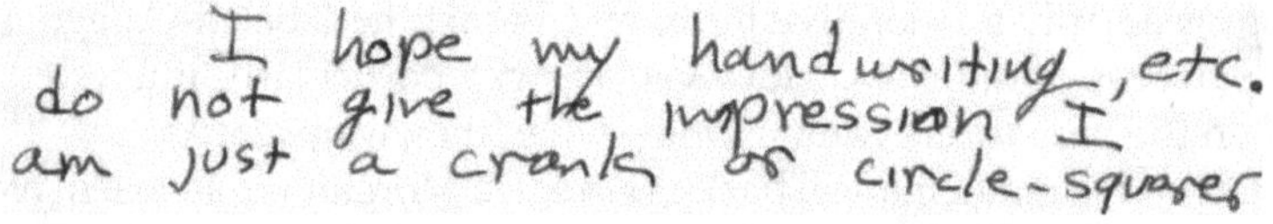

I hope my handwriting, etc. do not give the impression I am just a crank or circle-squarer

FIGURE T.1. Excerpt from a letter from John Nash to the NSA. (National Security Agency, https://www.nsa.gov/Portals/70/documents/news -features/declassified-documents/nash-letters/nash_letters1.pdf)

and a Critique of Metageometry in which he supposedly proved Euclid's parallel postulate—another impossible task that is a favorite of mathematical cranks—and attacked Einstein for his theories' reliance on non-Euclidean geometry.) However, Callahan refused to share his trisection technique, saying that he wanted to wait until he obtained a copyright. Presumably, he thought the proof so valuable that he dared not share it, lest someone else claim credit for it.

News of his trisection was widely reported throughout the country, including an announcement in *Time* magazine.[6] The *Pittsburgh Press* cited the mathematician Eric Temple Bell (1883–1960), who pointed out, correctly, that the problem was proved impossible in 1837. In response, Callahan was quoted as saying, "It is his privilege to think what he likes. This problem was just like many others that have been declared impossible of solution. A solution has been found."[7]

Eventually, Callahan shared his so-called proof. Instead of trisecting an arbitrary angle, he gave a convoluted method of tripling an angle. In other words, he began with an angle $\angle BDC$ and he constructed an angle $\angle BDE$ so that $\angle BDC = \frac{1}{3}\angle BDE$.[8]

Peruse the writings of mathematical cranks and we find many different, creatively incorrect methods of trisecting angles and squaring circles. The flaws in some proofs are immediately obvious to any mathematically trained reader—like Callahan's angle tripling. Other proofs are trickier to unravel—often because the writer presents a complicated mess of symbols, diagrams, and terminology. Also, sometimes the incorrect technique produces a good approximation. It is easy to be fooled by a convincing diagram.

Figure T.2 shows a common angle trisection argument. Draw a circle with its center at the vertex of the angle. Draw the chord determined by the angle. It is possible to trisect a line segment[9]—and hence, this chord—using a compass and straightedge. Then draw the segments from the vertex of the angle to the trisection points. If only angle

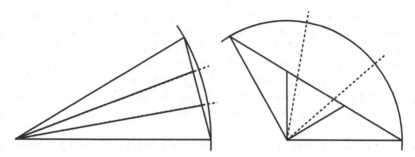

FIGURE T.2. Trisecting a chord does not trisect the angle.

trisection was that simple! As we see in the figure, if the angle is small—or even not so small—the procedure looks accurate (the dashed lines are the sought-after segments). But as the angle gets larger, it becomes apparent that trisecting the chord does not trisect the angle.

Many circle squarers are fixated on finding the "true" value of π. Among the candidates are 3 or 3.1 or 3.2 or 22/7 or $\sqrt{10}$ and so on. Through incorrect mathematics, deceptive figures, and approximations, they obtain rational or certain irrational values of π that can be constructed with a compass and straightedge.

Others mistake an example for a proof. It is possible to trisect *some* angles—45°, 90°, 180°, and so on. Thus, some cranks put forward these constructions as evidence that they can solve the general problem.

Unfortunately, many of these cranks do not have a good grasp of logical reasoning or of techniques for mathematical proof—they fail to understand syllogisms, they beg the question, they cannot give a proper reductio ad absurdum argument, and so on. Their solutions are often long and convoluted, using nonstandard terminology and notation, and riddled with mathematical errors.

In the eighteenth century, long before any of the problems were proved impossible, false proofs deluged the French Royal Academy of Sciences. They received approximately 150 articles on squaring the circle between 1741 and 1775.[10] The members of the academy believed—even without a rigorous proof—that the problems were impossible. As early as 1701 they wrote, "If the geometers dared to pronounce without absolute demonstrations, and were satisfied with the strongest probabilities, they would have come to the decision a long time ago with one voice that the quadrature of the circle is impossible."[11]

In 1740 Louis Castel (1688–1757) wrote, "It is not the famous geometers, the true geometers who seek the squaring of the circle: They know too much about it. It is the half-geometers who hardly know Euclid."[12] In fact, members of the academy were so tired of being inundated with quackery that in 1775 they passed a resolution not to accept solutions to the problems of circle squaring, angle trisection, or cube doubling.[13] (They also resolved not to accept proposals of perpetual motion machines.)

The mathematical physicist John Baez (1961–) proposed a "crackpot index" that was intended to provide "a simple method for rating potentially revolutionary contributions to physics."[14] The individual begins with a score of −5. Then Baez presented a list of 37 characteristics of a crackpot. Each time a criterion is met, a prescribed number of points is added to the index.

Mathematician Chris Caldwell was inspired by Baez's list and devised a mathematical version.[15] Some (lightly edited) examples from Caldwell's list are

- 1 point for each word in all capital letters;
- 5 points for every statement that is clearly vacuous, logically inconsistent, or widely known to be false;
- 10 points for each such statement that is adhered to despite careful correction;
- 10 points for not knowing (or not using) standard mathematical notation;
- 10 points for expressing fear that your ideas will be stolen;
- 10 points for each new term you invent or use without properly defining it;
- 10 points for stating that your ideas are of great financial, theoretical, or spiritual value;
- 10 points for beginning the description of your work by saying how long you have been working on it;
- 10 points for each favorable comparison of yourself to established experts;
- 10 points for citing an impressive sounding, but irrelevant, result;
- 20 points for naming something after yourself;
- 30 points for not knowing how or where to submit their major discovery for publication;

- 30 points for confusing examples or heuristics with mathematical proof;
- 40 points for claiming to have a "proof" of an important result but not knowing what established mathematicians have done on the problem.

Underwood Dudley (1937–) is the modern heir to De Morgan. He spent years collecting stories of mathematical cranks and wrote several humorous books and articles showcasing cranks he has encountered.[16]

As a joke, Dudley took values of π put forth by circle squarers from 1832 to 1879 (many from De Morgan's book) and fit a regression line to it. He concluded that π changed according to the function $0.0000056060t + 3.14281$, where t is years CE. He concluded that π was, well, π at 10:54 p.m. on November 10, 219 BCE.[17]

After his many years studying mathematical cranks, Dudley realized that they fit a pattern. In his book *The Trisectors*, he presented the following characteristics of the typical angle trisector (and presumably the circle squarers fit a similar mold):[18]

(1) They are male.
(2) They are old, often retired.
(3) They don't understand what it means for something to be mathematically impossible.
(4) Their mathematical background is minimal; it most likely ended with high-school geometry.
(5) They believe that the trisection of an angle is an important problem needing to be solved and that they will be richly rewarded with money or prestige for their work.
(6) Their proofs are always accompanied by dense, complicated figures.
(7) It is often impossible to convince them of their errors.
(8) They are prolific and persistent correspondents who will take up as much time as you give them.

Dudley concluded his description of these cranks by writing, "Now, will you know a trisector when you see one coming? And will you know what to do? Here is a hint: what you do involves your legs. No, you do not kick him."[19]

Alas, Dudley did not take his own advice. In the 1990s he was sued by one of the individuals that he featured in his book *Mathematical Cranks*[20]—one William Dilworth. The Federal District Court in Wisconsin threw the case out, but Dilworth appealed. The Seventh Circuit Court of Appeals found for Dudley. Dilworth then sued Dudley in a Wisconsin state court. Dilworth ended up losing and having to pay $7000 for the defense's legal expenses.[21]

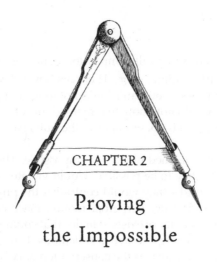

CHAPTER 2

Proving
the Impossible

No bid me run,
And I will strive with things impossible;
—William Shakespeare, *Julius Caesar*, Ligarius to Brutus[1]

THE MERRIAM-WEBSTER dictionary defines impossible as "incapable of being or of occurring,"[2] but we do not always use it that way.

We often use the term "impossible" as a substitute for "improbable." We use it to describe tasks that while not literally impossible, are highly improbable. If we hand a novice a scrambled Rubik's Cube, he will not solve it. The logical requirements are too complex to figure it out without help, and the probability of solving it by turning it at random is far too low. We'd say that it is impossible for him. Likewise, it is nearly impossible for an inexperienced bowler to roll a perfect 300 game[3] or for a monkey sitting at a keyboard to type the complete works of Shakespeare. These "needle in the haystack" scenarios aren't truly impossible, but the probabilities are so low, it is as if they are impossible. (We should note that some of these—like winning the lottery—are hopeless for any single person, but are likely to happen to *someone*.)

Other things are practically impossible, such as writing out, by hand, the first $10^{10^{10}}$ digits of π. There are several reasons we cannot carry out this task: a human's life is not long enough to write this many numbers, we do not know that many digits of π, and even if we did, there is not enough ink or paper in the universe to carry this out. It is impossible.

Other things are deemed physically impossible. These are ideas or actions that, were they to be possible, would violate our basic understanding of the world—they would contradict the physical laws that we have accepted as true. Perpetual motion machines are a prime example. The idea that there exists a device that could run indefinitely without an exterior energy source is ridiculous. It would violate one or more physical laws, such as the conservation of energy.

Of course, we have had a long history of misunderstanding the physical or biological world. Running a mile in under four minutes was once deemed impossible, but Roger Bannister achieved the impossible in 1954. Human flight was thought impossible, but the Wright brothers proved this wrong. The existence of a philosopher's stone or an alchemical process for turning lead into gold was labeled impossible when chemists discovered that lead and gold were different atomic elements. But now with the existence of particle accelerators, it is possible—although not practical—to carry out the alchemist's dream.[4]

Still other things are deemed impossible because they do not fit into our current scientific framework. In the late eighteenth century, scholars argued that it was impossible for stones to fall from the sky; they believed that there were no small celestial bodies other than the moon. They discredited as folklore eyewitness reports of meteorites ("thunderstones"). In 1768 a three-person team, including the young Antoine-Laurent Lavoisier (who would eventually become one of history's greatest chemists), investigated a meteorite using modern chemical methods. They concluded it was created when a bolt of lightning struck the pyrite-rich sandstone.[5] In 1807 two professors from Yale University published an article about a meteorite that landed in Weston, Connecticut. In disbelief, (the scientifically literate) President Thomas Jefferson exclaimed, "They may be right, but it is easier for me to believe that two Yankee professors would lie than to believe that stones would fall from heaven."[6] This anecdote may be apocryphal—or at least embellished—but it illustrates the beliefs of the time.[7]

Other things are deemed impossible because people aren't creative enough or far-sighted enough to imagine how they could happen. If we had described our modern computing technology to people in the nineteenth century, they would have said that such devices were impossible. Even if we had described the powerful computers that we have in our pockets and on our wrists to people in the 1950s—a time in which simple (by today's standards) computers filled an entire room—they would have shaken their heads in disbelief. Impossible.

Mathematically Impossible

What does it mean that something is mathematically impossible? And how do we *prove* that it is impossible?

Let's look at a simple example of an impossibility theorem—one that involves even numbers: 0, 8, −102, and so on. We all know what an even number is, but to work with them mathematically we need a clear, unambiguous definition: n is *even* if there exists an integer k such that $n = 2k$. So $0 = 2 \cdot 0$, $8 = 2 \cdot 4$, and $-102 = 2(-51)$ are even.

We can use this definition and properties of the integers to prove a theorem that should be a surprise to no one: it is impossible to find two even numbers whose sum is odd. Here's the proof: Suppose m and n are even numbers. Then there exist integers j and k such that $m = 2j$ and $n = 2k$. It follows that $m + n = 2j + 2k = 2(j + k)$. Because the sum of two integers is an integer, $j + k$ is an integer. Thus $m + n$ is even. An integer cannot be both even and odd, so we have proved our impossibility theorem.[8]

There are a few things we can learn from this example. Just as there are infinitely many even integers, there are infinitely many moves we can make with a compass and straightedge. We do not have to check all possible sums to prove this impossibility theorem. We can use general properties about integers and the even integers to prove it. In the same way, we will be able to use general properties of lines and circles to prove our impossibility theorems.

Also, if we begin with only even integers and apply our addition operation, the results will always be even. No matter how many numbers we add and in what order, we will never "get out" of the set of even numbers and obtain an odd one. Our sums will never yield 257 or 1301; it is impossible. As we shall see, this is analogous to the set of constructible numbers that we mentioned in chapter 1; when we apply

FIGURE 2.1. Sam Loyd's 15-puzzle. (S. Loyd, 1914, *Sam Loyd's Cyclopedia of Puzzles*, New York: Lamb Publishing)

certain arithmetic operations to the constructible numbers, we obtain other constructible numbers.

This example of adding even numbers may seem overly simplistic (although it isn't) and perhaps a little too contrived. Thus, we now show a more interesting example of an impossibility result. And again, the proof ends up with the sets of even and odd numbers.

Sam Loyd's Impossible Puzzle

In 1880 a mechanical puzzle swept across America, much as the Rubik's Cube did a century later. The mechanical toy, still available today, is the 15-puzzle. The objective is to slide the 15 numbered tiles left, right, up, and down in a 4 × 4 grid to place them in numerical order.

Around this time, the famous American puzzler Sam Loyd offered a $1000 prize (approximately $25,000 in today's dollars) for anyone able to solve *his* 15-puzzle. Loyd's puzzle looked like an ordinary one, but it began in a very specific configuration: the tiles were placed in numerical order, but 14 and 15 were transposed (see figure 2.1).[9]

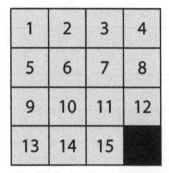 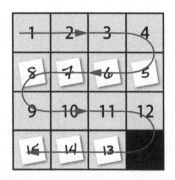

FIGURE 2.2. The 15-puzzle with the numbers relabeled.

FIGURE 2.3. A mixed-up 15-puzzle.

Loyd was not acting carelessly with his money. He knew that he would never pay out, for in 1879 two mathematicians had proved that such a starting configuration is impossible to solve.[10] Let's see why.

To solve the 15-puzzle, we must line the numbers up in order left-to-right, top-to-bottom. It turns out that for the proof it is easier to work with a winning solution in which the numbers follow a snake-like path: to the right on the first line, to the left on the second, to the right on the third, and to the left on the fourth. Instead of changing the rules of the game, imagine we tape new numbers on the old ones—cover the 5 with an 8, the 6 with a 7, and so on (see figure 2.2).

Now, suppose we are given a mixed-up 15-puzzle, as in figure 2.3. Take the numbers in the puzzle and list them in sequence, following the snake-like order and skipping over the blank. In our example, the list is 2, 13, 5, 1, 4, 12, 11, 10, 3, 14, 15, 6, 9, 7, and 8. For each number

TABLE 2.1. Inversions for our 15-puzzle

Sequence	2	13	5	1	4	12	11	10	3	14	15	6	9	7	8
Inversions	1	11	3	0	1	7	6	5	0	4	4	0	2	0	0

in the list, count how many numbers to its right are less than it. There is only one number less than 2 to its right, there are 11 numbers less than 13 to its right, and so on. Then add these numbers; our total is 44. These are the number of pairs of tiles that are in the wrong order—they are called *inversions*. In the winning configuration there are no inversions. Table 2.1 shows the inversions for our example.

Now slide a tile into the blank and see what happens to the inversions. If we were to slide a tile left or right into the blank—like the 14 or 15 in our example—then the sequence, and hence the number of inversions, does not change. The sequence is also unchanged when we slide a tile up or down where the snake changes rows. If we slide a tile up or down elsewhere, the number of inversions changes. But it doesn't affect all of the tiles; this move changes the order of three, five, or seven tiles, depending on the location of the blank and which tile moves into it. These are the only tiles whose inversions can change.

If we slide 12 down, it is inserted between 14 and 15. So it affects only 12, 11, 10, 3, and 14. Notice that 12 is larger than 11, 10, and 3, and smaller than 14. So, when it moves to the end, its inversions drop by 3, but 14's inversions go up by 1. The total number of inversions decreases by 2 to 42. Likewise, if we had moved 9 up, it would have changed 15, 6, 9 to 9, 15, 6, and because 9 is larger than 6 and smaller than 15 its inversions increase by 1, and 15's decreases by 1. Thus the total number of inversions is unchanged. See table 2.2.

In general, a vertical move affects $k + 1$ tiles, where k is 2, 4, or 6. The tile we slide is less than n of the remaining k tiles and greater than $k - n$. If we slide the tile down, the total number of inversions changes by $(k - n) - n = k - 2n$ and if we slide it up it changes by $n - (k - n) = 2n - k$. The exact change is unimportant; what *is* important is that these numbers are even. So, after every slide, the parity of the sum stays the same—an odd sum remains odd and an even sum remains even. If the starting configuration has an even sum, then the sum will remain even for the rest of the game. There is no way to move the tiles to turn the sum odd. Likewise, if the sum starts out odd, it will remain odd.

TABLE 2.2. Inversions after we slide tile 12 or tile 9

Original sequence	2	13	5	1	4	12	11	10	3	14	15	6	9	7	8
Inversions	1	11	3	0	1	7	6	5	0	4	4	0	2	0	0
Slide tile 12 down	2	13	5	1	4	11	10	3	14	12	15	6	9	7	8
Inversions	1	11	3	0	1	6	5	0	5	4	4	0	2	0	0
Slide tile 9 up	2	13	5	1	4	12	11	10	3	14	9	15	6	7	8
Inversions	1	11	3	0	1	7	6	5	0	4	3	3	0	0	0

TABLE 2.3. The number of inversions for Loyd's puzzle has a different parity from the target solution

Sam Loyd's puzzle	1	2	3	4	5	6	7	8	9	10	11	12	14	13	15
Inversions	0	0	0	0	0	0	0	0	0	0	0	0	1	0	0
Target	1	2	3	4	5	6	7	8	9	10	11	12	13	14	15
Inversions	0	0	0	0	0	0	0	0	0	0	0	0	0	0	0

Now let's return to Sam Loyd's challenge. He switched tiles 14 and 15. In our renumbered example, the switched tiles are 13 and 14. As we see in table 2.3, the total number of inversions for Loyd's puzzle is 1—an odd number. But the goal of the puzzle is to get the tiles in numerical order, and the total for this target solution is 0—an even number. Because the parity of the sum remains unchanged throughout the game, it is impossible to solve Loyd's puzzle and collect the $1000!

The Importance of Ground Rules

The rules are everything. If we do not enforce the rules, then the impossible can become possible. Just ask De Morgan and Dudley about the circle squarers and angle trisectors they encountered! In mathematics, the ground rules are the axioms and the definitions. They include the hypotheses in the specific problem or in the statement of the theorem. And they include the rules of logic that allow us to

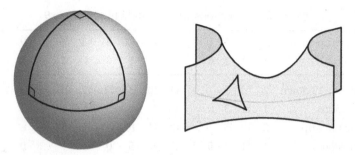

FIGURE 2.4. The interior angles of triangles may sum to more (left) or less (right) than 180°.

create solid mathematical proofs. If we ignore or change any of these, then we may be able to accomplish what was previously impossible.

If we allowed division in our even numbers example, then we can get an odd number from even numbers ($14 \div 2 = 7$ is odd). The impossible becomes possible. Likewise, Loyd's 15-puzzle is impossible to solve under the given rules. But if we are allowed to pop the pieces out and reassemble the puzzle, then it is solvable. Many children (and their parents!) have "solved" the Rubik's Cube using this method. There are mathematical examples of this type. Euclid proved that the sum of the interior angles of any triangle is 180°. Thus, it is impossible to construct a triangle whose interior angles sum to any other value. However, in the nineteenth century, mathematicians realized that if they changed the rules and modified Euclid's postulates, they could create new, fully consistent non-Euclidean geometries. These geometries have strange behaviors. For instance, the sum of the interior angles of a triangle may not equal 180°. In figure 2.4, we see a triangle on a sphere (the sides are arcs of great circles) in which all three angles are 90°, so the angle sum is 270°, which is greater than 180°. On the right we see a triangle sitting on a saddle-shaped surface. In this case, the interior angles sum to less than 180°. Thus, by changing the rules we can make the impossible possible.

Much of this book asks what happens if we change the rules—either by adding more tools to our toolkit, or taking some away, or doing something entirely different. Then we ask what we can construct. In particular, what does it take to solve the problems of antiquity?

Finally, a cautionary warning: we should not become overly confident. While we can say with certainty that some things are mathematically impossible, we should not fool ourselves into thinking that this reasoning carries over to other parts of life. Simon Newcomb (1835–1909), professor of mathematics at Johns Hopkins University, wrote the following on October 22, 1903, less than two months before the Wright brothers achieved flight in Kitty Hawk, North Carolina:[11]

> The mathematician of to-day admits that he can neither square the circle, duplicate the cube or trisect the angle. May not our mechanicians, in like manner, be ultimately forced to admit that aerial flight is one of that great class of problems with which man can never cope, and give up all attempts to grapple with it?

TANGENT
Nine Impossibility Theorems

> With willing hearts and skillful hands, the difficult we do at once;
> the impossible takes a bit longer.
> —Inscription on the memorial to the Seabees (US Naval Construction
> Battalions) outside Arlington National Cemetery

SOME OF THE greatest theorems in mathematics are impossibility theorems. Here we give nine of the most famous.

(1) **Irrationality of $\sqrt{2}$.** According to legend, Hippasus of Metapontum (fl. fifth century BCE), a follower of Pythagoras (ca. 570–ca. 495 BCE), shocked and angered his colleagues by proving that a side and a diagonal of a square do not have a common measure. Stated in today's terminology, he proved that $\sqrt{2}$ is an irrational number. That is, it is impossible to find integers m and n such that $\sqrt{2} = m/n$. We will learn more about this discovery in chapter 4.

(2) **Fermat's last theorem.** In 1637 Pierre de Fermat (1601 or 1607[1]–1665) famously wrote the following statement in the margin of one of his books: "It is impossible to separate a cube into two cubes, or a fourth power into two fourth powers, or in general, any power higher than the second, into two like powers. I have discovered a truly marvelous proof of this, which this margin is too narrow to contain." In other words, if $n > 2$ is an integer, then it is impossible to find positive integers a, b, and c such that $a^n + b^n = c^n$. The result, known as Fermat's last theorem, stubbornly resisted all proofs for over three and a half centuries. It was proved by Andrew Wiles in 1994 who had worked on the problem secretly for seven years.

(3) **The bridges of Königsberg.** In the middle of the eighteenth century, the Prussian city of Königsberg had seven bridges that crossed the River Pregel (see figure T.3). In their leisure, the residents looked

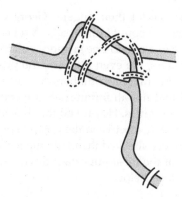

FIGURE T.3. A walking path that does not cross all seven bridges of Königsberg.

for a walking route that would take them across all the bridges once, and only once, preferably returning them to where they began. The puzzle made its way to Leonhard Euler (1707–1783) who proved in 1735 that it is impossible to find such a path. Euler's approach is now recognized as the beginning of the field of graph theory.

(4) **Insolvability of the quintic.** The quadratic formula is the high point of a high-school algebra class. It gives a simple expression to find the two x-values that satisfy $ax^2 + bx + c = 0$, namely

$$\frac{-b \pm \sqrt{b^2 - 4ac}}{2a}.$$

There are similar—although much more complicated—ways of specifying the roots of cubic (degree 3) and quartic (degree 4) equations. However, there is no such formula for polynomials of degree 5 (quintics) or higher. In particular, the polynomial $x^5 - x + 1$ has one real root that is approximately -1.67304, but it cannot be expressed using the integers, the four arithmetic operations, and the extraction of nth roots. Niels Abel (1802–1829) gave the first full proof of this impossibility theorem in 1824.

(5) **The nondenumerability of the continuum.** We have 10 fingers. We know that because we can match up the elements of the set $\{1, 2, 3, 4, 5, 6, 7, 8, 9, 10\}$ and our fingers in a one-to-one fashion. A child

does that when they count their fingers. Georg Cantor (1845–1918) took this idea further—to count infinite sets. A set is *countably infinite* if it can be matched up in a one-to-one fashion with the positive integers $\{1, 2, 3, \ldots\}$. The integers, the even integers, the prime numbers, and, most surprisingly, the rational numbers are all countably infinite. But Cantor proved that not all infinite sets are countable; there exist larger, uncountable, infinities. He proved that it is impossible to find a one-to-one correspondence between the positive integers and the real numbers. This discovery shocked the mathematical community and is now viewed as one of the most important discoveries of all time. It is at the heart of (6) and (9) below.

(6) **The halting problem.** Anyone who has written even the simplest computer program knows about programs that loop forever without stopping. It could be a simple program that prints the same text repeatedly ("Hello world! Hello world! Hello world!...") or it could be a subtle bug in the program that causes the program to enter an unending loop when the user enters some unexpected input. Wouldn't it be nice to have a computer program that could tell if a computer program will loop forever for a given input? Unfortunately, no such program exists. In 1936 Alan Turing (1912–1954) proved this impossibility theorem, now known as the halting problem.[2]

(7) **Arrow's impossibility theorem.** There are many famous examples of elections with "third-party spoilers." If candidates A and B were to run head-to-head A would win—perhaps in a landslide. But because candidate C, whose politics are similar to A's, enters the race, some would-be A-voters choose C instead and B wins the election. Thus B wins not because of the voters' preferences but because the plurality voting system doesn't handle three candidates well. There are other voting methods—approval voting, instant run-off, and so on—and each has its pros and cons. But no voting method is perfect. In 1950 the economist Kenneth Arrow (1921–2017) looked at *ranked voting systems* in which each voter has a personal ranking of the candidates and the voting system produces a societal ranking of the candidates. Arrow gave several commonsense criteria for what a fair voting method should satisfy. Then he proved that it is impossible to create a voting method that satisfies them all.

(8) **The parallel postulate.** From several definitions, five common notions, and five postulates, Euclid proved all the theorems in *Elements*. The fifth postulate, now known as the parallel postulate, is a mouthful and is somewhat difficult to grasp. John Playfair (1748–1819) gave the following more intuitive, equivalent version:

> There is exactly one line parallel to a given line that passes through a given point not on the line.

For centuries, mathematicians thought that this fifth postulate was redundant and could be proved from the other four. We now know that this is impossible. In the nineteenth century, mathematicians discovered non-Euclidean geometries—geometries that satisfy the first four postulates, but not the fifth. In these cases, Playfair's axiom—and hence Euclid's fifth postulate—fails. On a saddle, given a line and a point not on the line, there are infinitely many lines through the point that do not intersect the given line. And on the sphere all lines (great circles) intersect; so there are no lines through a point that are disjoint from a given line. Because of the existence of these non-Euclidean geometries, we know that it is impossible to prove the fifth postulate from the first four.

(9) **Gödel's incompleteness theorems.** This last impossibility theorem is subtle, deep, and shocking: There are theorems—true mathematical statements—that are impossible to prove. Mathematicians are familiar with conjectures that *seem* impossible to prove—with names like the twin primes conjecture, the Goldbach conjecture, and the Riemann hypothesis. Mathematical optimists say they will be proved eventually, but even if they are never proved, does that mean that proof is impossible? Maybe. There's a chance that proof is possible, but that mathematicians are not creative enough to discover it. But there is also the chance that it is true *and* unprovable. At the turn of the nineteenth century, mathematicians were trying to build a solid foundation for all of mathematics—to produce a set of definitions and axioms from which all mathematics could be derived. It was a noble, but ultimately hopeless pursuit. In 1931 Kurt Gödel (1906–1978) proved his incompleteness theorems, the first of which stated that in any sufficiently sophisticated axiomatic system there must be a true but unprovable statement. In a sense, this was the ultimate impossibility proof!

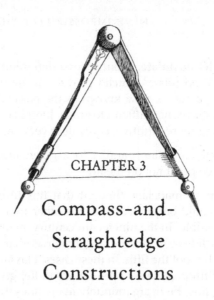

CHAPTER 3

Compass-and-Straightedge Constructions

No man can talk well unless he is able first of all to define to himself what he is talking about. Euclid, well studied, would free the world of half its calamities, by banishing half the nonsense which now deludes and curses it.
—Abraham Lincoln, 1860[1]

SQUARING THE CIRCLE, trisecting the angle, doubling the cube, and constructing regular polygons—the four most famous problems in the history of mathematics—ask the geometer to perform constructions using only a compass and straightedge. A compass. And a straightedge. And nothing else. Why?!? At first glance these requirements seem arbitrary and needlessly restrictive. Who set the rules? And what are they exactly?

Before answering these questions, we need to understand how the Greeks viewed and practiced geometry. To do so, we discuss one of the most influential books of all time—mathematical or otherwise.

Euclid was not the greatest Greek mathematician, but he is undoubtedly the most famous. His fame is not tied to any discoveries he made, but to the geometry text that he wrote, *Elements*. As we present the history of geometry in the next few hundred pages we will repeatedly refer to Euclid's propositions and his proofs—just as centuries of mathematicians have done. And thus, to understand the role of the compass and straightedge, we will begin with *Elements*.

Euclid's *Elements*

In 331 BCE, after conquering Egypt, the 25-year-old Macedonian king Alexander the Great created a new city on the Mediterranean Sea, west of the Nile Delta. He named this new capital city Alexandria. It was destined to become the bustling center of scholarship and Greek culture in Egypt. A century later it was the largest city in the world.

Alexander left Alexandria a few months after the founding and died seven years later, never returning to the city that bore his name. At the time of his death, there was no clear successor to his massive kingdom. His generals clashed for the next several decades. One of his most trusted generals, Ptolemy (ca. 323–ca. 283 BCE), ruled as satrap (governor) of Egypt for more than two decades after Alexander's death, and he became King Ptolemy I Soter ("the savior") in 305 BCE.

During his reign Ptolemy founded the Musaeum (the "Institution of the Muses") in Alexandria, which was a state-funded gathering place of more than a thousand great Hellenic thinkers. They conducted research, gave lectures, wrote, and collected knowledge, much like at Plato's Academy, today's Institute for Advanced Studies, or modern universities. It was not a museum in the modern sense—with paintings and sculptures—but an institution showcasing music, mathematics, poetry, astronomy, philosophy, and so on. The Musaeum was also home to the famous library of Alexandria, which contained many thousands of papyrus scrolls. The aim of the library was to house all the world's knowledge.

Thanks to the Musaeum, Alexandria was a center of mathematical activity for more than 700 years, from Euclid's arrival some time around 300 BCE until the tragic death of the philosopher, mathematician, and astronomer Hypatia in 415 CE (although the mathematical output was not always produced at the same consistently high level). Many mathematicians studied or taught in Alexandria—Euclid, Archimedes, Eratosthenes, Apollonius, Diophantus, Claudius Ptolemy, Pappus, Menelaus, Hypatia, and Proclus to name a few.

Unfortunately, we know very little about Euclid's life. Not only do we not know when he lived in Alexandria, we do not know when or where he was born or died. His mathematics helps us place him in time—we know who influenced him and whom he influenced. For instance, he came after Plato and the mathematicians in his circle, such as Theaetetus (ca. 417–369 BCE) and Eudoxus of Cnidus (ca. 400–ca. 347 BCE) (both of whom we shall encounter in chapter 4), after Aristotle (384–322 BCE), but before Archimedes. We believe his mathematical contributions occurred in Alexandria some time between 320 and 260 BCE—a surprisingly large window! Scholars have settled on the nice round date of 300 BCE as the middle of Euclid's working life.[2]

We believe Euclid was trained at Plato's Academy where he learned the mathematics of Theaetetus and Eudoxus; their work would form a large part of Euclid's *Elements*. And after gaining a reputation as a good mathematician and teacher, he was invited to relocate to Alexandria where he helped build the strong mathematics program at the Musaeum.

Elements is far and away Euclid's most famous book. Euclid is so closely tied to the work that we do not even have to say its name. We can simply say, "It is in Euclid." That gives us the source and seals it with a stamp of approval. However, he was the author of a number of other works—some of which survive to this day. His writings focused on topics such as geometry, the conic sections, ratios, number theory, astronomy, optics, mechanics, and logic.

When long works like *Elements* and the Bible were written on papyrus, they were broken into multiple scrolls called "books." *Elements* comprises 13 books, which we can think of as chapters. Books 1, 2, 3, 4, and 6 contain what we typically think of as Euclidean planar geometry—the Pythagorean theorem, angles, circles, similar polygons, and so on. Books 7 through 9 focus on number theory—divisibility, prime numbers, geometric sequences, the infinitude of the primes. Books 5 and 10 cover ratios and proportions, including topics such as incommensurable magnitudes (a term we will define in chapter 4) and the method of exhaustion, which were the precursors to irrational numbers and integral calculus. Books 11 to 13 contain solid geometry—cones, pyramids, spheres, and the Platonic solids. We will use the common shorthand I.47 to refer to proposition 47 in book I.

Many of the propositions were first proved by other mathematicians; we do not know which, if any, were original to Euclid. His main contribution was in bringing together all these known results and, beginning from a small number of definitions, postulates, and common notions, proving them in a logically sound way.

Elements was not intended to be an encyclopedic account of all current mathematical scholarship. It was just the elements—the basic definitions, axioms, and propositions on which all mathematics depends. Proclus Lycaeus (412–485 CE), one of the last classical philosophers, who is an important source of information about the history of Greek geometry, wrote,[3]

> [Euclid] included not everything which he could have said, but only such things as were suitable for the building up of the elements. He used all the various forms of deductive arguments, some getting their plausibility from first principles, some starting from demonstrations, but all irrefutable and accurate and in harmony with science.

Elements has a long and complicated history, including translations into various languages, additions, deletions, and commentaries.[4] Johannes Gutenberg invented the printing press around 1440, and in 1482 the first edition of *Elements* appeared in Venice. Since then it has been translated into many languages and reprinted thousands of times.

It is difficult to overstate the importance and popularity of *Elements*. It was a pillar of mathematical education for thousands of years. Although today's students do not study *Elements* itself, the theorems and proofs contained in it would be instantly familiar to any geometry student. It was the mathematical gold standard for Greek, Islamic, European, and American mathematics. The deductive structure has forever been the model of how to present mathematics.

The respect for *Elements* is strong and pervasive, and the deductive reasoning from definitions and axioms appealed to logicians, scientists, philosophers, and politicians. The logician Bertrand Russell (1872–1970) wrote, "At the age of eleven, I began Euclid, with my brother as my tutor. This was one of the great events of my life, as dazzling as first love. I had not imagined there was anything so delicious in the world."[5]

Thomas Jefferson had a lifelong love of mathematics. Several years after leaving office he wrote, "When I was young, mathematics was the passion of my life."[6] And in a letter to John Adams, he wrote, "I have given up newspapers in exchange for Tacitus and Thucydides, for Newton and Euclid; and I find myself much the happier."[7] His love of Euclid's style of deductive reasoning is visible in the Declaration of Independence, in which he began with the postulates of the new nation: "We hold these truths to be self evident."

Abraham Lincoln was also influenced by *Elements*. In a conversation several months before the 1860 Republican convention he said,[8]

> In the course of my law-reading I constantly came upon the word *demonstrate*. I thought, at first, that I understood its meaning, but soon became satisfied that I did not. . . . I consulted all the dictionaries and books of reference I could find, but with no better results. . . . At last I said, "Lincoln, you never can make a lawyer if you do not understand what *demonstrate* means;" and I left my situation in Springfield, went home to my father's house, and stayed there till I could give any proposition in the six books of Euclid at sight. I then found out what "demonstrate" means, and went back to my law studies.

Theorems vs. Problems

Before we discuss the compass and the straightedge, it is important that we understand the difference between the two types of Greek geometric propositions we will encounter: theorems and problems.

A theorem is typically a general statement that applies to a family of geometric objects. So, for instance, we may begin with a generic right triangle—with no assumption about the lengths of the sides or the measures of the non-right angles—or a central angle of a circle, or a pair of parallel lines and a transversal. Then we must prove that a certain property holds for the object or objects.

The Pythagorean theorem is a perfect example. It says that for *any* right triangle, the sum of the areas of the squares on the legs equals the area of the square on the hypotenuse (see figure 3.1). (Most readers are probably more familiar with the modern version, which is expressed in terms of the lengths of the legs, a and b, and the hypotenuse c: $a^2 + b^2 = c^2$.)

A problem, on the other hand, begins with some configuration of points, lines, and circles (and perhaps other shapes such as parabolas, hyperbolas, or ellipses), and requires the construction of another geometric object that satisfies some criteria. To solve a problem we must do two things. First, we must provide detailed step-by-step instructions for the construction. In particular, we are allowed to draw

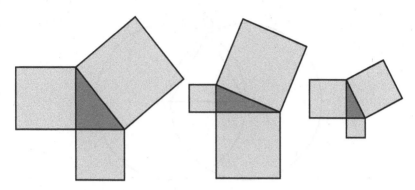

FIGURE 3.1. The Pythagorean theorem: the area of the square on the hypotenuse of a right triangle equals the sum of the squares on the legs.

the line or line segment joining two points (which is equivalent to using the straightedge) and to draw a circle with one point at its center and the other on its circumference (we imagine using a compass). Of course, as we build up our collection of solved problems we can use the results of previous constructions rather than starting from scratch each time. The second part requires proving that the construction achieves what it claims to have achieved. Thus, each problem has a theorem embedded in it implicitly: that the construction is valid.

The first proposition of Euclid's *Elements* is a prime example of a geometric problem. We are given a line segment, and we must construct an equilateral triangle with this segment as a side. Euclid began with a segment *AB* (see figure 3.2). He constructed a circle with center *A* and radius *AB* and another circle with center *B* and radius *AB*. The two circles intersect at a point *C*. He then constructed segments *AC* and *BC*. At this point the construction is finished, and Euclid turned his attention to justifying that *ABC* is equilateral. Here *AB* and *AC* are radii of the same circle, so *AB* = *AC*. Similarly, *AB* and *BC* are radii of the other circle, so *AB* = *BC*. Hence *AB* = *AC* = *BC*.

Problems were valuable to geometers because they functioned as proofs of existence. If an equilateral triangle or a perpendicular bisector is needed, a geometer can construct one because Euclid proved that it is possible to do with the tools available—lines and circles, or equivalently, a compass and straightedge.

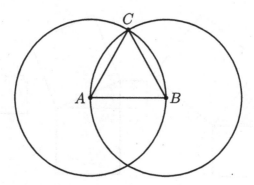

FIGURE 3.2. The first proposition of Euclid's *Elements* is the problem of constructing an equilateral triangle.

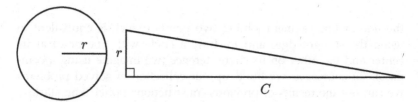

FIGURE 3.3. The area of a circle of radius r and circumference C is equal to the area of a right triangle with legs r and C.

To see the difference between problems and theorems, consider the following proposition of Archimedes. He proved that the area of a circle is equal to the area of a right triangle in which one leg is the radius of the circle and the other leg is the length of the circumference (see figure 3.3). This relationship is easy to verify with the area and circumference formulas that we all know.[9] The area of the circle of radius r is πr^2, and the area of the triangle is

$$\frac{1}{2} \cdot \text{base} \cdot \text{height} = \frac{1}{2} \cdot C \cdot r = \frac{1}{2} \cdot 2\pi r \cdot r = \pi r^2,$$

where $C = 2\pi r$ is the circumference of the circle. Figure 3.4 gives a geometric feeling for why this is true. Divide the circle into a large number of equal pie-piece shapes and reassemble them with their tips up. Then

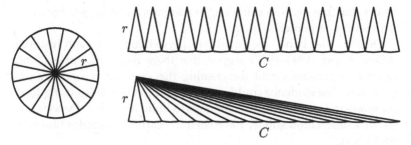

FIGURE 3.4. The area of the circle is $\frac{1}{2}rC$.

bring all their tips together as shown. The area of each piece does not change when we do this, but as we do this for increasingly more slices, it comes to resemble a right triangle with height r and base C. We will see Archimedes's proof in chapter 8.

Archimedes's proposition is a theorem, not a problem. Archimedes did not give a technique for constructing the triangle. Indeed, he can't: to do so requires constructing a line segment of length equal to the circumference, which is the problem of rectifying the circle.

The classical problems in this book are problems in the sense described here. To solve them, we need to be able to use lines and circles to construct the desired object—the square, the angle trisector, the regular heptagon, and so on—and if such a construction were possible, we would need to prove that it solves the problem.

When Did the Tradition Begin?

We do not know for certain who decided that compass-and-straightedge constructions were the gold standard for geometry. Heath put forth that it may have been Oenopides of Chios, who lived in the fifth century BCE. We do not know much about Oenopides. He was primarily an astronomer who may have visited Athens. According to Proclus (who cites Eudemus of Rhodes [fourth century BCE]), we can thank Oenopides for two compass-and-straightedge constructions: drawing a line perpendicular to a given line through a given point and copying a given angle. These constructions are straightforward, so Oenopides may have his name attached to them because he was the first to do so using only a compass and straightedge rather than with a carpenter's square.

On the other hand, the equally respected Stanford math historian Wilbur Knorr (1945–1997) argued that there is insufficient evidence to give Oenopides credit for creating the axiomatic foundations of geometry. The evidence could simply point to Oenopides using these compass-and-straightedge techniques to build a sundial. Oenopides's constructions could simply have been the earliest examples Eudemus could find.[10]

Regardless of how the tradition began, it was strongly reinforced by Euclid's *Elements*. When Euclid settled on his five postulates as the basis for his geometry and was able to build an impressive structure of mathematics, the compass and straightedge were firmly set as the geometric tools of choice. Greek geometers studied other curves, but when they did, they knew they were leaving the safe confines of Euclid's theory.

The Practical and the Theoretical

We encourage the reader to attempt the basic compass-and-straightedge constructions, especially if it has been years or decades since using these drawing tools. On the one hand, the constructions are satisfying and magical. It is a rewarding experience every time the familiar moves bisect an angle or construct a perpendicular bisector. The tactile experience of performing these constructions makes the mathematics that much more real. On the other hand, we quickly realize that the devices are not perfect. Maybe our straightedge is slightly out of position and our pencil barely misses a point that we need the line to pass through. Or perhaps the hinge on the dime-store compass isn't as tight as it should be, and the circle doesn't close up after turning 360°.

The circles and lines that we study in geometry are idealized versions of the circles and line segments that we can draw with a straightedge and compass. Points are zero-dimensional atoms and lines and circles are one-dimensional paths that have no thickness, unlike the points, lines, and circles we draw with a pencil.[11] Lines have infinite length and circles can be as large as we need them to be. No drawing tools have the ability to draw arbitrarily large—or infinite—figures. As Simplicius of Cilicia (ca. 490–ca. 560) observed 1500 years ago, "He would be a rash person who, taking things as they actually are, should postulate the drawing of a straight line from Aries to Libra."[12]

Some of the constructions that can be described algorithmically would have so many steps that they could not be carried out in any reasonable amount of time or space, and due to the imprecision of the drawing equipment and the fallibility of the human geometers, they could not be carried out with the required accuracy.

In 1833 the Swiss geometer Jakob Steiner (1796–1863) wrote,[13]

> Thus it happens that often in this way constructions are given, which, if it were necessary to carry out *actually* and *exactly* all that they include, would soon be given up, since thereby one would speedily be convinced that it is a very different matter actually to carry out the constructions, i.e., with the instruments in the hand, than it is to carry them through, if I may use the expression, simply by means of the tongue.

For instance, as we shall see, the regular 65,537-sided polygon is one that can be constructed using the Euclidean tools. At the end of the nineteenth century, Hermes of Lingen spent a decade figuring out how to do so. Clearly, we could not carry out his algorithm in practice. It would take far too long. Moreover, even if we sharpened the pencil before each use, its trail would still have a thickness, which introduces some inexactness to the construction. Add to this the errors that would creep in from using real tools, and before long we would lose all precision. Even if we could have thin enough lines and carefully calibrated tools, the paper would be an impressionistic, unintelligible mess of arcs and segments before we got very far into the process. The final polygon would be indistinguishable from a circle; drawn on a piece of standard notebook paper, the length of each of the 65,537 sides of the polygon would be virtually invisible![14]

In his 1913 book on squaring the circle, Hobson observed that geometry has two sides, the practical or physical—which includes spatial relations between objects and the drawing of these objects with a compass and straightedge—and the abstract or rational—which deals with ideal objects such as dimensionless points, lines, and circles:[15]

> The ordinary obliteration of the distinction between abstract and physical Geometry is furthered by the fact that we all of us, habitually and almost necessarily, consider both aspects of the

subject at the same time. We may be thinking out of a chain of reasoning in abstract Geometry, but if we draw a figure, as we usually must do in order to fix our ideas and prevent our attention from wandering owing to the difficulty of keeping a long chain of syllogisms in our minds, it is excusable if we are apt to forget that we are not in reality reasoning about the objects in the figure, but about objects which are their idealizations, and of which the objects in the figure are only an imperfect representation.

These observations are by no means new. They are reminiscent of Plato's allegory of the cave, in which prisoners are chained with their backs to a fire. All they can see are shadows cast on the wall, and they mistake these distorted images for reality. Plato argued that mathematics is true and real; he wrote that "the knowledge at which geometry aims is knowledge of the eternal, and not of aught perishing and transient."[16] Geometry is conducted through the study of ideal forms, and these ideal forms—the ideal circle, the ideal line, the ideal triangle, and so on—exist only in our minds, not as drawings on paper. The drawn figures are—like the shadows on the cave wall—imperfect representations of reality.

In Plato's *Republic*, Socrates said of mathematicians,[17]

And do you not know also that although they make use of the visible forms and reason about them, they are thinking not of these, but of the ideals which they resemble; not of the figures which they draw, but of the absolute square and the absolute diameter, and so on—the forms which they draw or make, and which have shadows and reflections in water of their own, are converted by them into images, but they are really seeking to behold the things themselves, which can only be seen with the eye of the mind?

Much later, Henri Poincaré (1854–1912), who was known for his poor drawing skills, wrote that "geometry is the art of reasoning well on badly made figures."[18]

Thus we are not truly discussing compasses and straightedges. We are discussing pure geometry—a geometry of ideal points, lines, and circles. But we shall still point at the wall of the cave and discuss our physical drawing tools.

The Rules of the Game

Let us be precise about the requirements for solving compass-and-straightedge problems. Every problem begins with a geometric object—a line segment for which we must construct a perpendicular bisector, an angle that we must bisect, a circle inside which we must draw a hexagon, and so on. But to use a compass or a straightedge we need points—we line up the straightedge between two points before we trace along its edge, and we place the tips of the compass on two points before sweeping out the circle. As we draw lines and circles we create new points. New points are produced in only one way: they are the points of intersection of previously drawn lines and circles.

Why must we be so restrictive? What prevents us from drawing an arbitrary new point with our pencil, or setting our straightedge on the page at random and drawing along its edge, or placing the tip of our compass anywhere we wish, opening the arms an arbitrary amount and drawing a circle? These are not radical moves unknown to crafts-men who have used such devices for millennia. The problem is that these arbitrary actions are not replicable. If Sally were to construct a geometric shape using such techniques, then Bob would be unable to recreate her figure. He may be able to construct a figure that resembles Sally's, but it would be impossible for him to duplicate it exactly. Such constructions may be fine for an artist or a craftsman, but not for a geometer. A geometric construction must be repeatable by anyone, anywhere, anytime—by the Greeks 2500 years ago, by us today, and by future geometry students—provided they have the same tools.[19]

Thus, we have the following strict rules.

1. Straightedge: Given points A and B, we can draw the line segment AB or the line AB.
2. Compass: Given points A and B, we can draw the circle with center A through B.
3. Points: New points are created as intersections of two lines, two circles, or a line and a circle.

These rules seem reasonable at first glance—given our prohibition of drawing random figures. There are no surprises, except perhaps that we can draw lines that are infinitely long. Again, recall that we aren't *really* working with a compass and straightedge. But if pressed to enforce that analogy, a healthy way to imagine our toolkit is that we have an infinitely extendable straightedge and an infinitely extend-able compass that can draw line segments and circles as large or as small as we need. Moreover, our desk contains enough paper to extend our working area as far as we need to accommodate our geometric drawing.

Perhaps more surprising than the things we are allowed to do with our compass and straightedge are the things we are not allowed to do. For instance, the rules seem to imply that we can't use a *locking compass*. That is, we are unable to open the compass to a fixed distance, like a pair of dividers, move them to another location on the paper, and draw a circle with that radius. Instead, Euclid's compass is a *collapsing compass* or a *snap compass*. When we lift the tips of the compass off the paper, the compass collapses (or snaps closed). In 1849 De Morgan wrote that "we do not allow a circle to be drawn with a compass-carried distance; suppose the compasses to close of themselves the moment they cease to touch the paper."[20]

This is troubling. Any high-quality compass purchased at a store that sells office supplies or drafting tools would certainly lock in place. Have geometry students (and their teachers!) been cheating all these years? Have they been doing something non-Euclidean? No, and using a locking compass isn't actually forbidden. We can prove that locking compasses are allowed. It turns out that the seemingly less powerful collapsing compass is just as powerful as the locking compass. Anything we can do with one, we can do with the other.

Euclid wasted no time in showing this. The second proposition in the first book of *Elements* states that given a line segment AB and a point C it is possible to construct a point D so that $CD = AB$. Thus, we can draw a circle of radius AB at C. Figure 3.5 shows Euclid's construction. First, construct the point E so that BCE is an equilateral triangle. Extend BE and CE beyond B and C. Draw a circle with center B and radius AB. It intersects BE at F. Finally, construct a circle with center E and radius EF. It intersects CE at D. It follows that $AB = CD$.[21]

Thus, although locking compasses aren't explicitly allowed, we may use one. The hypothesis of the collapsible compass has the benefit of simplicity, but when it comes to constructing figures, the locking compass is a major time saver. Euclid followed Aristotle's advice when he chose this seemingly weaker postulate. Aristotle wrote, "All other things being equal, that proof is the better that proceeds from the fewer postulates or hypotheses or propositions."[22] The simpler, the better; don't assume a locking compass when a collapsing one will suffice.

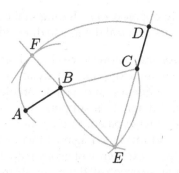

FIGURE 3.5. Transferring a length: given AB and C, we construct D so that $CD = AB$.

Another thing missing from the rules is the use of a *ruler*. It is very common for mathematicians and mathematics books to use the words straightedge and ruler interchangeably. Indeed, most geometry students use rulers as straightedges. But not only is the Euclidean straightedge not a ruler marked off with inches or centimeters, it is completely free of blemishes. A geometer using a ruler must not use the markings. As we shall see in chapter 10, Archimedes and other mathematicians investigated how geometry would change if we relaxed this criterion. They proved that if we have a straightedge that has two marks on it—so it can measure off one, and only one, distance—then we can solve more construction problems than we could with an unmarked straightedge. In particular, some problems of antiquity can be solved with this modified drawing tool.

Yet another item missing from this list is the ability to construct curves other than lines and circles. Although circles and lines dominated Greek geometry, they were by no means the only curves that were studied. Greek geometers defined and explored the conic sections—ellipses (of which the circle is a special case), parabolas, and hyperbolas. Although not as natural as the circle and the line, these were easy to define, and in the years to come they played a major role in mathematics and physics and in the story of the impossibility problems. The Greeks also took things to a further extreme and invented curves—and clever tools to draw them. For the most part these curves, with names like the quadratrix, the spiral, the conchoid, and the cissoid, did not live on to be standard geometric objects that had a life beyond their original purpose—to solve the problems of antiquity.

Plato was strongly opposed to including in the study of geometry curves that required mechanical devices to draw. Plutarch said,[23]

> For the art of mechanics, now so celebrated and admired, was first originated by Eudoxus and Archytas, who embellished geometry with its subtleties, and gave to problems incapable of proof by word and diagram, a support derived from mechanical illustrations that were patent to the senses.... But Plato was incensed at this, and inveighed against them as corrupters and destroyers of the pure excellence of geometry, which thus turned her back upon the incorporeal things of abstract thought and descended to the things of sense, making use, moreover, of objects which required much mean and manual labour. For this reason mechanics was made entirely distinct from geometry, and being for a long time ignored by philosophers, came to be regarded as one of the military arts.

In terms of Plato's cave, "We cannot use the material to study the immaterial because then we would be using shadows to study reality."[24]

The Greeks created a hierarchy for their curves. The line and circle were best, next came the conic sections, and all other curves came last. Indeed, they classified problems according to what geometric objects were required for the solution. *Plane* problems required only lines and circles, *solid* problems required one or more conic sections, and *linear* problems required some other curve (this is not the way we use the term linear today!). Of course, proving that a problem was one type and not another was beyond the capabilities of the Greek mathematicians.

In the fourth century CE, Pappus (ca. 290–ca. 350), one of the last great mathematicians in Alexandria, described this classification:[25]

> Those [problems] that can be solved by straight lines and the circumference of a circle are rightly called plane because the lines by means of which these problems are solved have their origin in the plane. But such problems that must be solved by assuming one or more conic sections in the construction, are called solid because for their construction it is necessary to use the surfaces of solid figures, namely cones. There remains a third kind that is called line-like. For in their construction other lines than the ones just mentioned are assumed, having an inconstant and changeable

origin, such as spirals, and the curves that the Greeks call [tetrag-onizousas], and which we call "quadrantes," [quadratrix] and conchoids, and cissoids, which have many amazing properties.

Pappus argued that it was necessary to use the simplest techniques available. One should not employ conic sections to solve a problem that is solvable with lines and circles. One should not solve solid problems using linear methods. He wrote, "It seems to be a grave error into which geometers fall whenever any one discovers the solution of a plane problem by means of conics or linear (higher) curves, or generally solves it by means of a foreign kind."[26] Federico Commandino (1509–1575) used the word "sin" (*peccatum*) in his translation instead of "grave error."[27]

The conversation over what curves should be allowed in geometry continued for hundreds of years. In the seventeenth century, Descartes expanded this classification of curves in light of his achievement in bringing algebra into geometry. Like Pappus, he called it a "geometric error" to use curves that are more complicated than needed. (He also added that "it would be a blunder to try vainly to construct a problem by means of a class of lines simpler than nature allows."[28])

Johannes Kepler (1571–1630) thought that geometric objects were knowable only if they were constructible by Euclidean rules. Thus he believed (although did not have a proof) that the heptagon was not knowable. The fact that some figures were knowable and some were not was part of God's plan. He wrote,[29]

> So no Regular Heptagon has ever been constructed by anyone knowingly and deliberately, and working as proposed; nor can it be constructed as proposed; but it can well be constructed fortuitously; yet it is, all the same [logically] necessary that it cannot be known whether the figure has been constructed or no.

Isaac Newton (1643–1727) pushed back on Descartes's algebraic classification, observing that simplicity in algebra, measured by the complexity of an equation, and simplicity in geometry, measured by the ease of construction, may not coincide. For example, the equation of a parabola is simpler than that of the geometrically simpler circle:[30]

> In Constructions that are equally Geometrical, the most simple are always to be preferred. This Law is beyond all Exception. But Algebraick Expressions add nothing to the Simplicity of the Construction. . . . Wherefore that is *Arithmetically* more simple

which is determined by the more simple Equations, but that is *Geometrically* more simple which is determined by the more simple drawing of Lines; and in Geometry that ought to be reckoned best which is Geometrically most simple.

Things are not too different today. Mathematicians are always looking for better, simpler, more elegant proofs. There is prestige bestowed upon the mathematician who proves a theorem first, but there is also value to improving a mathematical demonstration. It would be a waste to use calculus to solve a problem when there is a simple algebraic solution. Simplicity and elegance are highly valued in the mathematical community.

There was a long and occasionally intense debate about what curves should be included in geometry. Lines and circles? Definitely. Conic sections? Probably. Other more complex "mechanical" curves? At first, definitely not, then eventually, grudgingly, yes. Much later, mathematicians would have to consider curves defined in terms of algebraic expressions and after that transcendental curves.

TANGENT
The Tomahawk

What is life but the angle of vision? A man is measured by the angle
at which he looks at objects. What is life but what a man is
thinking of all day? This is his fate and his employer. Knowing is
the measure of the man. By how much we know, so much we are.
—Ralph Waldo Emerson, "Natural History of Intellect," 1893[1]

THE CRAFTY MATHEMATICIAN who wants to make an angle trisection device should consider the *tomahawk* in figure T.4, which dates
back to at least 1835. It is easy to make and use.[2] The top of the tool
is divided into thirds. The diameter of a semicircle sits on two-thirds
of the edge, and the perpendicular side of the handle is tangent to the
semicircle.

Given ∠ABC, align the tomahawk so that B is on the handle, AB
meets the butt of the tomahawk at D, and BC is tangent to the semicircular blade at E. Mark off points F and G using the marks on the top
of the T. Then the lines BF and BG trisect the angle.[3]

The proof that this is a valid trisection of ∠ABC is straightforward;
it follows from the fact that the right triangles BDF, BFG, and BEG are
congruent.

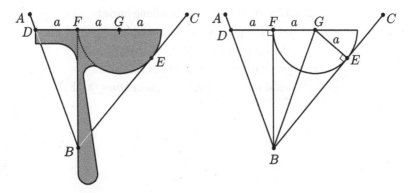

FIGURE T.4. We can use a tomahawk to trisect an angle.

CHAPTER 4

The First
Mathematical Crisis

For, as it happened, the inventor's sister,
quite unaware of what the Fates intended,
entrusted her own son to his [Daedalus's] instruction,
a likely lad of twelve, who had a mind
with the capacity for principles and precepts;
and from his observation of the spines
of fishes, which he'd taken as his model,
incised a row of teeth in an iron strip
and thereby managed to invent the saw.
Likewise, he was the first to bind two arms
of iron at a joint, so one is fixed
and the other, as it moves, inscribes a circle.

Daedalus envied him, and headlong hurled
this lad of precepts from a precipice,
the steep acropolis Minerva loves,
and lying, said the lad had slipped and fallen.
—Ovid, "Daedalus and Perdix," *Metamorphoses* (8 CE)[1]

ONE OF THE greatest dramas in all mathematics—the shocking discovery in the mid- to late fifth century BCE of what today we call irrational numbers[2]—came not from the study of numbers, but from geometry.[3]

Hippasus of Metapontum was a Pythagorean—a follower of the mystical cult leader and mathematician Pythagoras of Samos who had been dead a half century or more.[4] The Pythagoreans held the sacred belief that the universe was described by integers and their ratios. Thus we can imagine their disbelief when Hippasus made his discovery. But the Pythagoreans' deeply held certainty was about to crumble when faced with the undeniable truth.

To understand Hippasus's geometrical discovery, to see why it was so troubling to the Pythagoreans, and to see what it has to do with irrational numbers, we must first understand the Greek's view of numbers and their approach to ratios and proportions.

Numbers, Magnitudes, and Ratios

The Greeks had a very restricted view of what a number is. Numbers were the natural numbers: 1, 2, 3, and so on. In truth, Euclid didn't even consider 1 a number; it was the unit from which numbers were formed. Negative numbers and zero were not numbers, nor were nonintegral rational numbers like 1/2 and irrational numbers like $\sqrt{2}$. *Magnitudes* took the place of positive quantities that were infinitely divisible. Line segments, two-dimensional regions, three-dimensional solids, and angles are examples of magnitudes.

Greek geometry didn't contain numerical measurements—lengths of segments, numerical areas, measures of angles, and so on. Where we would say that two line segments are the same length, they would simply say that the segments were equal. Likewise, if two regions have equal areas, they would say that they were equal. For example, Euclid said that two triangles having the same base and height are equal—even if they are not congruent. Euclid did not define area, and he certainly did not assign a number to it.[5]

Nevertheless, the Greeks treated magnitudes in number-like ways. Magnitudes of the same kind can be added—geometrically joining the two segments, regions, and so on. They can be ordered: $a < b$ provided there exists a magnitude c such that $a + c = b$. For any two magnitudes a and b of the same type, either $a = b$, $a < b$, or $b < a$. They defined ma to be the sum of m a's.

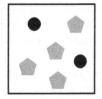 

FIGURE 4.1. The ratio of circles to pentagons is the same as the ratio of stars to triangles.

The Greeks had the notion of the ratio of two numbers. In figure 4.1 the ratio of the numbers of circles to pentagons is 2:4, and the ratio of stars to triangles is 3:6. When we look at these ratios today, we immediately think of the rational numbers 2/4, 3/6, and 1/2, because there are half as many circles as pentagons and half as many stars as triangles. But to the Greeks, a ratio of numbers was not a number. They did not place ratios on a number line, they did not make measurements with ratios, and they did not add, subtract, multiply, or divide ratios. On a more philosophical level, it is not clear that a ratio was a "thing." To us, 2/4 is a mathematical object, but to the Greeks 2:4 may have been only a relation between numbers.

In Plato's *Republic*, Socrates said, "You know how steadily the masters of the art repel and ridicule anyone who attempts to divide absolute unity when he is calculating, and if you divide, they multiply, taking care that one shall continue one and not become lost in fractions."[6] That is, they would not think of 1:3 as dividing 1 into 3 parts (1/3) because 1 is indivisible. Instead, it would mean 1 part of 3.

Yet they endowed ratios with some number-like properties. They knew what it meant for two ratios $a{:}b$ and $c{:}d$ to be the same,[7] which we write as $a{:}b :: c{:}d$. They are the same if there is some multiple of a and b, say n, and some multiple of c and d, say m, such that $an = cm$ and $bn = dm$. Returning to figure 4.1, if we triple the numbers of circles and pentagons and double the numbers of stars and triangles, then the numbers of circles and stars are the same and the numbers of pentagons and triangles are the same (as in figure 4.2). Written as fractions, this is simply

$$\frac{3 \cdot 2}{3 \cdot 4} = \frac{6}{12} = \frac{2 \cdot 3}{2 \cdot 6}.$$

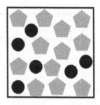 

FIGURE 4.2. Triple the numbers of circles and pentagons and double the numbers of stars and triangles.

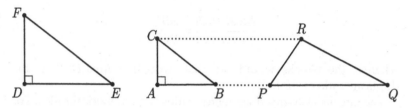

FIGURE 4.3. Triangles *ABC* and *DEF* are similar. Triangle *PQR* has the same height as triangle *ABC*.

They also had the notion of a ratio being greater than or less than another ratio. They had the notions of *duplicate* and *triplicate* ratios, which today we might view as squaring and cubing the ratios. They had a notion that can be seen as multiplying and canceling: the ratios *a*:*b* and *b*:*c* are *compounded* to form the ratio *a*:*c* (think $\frac{a}{b} \cdot \frac{b}{c} = \frac{a}{c}$).

The Greeks also considered ratios of magnitudes, such as lengths of line segments, areas of regions, volumes of solids, measures of angles, and so on. They did not consider mixed ratios, such as the ratio of an area to a length, but they did compare ratios of different kinds.

To get an idea of how the Greeks used ratios of magnitudes, consider the triangles in figure 4.3. Triangles *ABC* and *DEF* are similar, so *AB*:*BC* :: *DE*:*EF*. In this expression, all four magnitudes are of the same type—they are all line segments.

But ratios of different magnitudes can also be the same, such as

segment:segment :: area:area.

For instance, because the heights of triangles *ABC* and *PQR* in figure 4.3 are the same, the areas of the triangles are proportional

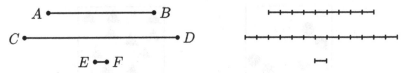

FIGURE 4.4. Segment *EF* measures *AB* and *CD*.

to the bases: $AB:PQ :: \text{Area}(ABC):\text{Area}(PQR)$. And because the areas of similar triangles are proportional to the squares of the lengths of corresponding sides, we'd say

$$\frac{\text{Area}(ABC)}{\text{Area}(DEF)} = \frac{AB^2}{DE^2},$$

whereas the Greeks would say that $\text{Area}(ABC):\text{Area}(DEF)$ is the duplicate ratio of $AB:DE$.

The Greeks also mixed geometric ratios with numerical ratios. For example, in figure 4.3 the two similar triangles happen to be 3-4-5 right triangles. So $AB:BC :: 4:5$. Moreover, it happens that PQ is twice as long as AB, so $\text{Area}(ABC):\text{Area}(PQR) :: 1:2$.

We are almost ready to define commensurable and incommensurable magnitudes. Let's take a closer look at numerical ratios of line segments. In figure 4.4, line segment AB is 9/13ths segment CD, so $AB:CD :: 9:13$. There are several ways of thinking about this. A modern conception is $AB/CD = 9/13$, where AB and CD are the lengths of the segments. Another way is that $13\,AB = 9\,CD$: if we lined up 13 copies of AB end to end, it would be as long as 9 copies of CD. Yet another way is that if we divide AB into 9 equal parts and CD into 13 equal parts, the parts would be of equal length. That is, there is a segment EF such that $9\,EF = AB$ and $13\,EF = CD$. We say that EF is a *common measure* of, or that it *measures*, AB and CD.

Naturally, we may ask: Given segments PQ and RS, is there a common measure? That is, does there exist a segment TU—perhaps one that is very, very short—and integers m and n such that $PQ = m \cdot TU$ and $RS = n \cdot TU$? Equivalently, do PQ and RS have a numerical ratio, $PQ:RS :: m:n$? Or, using today's terminology, is PQ/RS a rational number? If so, PQ and RS are *commensurable*; if there does not exist a common measure, they are *incommensurable*.

Pythagoras and his followers believed the answer was yes—that any two segments, any two areas, any two volumes, and so on are commensurable. They believed it until Hippasus discovered two incommensurable magnitudes.

The Pythagoreans and Hippasus of Metapontum

We know surprisingly little about Pythagoras—or perhaps it is not surprising. He was one of the earliest great Greek thinkers, born in approximately 560 BCE, or about two and a half centuries before Euclid. He was also the head of a secret society. Thus, almost by design, details of his life are hard to come by. Much of what we know about Pythagoras was written 1000 years after he lived, and much of it is contradictory or too fabulous to believe.

Pythagoras was born on Samos, a commercially prosperous Greek island in the eastern Aegean Sea, known at the time for its fine wines and red pottery. In his youth, Pythagoras traveled to Egypt and Babylonia. Later, he settled in the southern Italian city of Croton, where he attracted followers who viewed him as a religious and philosophical leader. In about 500 BCE he was forced to leave Croton, and he resettled in Metapontum, where he lived the rest of his days. He died in approximately 480 BCE.

Although Pythagoras predated Jesus by five and a half centuries, he was a contemporary of other mystical leaders such as Zoroaster, Buddha, Confucius, and Lao-Tzu. However, Pythagoras taught a philosophy that was unique among these founders of religions. He asserted that the way to transcendence was through mathematics. In particular, he asserted that the universe was built on numbers and their ratios. He taught that the ratios of numbers described and connected number theory, geometry, music, and astronomy.

If Pythagoras strummed one string of his lyre and then pressed his finger on the string making it half its length and strummed again, he'd hear sounds an octave apart. If instead he made the string 2/3 its original length then the sounds would form a pleasing "perfect fifth."

In geometry, the Pythagoreans knew that any triangle whose sides have a ratio of 3:4:5 is a right triangle. This, of course, can be shown using the theorem that bears his name:[8] $3^2 + 4^2 = 5^2$.

They knew the planets were closer to Earth than were the stars, and they believed that their motion could be predicted using numbers. We do not know their planetary theory, but the seven known heavenly bodies were likely paired with the seven strings of the lyre.

And a celestial harmony—the music of the spheres—was audible only to Pythagoras.

We do not know much about Hippasus—not even when he lived.[9] As with Pythagoras, much of what we know is sketchy, contradictory, and from sources written centuries after he died.[10] Hippasus was from Metapontum, which is located in the bay between the heel and sole of the boot-shaped Italy. He studied mathematics and music theory. He investigated the way the pitch changes for metal disks of varying thicknesses and for tumblers filled with varying amounts of water. He made mathematical contributions to the study of ratios and proportions. He may have been the first to construct a dodecahedron—the Platonic solid with 12 regular pentagons as faces—in a sphere.[11] We also believe he discovered incommensurable magnitudes.

Plutarch, writing about this discovery, used a certain Greek word (transliterated as *arretos*) to describe the incommensurability of the lengths. The word has a clever double meaning: "ineffable because irrational" and "unspeakable because secret."[12] The former seems to imply that there are no words to describe the relationship between the lengths. The latter implies that speaking about the discovery would be forbidden because of the secretive nature of the Pythagorean sect. The legend states that Hippasus was severely punished by the Pythagoreans. His offense was either that he discovered incommensurable magnitudes, which contradicted the Pythagorean belief that "all is number," or that he took credit for one of his mathematical discoveries rather than giving credit to Pythagoras. The exact punishment also varies from telling to telling. Either he was drowned at sea, or he was expelled from the Pythagoreans, who held a funeral for him. Of course, these stories are likely only that—good stories. Nevertheless, the existence of incommensurable magnitudes was a blow to the Pythagoreans and was later a motivation for Plato to encourage members of his academy to firm up the logical foundations of mathematics.[13]

We do not know how Hippasus discovered incommensurable magnitudes nor even which pair of magnitudes he found to be incommensurable. Most scholars believe Hippasus showed that the side and diagonal of a square are incommensurable (see figure 4.5). This fact is equivalent to $\sqrt{2}$ being an irrational number.

Others suggest that he discovered that the side and diagonal of a regular pentagon are incommensurable (see figure 4.5).[14] This result implies that the golden ratio, $\phi = (1 + \sqrt{5})/2$, is irrational. In addition

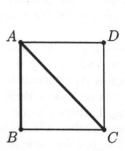 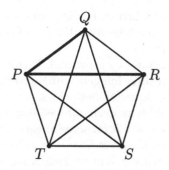

FIGURE 4.5. The side and diagonal of a square (AB and AC) have no common measure and neither do the side and diagonal of a regular pentagon (PQ and PR).

to technical mathematical reasons why the pentagon would be the natural first example, we know Hippasus was interested in the dodecahedron, which has regular pentagons as faces. Moreover, the pentagram star, which is formed by the diagonals of a pentagon, was the Pythagorean symbol of health and was used as an identification symbol.

Before we give the proof that the side and diagonal of a square are incommensurable, let us see how this result is related to $\sqrt{2}$. Consider the square in figure 4.5. Because $AB = BC$ and because ABC is a right triangle, the Pythagorean theorem tells us that $AC^2 = AB^2 + BC^2 = 2AB^2$. So

$$\frac{AC^2}{AB^2} = \frac{2AB^2}{AB^2} = 2.$$

Thus $AC/AB = \sqrt{2}$. If there were a common measure EF of AB and AC, say $AB = m \cdot EF$ and $AC = n \cdot EF$, then

$$\sqrt{2} = \frac{AC}{AB} = \frac{m \cdot EF}{n \cdot EF} = \frac{m}{n}$$

is a rational number.

The proof is a classic use of proof by contradiction, or reductio ad absurdum. We assume the negation of what we want to prove—in this case, that $\sqrt{2}$ is a rational number—and then show that this leads to a contradiction. Having done so, we conclude that the original statement must be true. We give the modern version, not the geometric version.

Suppose $\sqrt{2}$ is rational. Then we can write $\sqrt{2} = m/n$ where m and n are positive integers with no common factors. It follows that $m = \sqrt{2}n$, or, by squaring both sides, $m^2 = 2n^2$. From this expression it follows that m^2 is an even number. But because m^2 is even, so is m. Because m is even, $m = 2k$ for some integer k. Substituting that back into the earlier expression gives $(2k)^2 = m^2 = 2n^2$. But this implies that $2k^2 = n^2$. So n^2, and hence n, is even. Finally we reach our sought-after contradiction: because both m and n are even, they share a common factor of 2, which is impossible. Hence, $\sqrt{2}$ is irrational.

The Acceptance of Incommensurable Magnitudes

After Hippasus's discovery of incommensurable magnitudes, the mathematical community turned its attention to this interesting new problem. Knorr described the construction of the theory of incommensurable magnitudes as "a massive project which engaged the best efforts of the most notable fourth-century mathematicians: Theodorus, Theaetetus, Archytas and Eudoxus."[15]

In his dialogue *Theaetetus*, Plato wrote the following of his mathematical tutor Theodorus of Cyrene (ca. 465–after 399 BCE):[16]

Theodorus here was demonstrating to us with the aid of diagrams a point about powers. He was showing us that the power of 3 square feet and the power of 5 square feet are not commensurable in length with the power of 1 square foot; and he went on in this way, taking each case in turn till he came to the power of 17 square feet; there for some reason he stopped.

In other words, Theodorus proved that $\sqrt{3}$, $\sqrt{5}$, $\sqrt{6}$, $\sqrt{7}$, $\sqrt{8}$, $\sqrt{10}$, $\sqrt{11}$, $\sqrt{12}$, $\sqrt{13}$, $\sqrt{14}$, $\sqrt{15}$, and perhaps $\sqrt{17}$, are irrational (or, more precisely, that the side of a square with area 3, 5, 6, and so on is incommensurable with the side of a square with area 1).[17]

According to Plato, Theaetetus—another student of Theodorus and a student, and later a teacher, at Plato's Academy—extended Theodorus's result, showing that if a positive integer n is not a perfect square, then $\sqrt{n}$ is irrational. In addition to his study of these and other irrationalities, he also constructed the five Platonic solids and proved that there were only five. Plato thought very highly of Theaetetus, featuring him in two dialogues.[18] He wrote, "This boy advances toward learning and investigation smoothly and surely and successfully, with perfect gentleness, like a stream of oil that flows without a sound, so that one marvels how he accomplishes all this at his age."[19] We believe that Theaetetus is responsible for most of the mathematics in books X and XIII of Euclid's *Elements*. In a commentary on book X of *Elements* Pappus wrote that the theory of incommensurable quantities[20]

> underwent an important development at the hands of the Athenian, Theaetetus, who is justly admired for his natural aptitude in this as in other branches of mathematics. One of the most gifted of men, he patiently pursued the investigation of the truth contained in these branches of science ... and was in my opinion the chief means of establishing exact distinctions and irrefutable proofs with respect to the above-mentioned quantities.

Archytas of Tarentum (ca. 428–ca. 350 BCE)—the third individual mentioned by Knorr—was a friend of Plato's, a mathematician, a statesman, a philosopher, and a military commander. Among his results was a proof that, in modern terminology and notation, certain values, such as $\sqrt{n(n+1)}$, are irrational. His motivation for this work came out of music theory.

However, the real hero of this story is Archytas's student, Eudoxus. Thanks to Eudoxus's work on ratios and proportions, Euclid gave two different definitions for proportionality—a simpler definition for proportionality of numbers and a more complicated one for geometric magnitudes—one that allowed the Greek mathematicians to work with incommensurable magnitudes.[21] The latter definition is essentially the modern definition of the real numbers given by Richard Dedekind in 1872.[22] The British mathematician Isaac Barrow (1630–1677) offered high praise for this work: "There is nothing in the whole body of the *Elements* of a more subtile invention, nothing more solidly established and more accurately handled, than the doctrine of proportionals."[23] Similarly, his countryman Arthur Cayley (1821–1895) wrote, "There is hardly anything in mathematics more beautiful than this wondrous fifth book."[24]

Eudoxus is probably the greatest Greek mathematician no one knows about—and arguably the second greatest to Archimedes. Not only did Eudoxus prove important mathematical theorems, such as ones about the volumes of the cone and pyramid, he did important work building the foundations of mathematics. We credit Eudoxus with devising the method of exhaustion, which is a limiting process used to compute an area and which we now recognize as a precursor to the ideas used in integral calculus; for formalizing the axiomatic method, which Euclid would later employ so successfully; and for making sense of ratios and proportions after the crisis of incommensurable magnitudes. Although none of his writings survive, we believe that the mathematics in books V, VI, and XII in *Elements* are due to Eudoxus.

Eudoxus was from the Greek city of Cnidus—a prosperous commercial center on the coast of the Aegean Sea in present-day Turkey. Although he had little money, he moved to a town near Athens and learned philosophy from Plato—walking two hours each way to hear Plato's lectures. When he returned to Cnidus, his friends raised the funds to send him to Alexandria, where he spent the next 16 months. Later, after setting up a successful school in Cyzicus, he returned to Athens, and Plato welcomed him back as a master teacher.

Plato respected Eudoxus even though they had differing views on the theory of forms and ideas. Also, Eudoxus was a proponent of hedonism, arguing that pleasure is the greatest good. Plato objected to this belief on intellectual grounds. On this, Aristotle wrote,[25]

> [Eudoxus's] arguments gained strength rather from the excellence of his own character than from any intrinsic worth of their own; for he had, of all men, the highest reputation for temperance, and was, consequently, believed to take up this position, not because he was any friend of pleasure, but because he was convinced of the truth of his assertions.

Eudoxus is known for his contributions to and lectures in mathematics, astronomy, cosmology, geography, philosophy, theology, meteorology, medicine, oration, and law. G. L. Huxley wrote, "It is greatly to be deplored that not a single work of Eudoxus is extant, for he was obviously a dominant figure in the intellectual life of Greece in an age of Plato and Aristotle."[26]

TANGENT
Toothpick Constructions

And finally, when I had trained my dull memory to treasure up an endless
array of soundings and crossing-marks, and keep fast hold of them, I judged
that my education was complete; so I got to tilting my cap to the side of my
head, and wearing a toothpick in my mouth at the wheel.
— Mark Twain, *Life on the Mississippi*[1]

IN 1939 T. R. DAWSON investigated the geometric constructions that
are possible using an unlimited supply of identical toothpicks.[2] As
usual, knowing the ground rules is crucial. A "point" in this geometric
system is either an endpoint of a toothpick or a point of intersection of
two toothpicks. (These toothpick constructions may be tricky to do
in reality because the toothpicks have a thickness, but, being good
mathematicians, we will ignore that pesky detail.)

We can lay down a toothpick so that one end is at a point and
the toothpick passes over another suitably close point. We can place
a toothpick so that one end is on a point and the other end is on a
nearby toothpick. Lastly, given two points less than two toothpicks
apart, we can lay down two toothpicks to make the legs of an isosceles
triangle with these points as the base vertices.

Although we forbid extending a line segment by sticking two toothpicks end-to-end or having two toothpicks overlap in a piggy-backing
fashion, it is straightforward to extend a line segment: In figure T.5 we
have two points A and B. Lay a toothpick BC through A and B. Then
extend BC by constructing three equilateral toothpick triangles.

Dawson showed how to perform some of the classic geometric constructions using toothpicks. For example, the right-hand diagram in
figure T.5 shows how to bisect an angle PQR. We do so by building
two equilateral triangles, QRS and PQT. The toothpicks RS and PT
cross at U. Then QU is the desired angle bisector.

Then Dawson proved that toothpick geometry is equivalent
to straightedge-and-compass geometry. This statement may sound

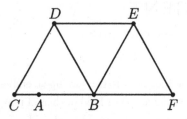
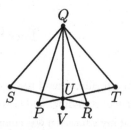

FIGURE T.5. Toothpick constructions of a line (left) and an angle bisector (right).

fishy—after all, we can draw a circle with a compass, but not with toothpicks. Here Dawson focused on points that are constructible with a compass and straightedge—points that are the intersections of the constructed lines and circles. Dawson proved that if we are given two points A and B and an unlimited supply of identical toothpicks whose lengths are at least as long as AB, then we can use the toothpicks to construct every compass-and-straightedge point, and vice versa—every point constructible by toothpicks is constructible with a compass and straightedge!

However, just because toothpick constructions are equivalent to compass-and-straightedge constructions, does not mean they are as easy. As an exercise, pick your favorite Euclidean construction and try to achieve the same end using toothpicks.

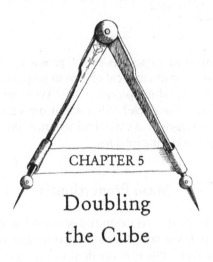

CHAPTER 5

Doubling the Cube

Den lieb ich, der Unmögliches begehrt. (I love those
who yearn for the impossible.)
— *Goethe*, Faust

TODAY THE PROBLEM of doubling the cube is probably the least known of the problems of antiquity, but that was not always the case. It garnered a lot of attention from the Greeks and later from European mathematicians. For instance, Eutocius (ca. 480 CE) gave 12 solutions in his commentary on Archimedes's *On the Sphere and Cylinder*.

In a way, this chapter is representative of this entire book and mathematics in general. In their attempts to solve the problem of doubling the cube, mathematicians attacked the problem in several ways. First, they generalized the problem: although their aim was to construct a line segment of length $\sqrt[3]{2}$ from a unit line segment, they realized that if they could solve a certain more general problem, they would be able to solve the simpler specific problem. Often this approach of generalizing a problem is fruitful, as the seemingly more difficult general problem is easier to solve. Of course, in this case the attempt was bound to fail (because the Delian problem is impossible) but, nevertheless, it was an important step that allowed mathematicians to understand the problem at a higher, more abstract level.

They also looked at variations of the problem. What if we were allowed to use other mathematical objects to help solve it? What if we could use certain three-dimensional solids? What about using curves besides the line and the circle? What about drawing tools besides a compass and straightedge? As we shall see, mathematicians took the same approach for all four problems.

Mean Proportionals

The *mean proportional* of two magnitudes a and b is the magnitude x such that $a{:}x :: x{:}b$. If we take the modern approach and think of a, b, and x as real numbers, this is equivalent to $a/x = x/b$. Solving for the mean proportional then amounts to solving the quadratic equation $x^2 = ab$. Geometrically, then, we must begin with segments of length a and b and construct a segment of length $x = \sqrt{ab}$. Euclid presented the construction in proposition VI.13.

His procedure is illustrated in figure 5.1. Join segments of length a and b to form a segment of length $a + b$, say AB and BC. Construct a semicircle with diameter AC. Construct a segment perpendicular to AC through B that meets the circle at D. A similar triangles argument shows that BD is the mean proportional of a and b.

The quantity $\sqrt{ab}$ is the *geometric mean* of a and b and can be viewed as a type of average. A nice way of viewing the geometric mean is to interpret it as a statement about areas:[1] a rectangle with sides a and b has the same area as a square with side length $\sqrt{ab}$.

Euclid introduced a special case of the mean proportional in definition VI.3. He wrote, "A straight line is said to have been cut in *extreme and mean ratio* when, as the whole line is to the greater segment, so is the greater to the less."[2] That is, to divide AB into extreme and mean ratio, we must find a point C on AB so that $AB{:}AC :: AC{:}BC$ or equivalently, find C so that AC is the mean proportional of AB and BC. If we let $AC = 1$ and denote AB using the traditional ϕ (see figure 5.2), then $\phi = 1/(\phi - 1)$, or $\phi^2 - \phi - 1 = 0$. The (positive) solution to this equation is $\phi = \left(1 + \sqrt{5}\right)/2 \approx 1.6180339\ldots$, the *golden ratio*.

In chapter 4 we pointed out that the golden ratio appears in a regular pentagon and that Hippasus may have proved that the side and diagonal of a pentagon are incommensurable. Let's fill in the details here. Figure 5.3 shows a pentagram inscribed in a regular pentagon. Observe that ADE and BCD are similar isosceles triangles.

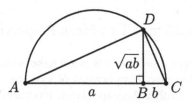

FIGURE 5.1. A construction of the mean proportional to segments of length a and b.

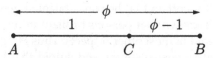

FIGURE 5.2. The point C divides AB into the extreme and mean ratio.

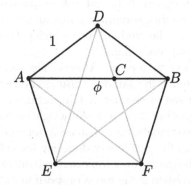

FIGURE 5.3. The point C divides the diagonal AB into the extreme and mean ratio.

So $DE:AE :: BD:BC$. But $AC = BD = AE$ and $DE = AB$. Thus $AB:AC ::$ $AC:BC$. So C divides AB into the extreme and mean ratio. If the side of the pentagon has length 1, then, because each side is the same length as AC, the diagonals have length ϕ.

The golden ratio has a long history and many fascinating mathematical properties (and many falsely attributed properties[3]). The idea of computing mean proportionals can point toward a new way of looking at the problem of doubling the cube.

Hippocrates's Two Mean Proportionals

The merchant Hippocrates of Chios is not to be confused with his contemporary, the physician Hippocrates of Cos (ca. 460–ca. 370 BCE), who was born on another Aegean island, 100 miles southeast of him, and whose ethical oath has been recited by generations of new doctors.

Hippocrates learned mathematics while he was in Athens trying to prosecute thieves who stole all his money and property.[4] It is unclear who robbed Hippocrates—it was either pirates or unscrupulous tax collectors—but it is certain that he was a talented geometer. Aristotle wrote, "It is well known that persons brilliant in one particular field may be quite stupid in most other respects. Thus Hippocrates, though skilled in geometry, was so supine and stupid that he let a customs collector of Byzantium swindle him out of a fortune."[5]

According to Proclus, Hippocrates wrote his own version of *Elements of Geometry*, well before Euclid did. In this work he used letters to represent points on the geometric figures, an important advance that we take for granted today and without which geometrical arguments would be much more confusing.

When Hippocrates arrived in Athens, the local mathematicians were already studying the famous compass-and-straightedge problems. As we shall see in chapter 7, Hippocrates attempted to solve the problem of squaring the circle. Although he was not successful, he was the first mathematician to square a nonrectilinear region. He also set his sights on the problem of doubling the cube. Again, he was unsuccessful, but he made some important advances in reframing and generalizing the problem using mean proportionals.

Rather than trying to double the cube, let's step down a dimension and try to double a square. That is, given an $a \times a$ square, construct a square with twice the area, $2a^2$. The side length of this larger square is $\sqrt{2}a$, which is precisely the length of the diagonal of the original square. So doubling the square is a straightforward task. But we ignore this easy solution and go in a different direction.

Let's take two copies of our original square and place them side by side so they form an $a \times 2a$ rectangle. Doubling the $a \times a$ square is equivalent to constructing a square with the same area as the $a \times 2a$ rectangle (see figure 5.4). Let's say that the side length of this larger square is x. When we view the problem this way, we see that x is the mean proportional of a and $2a$. That is, x satisfies $a{:}x :: x{:}2a$. So if we can

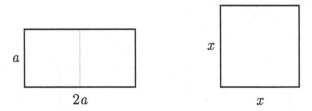

FIGURE 5.4. We wish to find an $x \times x$ square with the same area as the $a \times 2a$ rectangle.

find the mean proportional between a and $2a$—which we can with a compass and straightedge—we can double the square.

Hippocrates realized that mean proportionals were the key to understanding the problem of doubling the cube as well. However, unlike for the problem of doubling the square, doubling the cube required finding two mean proportionals. Because it was possible to construct mean proportionals with a compass and straightedge, Hippocrates was hopeful that constructing *two* mean proportionals would also be achievable.

The magnitudes x and y are the two mean proportionals between a and b provided $a{:}x :: x{:}y :: y{:}b$, or equivalently, $a/x = x/y = y/b$. How is this related to the doubling of the cube? A little algebraic manipulation shows that

$$x^3 = \frac{x}{a} \cdot \frac{x}{a} \cdot \frac{x}{a} \cdot a^3 = \frac{x}{a} \cdot \frac{y}{x} \cdot \frac{b}{y} \cdot a^3 = a^2 b.$$

In other words, an $a \times a \times b$ rectangular box has the same volume as an $x \times x \times x$ cube (see figure 5.5).

If we begin with an $a \times a \times a$ cube and double it, by stacking one atop another, we get an $a \times a \times 2a$ box. So if we can find the two mean proportionals between a and $b = 2a$, then x would satisfy $x^3 = 2a^3$. In other words, the cube with side length x would have double the volume of the $a \times a \times a$ cube.

As Eutocius wrote,[6]

After they [geometers] had all puzzled for a long time, Hippocrates of Chios was first to come up with the idea that if one could take two mean proportionals in continued proportion between two lines, of which the greater is double the smaller, then the cube will be doubled. Thus he turned one puzzle into another one, no less of a puzzle.

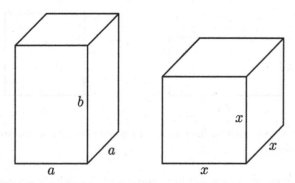

FIGURE 5.5. If $a{:}x :: x{:}y :: y{:}b$, then the $a \times a \times b$ and $x \times x \times x$ boxes have the same volume.

Viewed in terms of numbers, the problem of finding two mean proportionals is equivalent to the problem of computing cube roots. If we want to find the cube root of a number b, we can find the two mean proportionals x and y between 1 and b. Then $x = \sqrt[3]{b}$. Conversely, if we can find cube roots, then $x = \sqrt[3]{a^2b}$ and $y = \sqrt[3]{ab^2}$ are the two mean proportionals between a and b.

Hippocrates's discovery was a beautiful mathematical move—one that has been repeated over and over by generations of mathematicians. He took a very specific problem—in this case doubling the cube—and turned it into a more general one—the problem of finding two mean proportionals. If it is possible to solve this more general problem, then it is possible to solve the specific one. Sometimes the process of looking at the more general case makes the problem easier—if not easier to solve, then perhaps easier to see the solution.

Unfortunately, despite his good work, he was unable to construct the two mean proportionals or to double the cube.

Archytas's, Eudoxus's, and Menaechmus's Solutions

According to Eratosthenes, three members of Plato's Academy found methods to double the cube: Archytas, Eudoxus, and Menaechmus—of course, none used only a straightedge and compass.[7]

The first solution came in the first half of the fourth century BCE. Archytas gave a three-dimensional solution that is "a stunning *tour de force* of stereometric insight."[8] This solution relies on finding the point of intersection of three surfaces of revolution: a cylinder, a cone, and a degenerate torus. The solution was amazingly creative, but it did not give a convenient, planar technique for constructing the two mean proportionals between two segments.[9]

According to Eratosthenes, Eudoxus had a solution for doubling the cube that made use of "curved lines," but to the chagrin of scholars who came after, no one copied it down.[10]

Sometimes, mathematics created to solve one problem takes on a life of its own. One of the earliest examples is the conic sections—the parabola, ellipse, and hyperbola. Today they model planetary motion and the trajectory of a thrown ball; they are cross sections of reflective mirrors and nuclear cooling towers. But these applications came much later. The conic sections were discovered by Menaechmus (fl. ca. 350 BCE) in the fourth century BCE while trying to double the cube.

Menaechmus was a student of Eudoxus. In his summary of Greek mathematics, Proclus wrote, "Amyclas of Heraclea, one of the friends of Plato, and Menaechmus, a pupil of Eudoxus and associate of Plato, and his brother Dinostratus made the whole of geometry more perfect."[11] We believe that Menaechmus was a tutor for Alexander the Great. According to one story, Alexander asked Menaechmus to give him an easy method to learn geometry. Menaechmus replied, "O king, for traveling through the country there are private roads and royal roads, but in geometry there is one road for all."[12]

Today we know the conic sections by their equations, but as the name implies, they began as curves obtained by slicing a cone with a plane. Menaechmus's cones had varying angles at their apex, but he always sliced them with planes perpendicular to the generating line (the line that, when revolved about the axis, sweeps out the cone). Sliced in this way, an acute-angled cone produces an ellipse, a right-angled cone yields a parabola, and an obtuse-angled cone creates a hyperbola (see figure 5.6).

Apollonius of Perga (ca. 262–ca. 190 BCE) was the most significant contributor to the theory of conic sections. His masterpiece *Conics*, which contains 389 propositions in eight books, summarized all previous work, made new contributions, and systematized the study of conic sections. He also coined the terms ellipse, hyperbola, and parabola. His work was so highly respected that he acquired the title "Great Geometer," which is quite a pronouncement, following, as he did, after Euclid and Archimedes.

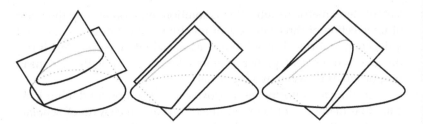

FIGURE 5.6. The conic sections.

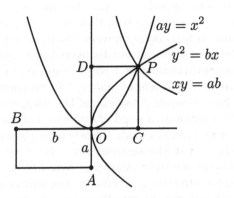

FIGURE 5.7. We can use the point of intersection of the three conic sections to find the two mean proportionals to a and b.

Let us look at how the conic sections help solve the problem of finding the two mean proportionals. We will use a modern approach that would have been alien to the Greeks: Cartesian coordinates and analytic geometry. We want to find x and y such that $a/x = x/y = y/b$. By cross-multiplying, we obtain three equations: $x^2 = ay$, $y^2 = bx$, and $xy = ab$, which are the equations of two parabolas and a hyperbola. The graphs of the three curves are shown in figure 5.7. We can use any pair of these curves to find the coordinates x and y of the point of intersection P, which are the two mean proportionals.

Menaechmus's precise argument is lost; all we have is Eutocius's account from 800 years later.[13] According to Eutocius, Menaechmus's solution is not too different from ours. The Greeks, of course, began with segments OA and OB of length a and b (such as in figure 5.7) and wanted the segments OC and OD. Menaechmus gave two methods for finding the point P; in the first, he found the intersection of one of the parabolas and the hyperbola, and in the second, he found the intersection of the two parabolas.[14]

To illustrate, here is how he could have described the hyperbola. He wanted to find the points C and D on the two perpendicular lines OA and OB so that OC and OD are the two mean proportionals for the segments OA and OB (again, see figure 5.7). Then the segments would satisfy $OA{:}OC :: OD{:}OB$ (that is, $a/x = y/b$), or equivalently $OA \cdot OB = OC \cdot OD$ ($ab = xy$). In other words, we have a fixed rectangular area formed by the segments OA and OB. Then there are various ways of situating C and D so that the area of rectangle $OCPD$ equals $OA \cdot OB$. The possible locations for P describe the hyperbola.

Recall that Pappus classified geometric problems into plane, solid, and linear depending on which curves were required to solve them. This solution of Menaechmus's shows that the problem of finding two mean proportionals is at worst a solid problem.

A Solution Using a Pair of Carpenter's Squares

According to Plutarch, Plato did not approve of Menaechmus's cube-doubling solution because it relied on a mechanical instrument. This implies that Menaechmus possessed a device that could draw conic sections or could be used to compute mean proportionals. There's no further information about this instrument, but one possibility is that the following mechanical procedure is due to Menaechmus.[15] Eutocius credits Plato with this discovery, but given Plato's disdain for mechanical constructions, this is unlikely.[16]

In their toolboxes, woodworkers have an L-shaped device, called a carpenter's square, that they use to draw accurate perpendicular lines. But Plato or Menaechmus or someone else used a pair of these tools to find the two mean proportionals between two lengths. The technique is shown in figure 5.8. Draw two segments OA and OB of length a and b, respectively, such that $\angle AOB = 90°$. Then extend the segments to lines. Now here's the tricky part: let the squares slide along each other, and maneuver them so that the other edges pass through the points A and B and the corners meet the lines OB and OA, as shown. Call the corner points C and D, and let x and y be the lengths of OC and OD, respectively. Then the triangles BDO, ACO, and CDO are similar, and from this it follows that $a/x = x/y = y/b$. Notice that the points A,

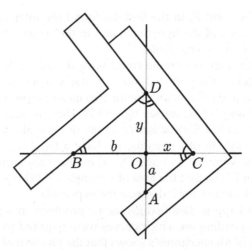

FIGURE 5.8. We can use two sliding carpenter's squares to find the two mean proportionals between a and b.

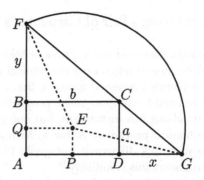

FIGURE 5.9. In this figure x and y are the two mean proportionals between a and b.

B, C, and D are exactly the same as the points with the same names in figure 5.7. Furthermore, in our diagram $b = 2a$, thus $x = \sqrt[3]{2}a$.

Apollonius's Solution

Eutocius gave the following construction for finding the two mean proportionals. He attributed it to several people, the oldest being Apollonius, who we thus may assume discovered it.[17] Construct a rectangle $ABCD$ with side lengths a and b and center E (see figure 5.9). The next step is the tricky part—the part that we cannot accomplish using a compass and straightedge. Using trial and error, draw a circle with center E passing through AB at F and AD at G in such a way that C, F, and G are collinear.[18] Having accomplished this, $x = DG$ and $y = BF$ are the two mean proportionals between a and b.

To see why this is the case, first observe that AFG, DCG, and BFC are similar triangles. So

$$\frac{y+a}{x+b} = \frac{a}{x} = \frac{y}{b}.$$

Next construct EF, EG, and perpendiculars from E to AG and AF meeting at P and Q, respectively. Then the hypotenuses of the two right triangles EGP and EFQ are radii of the circle, so they are equal. By the Pythagorean theorem, $FQ^2 + EQ^2 = EF^2 = EG^2 = EP^2 + GP^2$, or equivalently, $(y+a/2)^2 + (b/2)^2 = (x+b/2)^2 + (a/2)^2$. After a little algebra we conclude that

$$\frac{y+a}{x+b} = \frac{x}{y}.$$

So

$$\frac{a}{x} = \frac{x}{y} = \frac{y}{b}.$$

TANGENT
Eratosthenes's Mesolabe

We have heard much about the poetry of mathematics, but very little of it has as yet been sung. The ancients had a juster notion of their poetic value than we. The most distinct and beautiful statements of any truth must take at last the mathematical form.
—Henry David Thoreau[1]

IN THE LETTER in which Eratosthenes gave the colorful, if fictitious, origins of the cube doubling problem, he also described a mechanical invention called a *mesolabe* ("mean-taker"). This device is used to find the two mean proportionals between two given magnitudes.

The apparatus consists of three rigid congruent right triangles sitting in a rectangular frame (see figure T.6). Triangle I is fixed in place, but triangles II and III are free to slide left and right and can slide over one another. There is a rod attached to the upper left-hand corner that can pivot. If we arrange the parts so that the rod, the leg of I, and hypotenuse of II meet at one point, and the rod, the leg of II, and the hypotenuse of III meet at a point, then, by an argument using similar triangles, the lengths of the segments labeled a, b, x, and y satisfy $a/x = x/y = y/b$.

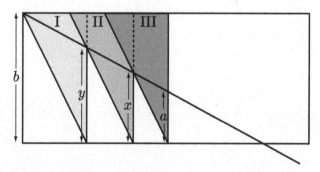

FIGURE T.6. Eratosthenes's mesolabe.

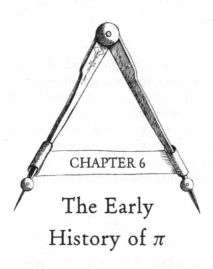

CHAPTER 6

The Early History of π

[The courtier] wanted to evaluate Giotto's work, finally asking
him for a small sketch to send to His Holiness. Giotto, who
was a most courteous man, took a sheet of paper and a brush dipped
in red, pressed his arm to his side to make a compass of it, and with
a turn of his hand made a circle so even in its shape and outline
that it was a marvel to behold. After he had completed the circle,
he said with an impudent grin to the courtier: "Here's your drawing."
The courtier, thinking he was being ridiculed, replied: "Am I
to have no other drawing than this one?" "It's more than sufficient,"
answered Giotto, "Send it along with the others and you will see
whether or not it will be understood."

Realizing that this was all he was going to obtain, the envoy left Giotto
rather dissatisfied, thinking he had been tricked. Nevertheless,
in sending the other drawings and the names of the artists who had
done them to the pope, he also included that of Giotto, recounting the
method he had used in making his circle without moving his arm
and without the use of a compass. As a result, the pope and many of his
knowledgeable courtiers realized just how far Giotto surpassed all
the other painters of his time in skill.
—Giorgio Vasari (1511–1574), *The Lives of the Artists*[1]

WE ALL KNOW and love π. Math enthusiasts memorize its digits and celebrate the number by eating pie on March 14 each year.[2] It is a ubiquitous number that shows up in the most unexpected places.

It relates the circumference of a circle to its diameter ($C = \pi d$, or equivalently $C = 2\pi r$, where r is the radius) and the area of a circle to the radius ($A = \pi r^2$). It relates the volume ($V = \frac{4}{3}\pi r^3$) and surface area ($S = 4\pi r^2$) of a sphere to its radius. It shows up in the normal distribution in probability (the so-called bell curve),[3]

$$f(x) = \frac{1}{\sigma\sqrt{2\pi}}e^{-(x-\mu)^2/(2\sigma^2)}.$$

It is one ingredient in Euler's identity, often cited as the most beautiful formula in mathematics, $e^{i\pi} + 1 = 0$. It appeared unexpectedly in Euler's solution to the Basel problem

$$\frac{1}{1^2} + \frac{1}{2^2} + \frac{1}{3^2} + \frac{1}{4^2} + \cdots = \frac{\pi^2}{6}.$$

It is found in the Gauss–Bonnet theorem, which relates the geometry of a surface to its topology,[4] and in Cauchy's integral formula, which allows us to compute the value of a complex function at a point by evaluating the function on a path around the point.[5] And on and on. As De Morgan wrote, "This mysterious 3.14159 ... comes in at every door and window, and down every chimney."[6] Most important to our story, π is intimately related to the problem of squaring the circle.

The first appearance of the symbol π with its modern meaning was in 1706 in Welsh mathematician William Jones's (1675–1749) *Synopsis palmariorum mathesios*. The historian of mathematics Florian Cajori (1859–1930) wrote,[7]

> William Jones made himself noted, without being aware that he was doing anything noteworthy, through his designation of the ratio of the length of the circle to its diameter by the letter π. He took this step without ostentation. No lengthy introduction prepares the reader for the bringing upon the stage of mathematical history this distinguished visitor from the field of Greek letters. It simply came, unheralded, in the following prosaic statement.

Cajori then quoted the following passage from Jones:[8]

There are various other ways of finding the *Lengths*, or *Areas* of particular *Curve Lines*, or *Planes*, which may very much facilitate the Practice; as for Instance, in the *Circle*, the Diameter is to the Circumference as 1 to $\overline{\frac{16}{5} - \frac{4}{239}} - \overline{\frac{1}{3}\frac{16}{5^3} - \frac{4}{239^3}} - \overline{\frac{1}{5}\frac{16}{5^5} - \frac{4}{239^5}}-$, &c. = 3.14159, &c.$= \pi$. This *Series* (among others for the same purpose, and drawn from the same Principle) I received from the Excellent Analyst, and my much Esteem'd Friend Mr. *John Machin*; and by means thereof, *Van Ceulen's* Number, or that in Art. 64.38 may be Examin'd with all desirable Ease and Dispatch.

In the next paragraph Jones wrote that $c = d \times \pi$. (We will meet John Machin [1680–1751] and van Ceulen [1540–1610] later in the book.)

Jones did not have the stature required to entice mathematicians around the world to adopt the symbol. However, Euler did, and in 1748 he used the symbol π for this same purpose.[9] He wrote,[10]

It has been found that the semicircumference [of the unit circle] is by approximation 3.141592653589793 . . . [Euler gives π to 127 decimal places[11]] for which number I would write π, so that π is the semicircumference of the circle of which the radius =1, or π is the length of the arc of 180 degrees.

Although we can date the first use of the symbol π to the eighteenth century, this is not the beginning of the story of this constant. Not even close. The circle constant was discovered and rediscovered by many cultures going back thousands of years. In what follows we give a whirlwind tour of the early history of π.

Mesopotamia

The plain between the Tigris and Euphrates rivers in Mesopotamia (currently Iraq and Kuwait and parts of Syria and Turkey) was a fertile location for both agriculture and mathematics. Humans had inhabited this "cradle of civilization" since at least 10,000 BCE. By the middle of the fourth millennium BCE, the Sumerians were firmly established in the region. They had an advanced civilization sporting a written language, schools, irrigated fields, the first cities, and a base-60 (sexagesimal) number system. We divide the hour into 60 minutes, the minute into 60 seconds, and the circle into 360 degrees thanks to this number system. In the eighteenth century BCE the Babylonians, led by Hammurabi (1792–1750 BCE), conquered Mesopotamia. The Babylonians adopted the sexagesimal number system and created some amazing mathematics.

Fortunately for us, some of their writing (and mathematics) has survived and is available for scholarly study. This is largely because of what they wrote on: they used a stylus to press wedge-shaped cuneiform symbols into clay tablets and let them dry in the sun. Remarkably, thousands of these durable tablets have been found intact.

One tablet from approximately 1800–1600 BCE shows a square with both diagonals. The cuneiform writing gives an approximation of $\sqrt{2}$ in sexagesimal that is accurate to an amazing six decimal places.[12] Another tablet from the same era gives a table showing 15 Pythagorean triples—that is, whole numbers a, b, and c such that $a^2 + b^2 = c^2$.[13]

Remarkably, the Babylonians knew of both versions of π—the circumference constant and the area constant. And they knew they were the same number; that is, they knew the relationship $A = \frac{1}{2}Cr$, although they stated it for a semicircle, not a circle. However, in many Babylonian tablets the value they took for π is extremely crude: 3. This gave the Babylonians the formulas $C = 3d$ and $A = C^2/12$.

Interestingly, the Babylonians possessed a better approximation for π. One of the Susa tablets, found in 1936, relates the circumference of a circle and the perimeter of the inscribed regular hexagon by[14] $P = (24/25)C$. If the radius of the circle is r, then the perimeter of the hexagon is $6r$, so this expression implies that $\pi = 3\frac{1}{8} = 3.125$. Seidenberg suggested that the Babylonians knew that 3 was a crude approximation for π, but worked with it anyway. He wrote, "Our view is that the Old-Babylonians didn't care!... There are many absurdities in Babylonian geometry... Once the 3 had been fixed in tradition, it would be difficult to change... The new value [$\pi = 3\frac{1}{8}$]... appears to have had no impact on mathematics itself."[15]

Egypt

Meanwhile, off to the southwest, along the Nile River, the Egyptians were creating their own mathematics. Unfortunately, the Egyptians wrote on papyrus, which is not as durable as the Babylonians' clay tablets. Thus, we have less information about their mathematical

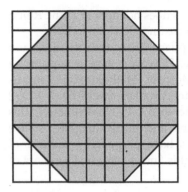
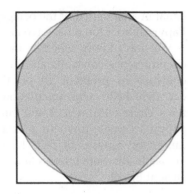

FIGURE 6.1. An octagon inscribed in a square appeared in problem 48 of the Rhind Papyrus.

accomplishments. Fortunately, the dry climate allowed some mathematical items to survive.

The Egyptians used an additive base-10 number system, and the numbers were expressed using either hieroglyphic or hieratic symbols. They expressed fractions as sums of unit fractions (for instance they would write $1/2 + 1/5 + 1/8$ rather than $33/40$). Several sets of mathematical problems and solutions have survived. They were problems in arithmetic, basic algebra, and planar and solid geometry.

Evidence of the Egyptians' knowledge of π can be found in a piece of papyrus from 1650 BCE that is 1 foot high and 18 feet long. It was copied by a scribe named Ahmes from a prototype from 2000–1800 BCE. Today it is called the Ahmes Papyrus or the Rhind Papyrus (for Henry Rhind who purchased it in 1858).

The Rhind Papyrus contains 84 problems and their solutions. Problem 50 states that the area of a circle is equal to that of a square whose side is the diameter diminished by 1/9. That is, a circle with diameter d has area $A = \left(\frac{8}{9}d\right)^2$. If this formula were true, it would imply that $\pi = 4\left(\frac{8}{9}\right)^2 = \frac{256}{81} \approx 3.1640\ldots$.

We do not know how the Egyptians arrived at this formula, but there is a possible clue in problem 48.[16] It shows an irregular octagon inscribed in a 9×9 square (see figure 6.1). The area of the octagon is 63 small squares. If the side length of the large square is d, then the octagon has area $\frac{63}{81}d^2$, which is approximately equal to $\frac{64}{81}d^2 = \left(\frac{8}{9}d\right)^2$. In figure 6.1 we see that a circle inscribed in the large square is almost equal in area to that of the octagon (this does not appear in the Rhind Papyrus); so it too must have area approximately equal to $\left(\frac{8}{9}d\right)^2$.

We don't know whether the Egyptians knew that the area and circumference constants are the same. However, there is supporting evidence in problem 10 on the Moscow Mathematical Papyrus (ca. 1850 BCE). One calculation in the problem could be $A = Cr/2$ for a circle with area A and circumference C.[17] Unfortunately, this papyrus comes down to us damaged and partly illegible, so we cannot be certain. In fact, historians have proposed a number of plausible alternate interpretations of this "most contested problem of ancient Egypt."[18]

India

The *Sulvasutras* are Hindu instruction manuals providing details on how to construct sacrificial fire altars of various sizes and shapes. *Sulvasutras* means "the rules of the cords" and refers to the process of doing constructive compass-and-straightedge-like geometry using rope.

The *Sulvasutras* are difficult to date. They were probably written during the first millennium BCE, although some of the geometric techniques are certainly older than that. So this mathematics occurred well before the more fruitful and familiar "golden age" of Indian mathematics (roughly 400–1600 CE) in which they introduced our decimal number system, negative numbers, the concept of zero, and so on.

Different altars were required for different occasions and for different schools, but, according to Seidenberg, "In these controversies, those concerning the altars at any rate, the area was understood to be constant, and this led, as we suggest, to the problems of squaring the circle and of turning the square into a circle. In the *Sulvasutras* there are attempts at solving these problems."[19]

The *Sulvasutras* presents the following method for creating a circle with the same area as the square $ABCD$ with center E (see figure 6.2).[20] Sweep an arc with center E and radius EA. The perpendicular bisector of AD intersects the arc at F and AD at G. Let H be the point on FG such that $GH = \frac{1}{3}FG$. Then the *Sulvasutras* claims that the circle with center E and radius EH has the same area as the square.

If these areas were indeed equal, it would have implied that $\pi = 54 - 36\sqrt{2} = 3.088\ldots$. The *Sulvasutras* also gives a technique for

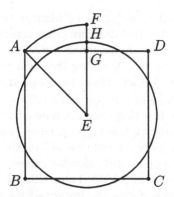

FIGURE 6.2. The *Sulvasutras* states that this circle and square have the same area.

squaring the circle.[21] It is essentially the reverse of this process, but using a rational approximation for $\sqrt{2}$.

China

This chapter considers the history of π before the Greek contributions of Euclid and Archimedes, but this section is an exception. Although it comes chronologically after Archimedes, the mathematics was developed independently. Archimedes's *Measurement of a Circle*, in which he gave the bounds 22/7 and 223/71 for π, did not appear in China until 1631.[22]

The most influential Chinese treatise on mathematics is *The Nine Chapters on the Mathematical Art (Jiuzhang suanshu)*. Some writers refer to it as the "Chinese *Elements*," but, although it was similarly influential, the contents and approaches of the two texts are quite different. *The Nine Chapters* was a practical handbook of mathematics featuring 246 problems and general methods for solving them, on topics such as engineering, surveying, trade, and taxation. Unlike *Elements*, it did not contain rigorous proofs of the mathematics contained within its covers. We do not know when *The Nine Chapters* was written, but most likely it was during the Han period, between 206 BCE and 221 CE. It may have been based on an earlier work that was destroyed in a mass burning ordered by the Qin dynasty ruler Shih Huang Ti (259–210 BCE).

In problems 31 and 32 of chapter 1 ("Mensuration of fields"), we see that they knew that the two different circle constants were the same. It states that the area of the sector of a circle is half of the product of the radius and the arc. Applying that to a full circle, we obtain Archimedes's result, $A = \frac{1}{2}rC$. However, it also states that $A = \frac{1}{4}d^2$ and that $A = \frac{1}{12}C^2$, both of which imply that $\pi = 3$.

The value $\pi = 3$ was widely used in Chinese mathematics. However, it was well known to be a crude approximation, and a number of scholars tried to obtain a more accurate value. Liu Xin (ca. 50 BCE–23 CE), a first-century astronomer and calendar expert, used several different values of π, such as 3.1547. Later, Zhang Heng (78–139 CE), studying the volumes of spheres, improved the traditional assumption that $\pi = 3$ by taking $\pi = \sqrt{10} = 3.162\ldots$. He also reportedly came up with the approximation $736/232 = 3.172\ldots$. Wang Fan (217–257) knew that Zhang Heng's values were too large and he obtained the better $142/45 = 3.155\ldots$.[23]

The most significant progress toward an approximation of π came from Liu Hui (ca. 220–ca. 280 CE). We do not know anything about the life of Liu other than that he was a mathematician in the Wie Kingdom after the fall of the Han dynasty. But his contributions were significant and exerted a lasting impact on Chinese mathematics.

In 263 he wrote a long and important commentary on *The Nine Chapters*. It is in these writings that we see his work on π. First, Liu justified the formula $A = \frac{1}{2}rC$. He began with a regular hexagon inscribed in a circle and then repeatedly doubled the number of sides to obtain regular polygons of 12, 24, and so on, sides. We will be more generic and just refer to n- and $2n$-gons. Suppose we have a regular n-gon and a regular $2n$-gon both inscribed in a circle of radius r with center O. Figure 6.3 shows one side of the n-gon, BC, and one side of the $2n$-gon, BD. Because OD is a perpendicular bisector of BC, the area of triangle OBD is $\frac{1}{2} \cdot OD \cdot BE = \frac{1}{2} \cdot r \cdot \frac{1}{2}s_n = \frac{1}{4}rs_n$, where s_n is the length of a side of the regular n-gon. The regular $2n$-gon is composed of $2n$ triangles congruent to OBD, so its area is $A_{2n} = 2n \cdot \frac{1}{4}rs_n = \frac{1}{2}rns_n = \frac{1}{2}rP_n$, where $P_n = ns_n$ is the perimeter of the regular n-gon. Liu (foreshadowing calculus) argued that as the number of sides increases, the perimeters of the polygons get closer to the circumference of the circle, and thus the lengths of the perimeters get closer to C. So, when there are infinitely many sides, the area formula becomes $A = \frac{1}{2}rC$.

Then Liu turned his attention to obtaining a good approximation of π. Again, he began with a regular hexagon and repeatedly doubled

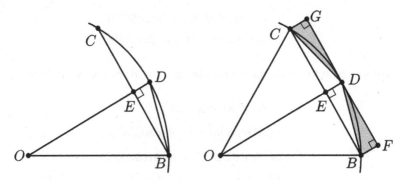

FIGURE 6.3. A side of a regular n-gon and a regular $2n$-gon inscribed in a circle (left). Liu Hui used the gray triangles to obtain an upper bound for the circular area.

the number of sides. The areas of the regular polygons approach the area of the circle, which, if we assume $r = 1$, is π.[24] He returned to the area formula $A_{2n} = \frac{1}{2}rns_n$ to find a method for obtaining the lengths of the sides, s_n. Turning again to figure 6.3 we see two right triangles: OBE and BDE. By applying the Pythagorean theorem (or the "Gougu theorem" as Liu knew it) to the two triangles and doing some algebra,[25] we find that

$$s_{2n} = \sqrt{r\left(2r - \sqrt{4r^2 - s_n^2}\right)}.$$

Thus, Liu could use these two formulas—for side length and area—to obtain a lower bound for π (because the polygon is a subset of the circle). But he was not content with this. He also showed how these ingredients could be used to obtain an upper bound. The key was to construct a rectangle on each side of the n-gon so that one side is tangent to the circle. Figure 6.3 shows one such rectangle ($BCGF$).

The circle is contained in the union of the regular $2n$-gon and these n rectangles. But the rectangles and the $2n$-gon intersect; so we are essentially adding the areas of $2n$ triangles, such as the two gray ones in figure 6.3. Moreover, triangles CDG, CDE, BDF, and BDE are all congruent. Thus, we have

Area(sector $OBDC$) < Area($OBFGC$)

$\qquad\qquad$ = Area($OBDC$) + Area(BDF) + Area(CDG)

$$= \text{Area}(OBDC) + \text{Area}(BCD)$$
$$= 2\,\text{Area}(OBDC) - \text{Area}(OBC).$$

Multiply by n to obtain the areas of the circle and the regular polygons:

$$A = n \cdot \text{Area}(\text{sector } OBDC)$$
$$< n\,(2\,\text{Area}(OBDC) - \text{Area}(OBC))$$
$$= 2n \cdot \text{Area}(OBDC) - n \cdot \text{Area}(OBC)$$
$$= 2A_{2n} - A_n.$$

Thus, Liu obtained the inequalities $A_{2n} < A < 2A_{2n} - A_n$, or, assuming $r = 1$, $A_{2n} < \pi < 2A_{2n} - A_n$.

Unlike Archimedes, who used a similar approach 500 years earlier, Liu had a decimal-based number system. So he was able to obtain the following bounds from an inscribed 192-gon:

$$\frac{98,157}{31,250} = 3.141024 < \pi < 3.142704 = \frac{196,419}{62,500}.$$

For simplicity, he gave the usable approximation $157/50 = 3.14$. Although we do not know for sure, he may also have computed the impressive approximation $3927/1250 = 3.1416$, which came from a 3072-gon!

Unlike his predecessors who used ad hoc techniques, Liu, like Archimedes, could obtain as precise an approximation as desired. All it required was doubling the number of sides enough times.

Later, a Chinese cartographer Zu Chongzhi (429–500), who had expertise in mathematics and astronomy, used Liu Hui's algorithm to compute π to eight digits: 3.1415926. He also proposed using $355/113 = 3.1415929\ldots$, which is accurate to a remarkable seven digits. He suggested the simpler $22/7$, which was Archimedes's bound.

The Biblical Value of π

The Old Testament of the Bible provides a detailed description of the temple that King Solomon built on the Temple Mount in Jerusalem in the tenth century BCE. The first book of Kings 7:23 describes

FIGURE 6.4. The brazen sea in King Solomon's temple. (I. Singer, ed., 1902, *The Jewish Encyclopedia, Vol. III*, New York: Funk and Wagnalls)

the "molten sea" or "brazen sea"—a metal basin that sat on the backs of 12 oxen and was used for the ablution of the priests (see figure 6.4):[26]

> And he [King Solomon] made the molten sea of 10 cubits from brim to brim, round in compass, and the height thereof was five cubits; and a line of 30 cubits compassed it round about.

In other words, this biblical verse gave the circumference of Solomon's vessel as 30 cubits and the diameter as 10. This implies that $\pi = C/d = 30/10 = 3$.

This verse was written between the tenth and sixth centuries BCE. Certainly the more accurate Babylonian and Egyptian values for π were available at the time. Even much later, the Talmud (a commentary on the Old Testament) stated, "Every [circle] whose circumference is three handbreadths, is one handbreadth wide."[27]

Religious believers defend the Bible's crude value of π in various ways.[28] Some point out that the passage is not a blueprint for Solomon's temple, so giving approximate lengths is perfectly reasonable; anyone needing a more precise value would use one. A mystical argument is that inside King Solomon's temple, the value of π *was* 3. A believer looking for a loophole argues that the measurements do not take into account the thickness of the brim: if the *inside* circumference was 30 cubits and the diameter of the container—brim and all—was 10 cubits, then the brim was 0.225 cubits (4 inches) thick.[29]

Still others look to numerology for evidence of a more exact value of π. Just as the Romans had some letters do double duty as numbers (I, V, X, L, C, D, and M), so did Hebrew letters double as numbers. Gematria is the process of assigning a numerical value to a Hebrew word and using these numerical values to draw additional meaning from a text. The word "line" in the quote from the first book of Kings above is written in a different way than it is read. Transliterating from the original Hebrew, "line" is written "qwh," but it is read "qw." The numerical equivalents of these letters are $q = 100$, $w = 6$, and $h = 5$, so qwh $= 111$ and qw $= 106$. If we multiply the given biblical value for π (which is 3) by $111/106$ we obtain a very good rational approximation for π: $333/106 = 3.141509\ldots$[30]

Martin Gardner (1914–2010), writing as "Dr. Matrix," gave this tongue-in-cheek numerological interpretation of π in the Bible:[31]

> One might suppose the Old Testament writers had no better estimate of pi than 3, but Dr. Matrix thinks otherwise. Consider the verse in which pi is first mentioned, 1 Kings 7:23. The initial 1, subtracted from the terminal 23, gives the ratio 7:22, and 22 divided by 7 is 3.14+, a fair approximation of pi. For a still better value, twice 7 is 14, half of 2 is 1, and twice 3 is 6. This gives 1416, the first four decimals of an excellent approximation of pi.

In a satirical response to a 1982 US District court case on creationism being taught in schools,[32] several professors at Emporia State University formed the Institute for Pi Research, which argued that the biblical value of π be given equal time in the classroom. Founder Samuel Dicks said, "To think that God in his infinite wisdom would create something as messy as this (3.14 and on) is a monstrous thought."[33]

Unfortunately, we will likely never know the true measurements of this vessel. As we find out in Jeremiah 52:17, in 586 BCE the brazen sea was broken down for scrap metal: "And the pillars of brass that were in the house of Jehovah, and the bases and the brazen sea that were in the house of Jehovah, did the Chaldeans break in pieces, and carried all the brass of them to Babylon."

TANGENT
The Great Pyramid

Cranks of the first kind say "I am right." Period. No reason necessary, no reason given. Cranks of the second kind say "I am right, because ...," and proceed to try to justify their work with mathematics. Cranks of the third kind recognize that what they have done conflicts with mathematics and therefore, since either they or mathematics are incorrect, they try to change mathematics to fit what they have accomplished. There are no cranks of the fourth kind.

—Underwood Dudley, *Mathematical Cranks*[1]

PYRAMIDOLOGISTS ASCRIBE ALL manner of biblical or mystical significance to the Egyptian pyramids. A favorite topic is that the design of the Great Pyramid of Giza, which was constructed some time around 2600 BCE, is closely related to π. This observation dates back to H. C. Agnew in 1838, but the popularity of the "π theory" is due to the sensational publications of John Taylor and Charles Piazzi Smyth (the Astronomer Royal of Scotland) in the 1850s and '60s.[2] Their conclusion was not that the ancient Egyptians were more mathematically advanced than previously thought, but rather that it proved that the pyramids were the result of God's handiwork. The pyramids concealed in their measurements deep secrets about mathematics and astronomy. Taylor's and Smyth's books are full of carefully chosen excerpts from the Bible that support their theories.

The observation is indeed striking. If we take the altitude of the pyramid and use it as the radius of a circle, then the circumference of the circle is very nearly the perimeter of the base of the pyramid. Said another way, if we built a hemisphere as tall as the pyramid, then the equator would be the same length as the perimeter of the base of the pyramid (see figure T.7). Indeed, the numbers check out—each side of the base pyramid measures $b = 230.36$ meters, so the perimeter is $4b = 921.46$ meters. The altitude of the pyramid is

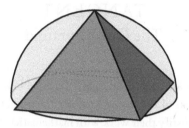

FIGURE T.7. If a hemisphere had the same height as a pyramid, they would have almost the same base perimeter.

$a = 146.64$ meters, so the circumference of the circle with that radius would be $2\pi a = 922.37$ meters![3] Moreover, if this was indeed the architect's intent, then we could use the altitude and the width of the pyramid to find his value of π. Solving $2\pi a = 4b$ for π, using the measured values of a and b, we obtain $\pi = 3.142$.

This is truly remarkable. About that there is no doubt. But this is simply a coincidence; the associated value of π is *too good*. There is no evidence that the ancient Egyptians possessed such an accurate value of π. Indeed, the famous Rhind Papyrus[4] (which is from 600 to 800 years after the pyramid's construction) has the value $\pi = 3.16$.

Of course, most likely this appearance of the value is an accidental by-product of another design decision. For example, when Herodotus traveled to Egypt in the fifth century BCE, he was told that the pyramid was designed so that the area of each face would equal the square of the altitude[5]—and the measurements support this claim. Suppose this was the design requirement for the pyramids and that the Egyptians knew nothing of π. Then suppose Taylor went in afterward to test out his "π theory." Such a pyramid would give an approximate value for π of $3.1446\ldots$, which is remarkably close, but not equal to, the true value of π—an amazing coincidence.[6] And that's all it is.

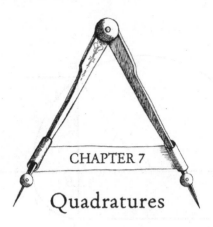

CHAPTER 7

Quadratures

Animals that draw chariots afford us a very simple demonstration of the
squaring of a circle, which is made by the wheels of these chariots by means
of the track of the circumference which forms a straight line.
—Leonardo da Vinci[1]

TO FULLY APPRECIATE the problem of squaring the circle, we
must understand how we can square *anything* using a compass
and straightedge. As we shall see, the Greeks were able to square
any polygon—regardless of its complexity—and certain regions with
curved boundaries.

The Quadrature of Polygons

To square a figure is to construct a square with the same area as the
figure using only a compass and straightedge. Through a string of
propositions in books I and II of *Elements*, Euclid showed how to
square a polygon. We present a sequence of constructions in a slightly
different order than Euclid did and give a slightly different method.
We break the task into three separate steps: first we square a rectangle,
then a triangle, then a general polygon.

Suppose we are given an arbitrary rectangle *ABCD*, as in figure 7.1.
If it is not already a square, then one side is longer than the other;

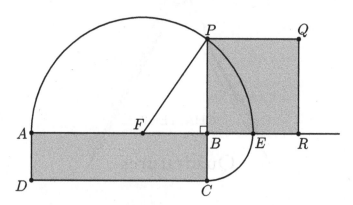

FIGURE 7.1. Squaring a rectangle.

let's say that AB is the longer side. Extend AB rightward. Construct a circle with center B through C; it intersects AB at E. Find the midpoint F of AE. Construct a circle with diameter AE and center F. Extend the line BC until it meets the circle at the point P. Construct a square on the segment BP, which we call $BPQR$. The following sequence of equalities proves that Area($ABCD$) = Area($BPQR$):

$$\begin{aligned}
\text{Area}(BPQR) &= BP \cdot BR \\
&= BP^2 \\
&= FP^2 - FB^2 && \text{by the Pythagorean theorem} \\
&= (FP - FB)(FP + FB) && \text{the difference of squares} \\
&= (FE - FB)(AF + FB) && \text{because } FP = FE = AF \\
&= BE \cdot AB \\
&= BC \cdot AB \\
&= \text{Area}(ABCD).
\end{aligned}$$

So we can square any rectangle. To square a triangle, it suffices to construct a rectangle with the same area as the triangle.[2] Consider triangle ABC. Drop a perpendicular from B to the point D on AC. (In figure 7.2, D is on the segment AC, but that need not be the case in general.) Find the midpoint E of BD. Construct a rectangle on AC with height equal to DE, which we call $APQC$. Because the area of a triangle is $\frac{1}{2} \cdot$ base $\cdot$ height, and BD is the height of ABC,

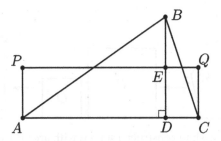

FIGURE 7.2. The rectangle *ACQP* has the same area as triangle *ABC*.

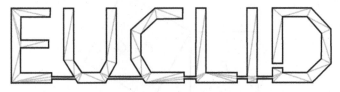

FIGURE 7.3. An 82-sided polygon.

$$\text{Area}(ABC) = \tfrac{1}{2} \cdot AC \cdot BD$$
$$= DE \cdot AC$$
$$= AP \cdot AC$$
$$= \text{Area}(ACQP).$$

We are now ready to attack a general polygon, even one that is extremely complicated, such as the 82-sided polygon in figure 7.3. We know we can divide an n-sided polygon into $n - 2$ triangular regions, and we can square all of these triangles. So we could take the 82-gon and create 80 squares, the areas of which sum to the area of the polygon. So then finally we ask: Can we turn a collection of squares into one large square with that same area?

It suffices to show that we can combine two squares of areas P and Q into one square of area $P + Q$, as shown in figure 7.4.[3] To do so, situate the squares corner to corner so that two sides meet at a right angle (as in figure 7.5). Connect the corners of the squares to make the hypotenuse of a right triangle, and construct a square on the hypotenuse. The Pythagorean theorem guarantees that the area of this square is $P + Q$.

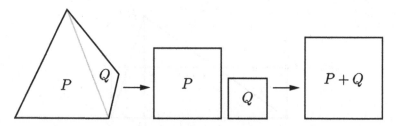

FIGURE 7.4. We wish to construct a square with area $P + Q$.

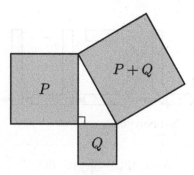

FIGURE 7.5. We may use the Pythagorean theorem to take squares of area P and Q and make a square of area $P + Q$.

Using this ingenious procedure, it's possible to square any rectilinear region. Proclus, after writing about being able to construct a parallelogram with area equal to a given rectilinear figure, wrote,[4]

It is my opinion that this problem is what led the ancients to attempt the squaring of the circle. For if a parallelogram can be found equal to any rectilinear figure, it is worth inquiring whether it is not possible to prove that a rectilinear figure is equal to a circular area.

But these techniques alone do not enable us to square a region with curved boundaries—not to mention a circle. To do so we need stronger geometric theorems.

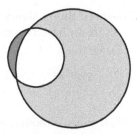

FIGURE 7.6. Two lunes (the shaded regions).

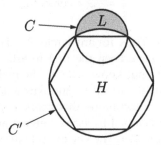

FIGURE 7.7. If we can square lune L, we can square circle C.

Hippocrates's Quadrature of Lunes

Today we'd say that a circle squarer is loony, but it may be that one of the first legitimate attempts to square the circle involved crescent-shaped *lunes*. A lune is the region bounded by two circular arcs both of which lie on the same side of their common chord (see figure 7.6).[5]

Hippocrates of Chios, who we recall from chapter 5 realized that the problem of doubling the cube was a special case of finding two mean proportionals, saw that he would be able to square a circle if he could square a certain lune.

Suppose we want to square the circle C in figure 7.7. Construct a hexagon H so one side of H is a diameter of C. Then circumscribe H with a circle C'. The circles determine a lune L. Hippocrates proved that if he could square L, then he could square C.

To prove this, he needed a result that related the area of a circle to the area of a square. This later appeared as proposition XII.2 of Euclid's *Elements*, "Circles are to one another as the squares on the diameters."[6] The Greeks viewed this statement geometrically, but we will express it arithmetically. If we have two circles with areas A_1 and A_2 and diameters d_1 and d_2, then

$$\frac{A_1}{A_2} = \frac{d_1^2}{d_2^2}.$$

This result is very familiar to us, although we are used to seeing it in a different form: if the circles have radii r_1 and r_2, then

$$\frac{A_1}{r_1^2} = \frac{A_2}{r_2^2} = \text{constant},$$

and this constant is π. Thus, for any circle, $A = \pi r^2$.

Hippocrates may have been the first to articulate this result, but even if he did, we do not know whether he had a rigorous proof of it. Archimedes credits the result to Eudoxus who proved it using his *method of exhaustion*—a technique that bears a strong resemblance to the calculus ideas developed by Isaac Newton and Gottfried Leibniz (1646–1716) 2000 years later. The key to Eudoxus's proof is that we can approximate a circle by many-sided inscribed and circumscribed polygons, and because an analogous result holds for them, the result holds for the circle. We will say more about this proof in chapter 8.[7]

Returning to Hippocrates's example, we recall that one side of the hexagon is the diameter of C. Moreover, if a hexagon is inscribed in a circle, as H is in C', then the side of the hexagon has the same length as the radius of the circle. So, by the previous proposition, Area(C') = 4 Area(C). Visually, as in figure 7.8, we think of four copies of C as six semicircles arranged along the sides of H together with one copy of C. If we subtract six segments of the circle (S in the diagram) from C' we obtain H. If we subtract them from the semicircles of C, we obtain six lunes (L). Putting all of this together,

$$\begin{aligned} \text{Area}(H) &= \text{Area}(C') - 6\,\text{Area}(S) \\ &= 4\,\text{Area}(C) - 6\,\text{Area}(S) \\ &= \text{Area}(C) + 6\left(\tfrac{1}{2}\,\text{Area}(C) - \text{Area}(S)\right) \\ &= \text{Area}(C) + 6\,\text{Area}(L). \end{aligned}$$

Equivalently, Area(C) = Area(H) − 6 Area(L).

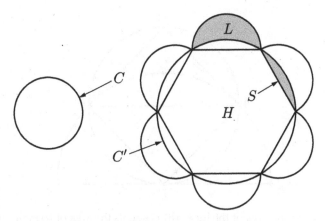

FIGURE 7.8. The area of the circle C equals the area of the hexagon H minus six copies of the lune L.

We know that H is squarable because it is a polygon. Thus, if—and this is a big if—the lune L is squarable, then so is the circle C.[8] One can imagine Hippocrates toiling day and night trying to square this particular lune. Alas, it was not to be. Although he was able to square three other lunes, this one remained stubbornly out of reach.

Let's look at one of his squarable lunes. Begin with a circle with inscribed square $ABCD$ (as in figure 7.9) and a circle with center D and radius AD. These circles determine a lune $ABCG$. Hippocrates proved that the area of the lune equals the area of triangle ACD and is thus squarable.

The first ingredient in the proof is the Pythagorean theorem, which implies that $AC^2 = AB^2 + BC^2$. Because $AB = BC = AD = ED = \frac{1}{2} EF$, this implies that $AC^2 = 2\left(\frac{1}{2} EF\right)^2 = \frac{1}{2} EF^2$.

The proposition on circular areas is true for semicircular areas, so

$$\frac{\text{Area(semicircle } ABC)}{\text{Area(semicircle } EGF)} = \frac{AC^2}{EF^2} = \frac{AC^2}{2\,AC^2} = \frac{1}{2}.$$

Thus Area(semicircle EGF) = 2 Area(semicircle ABC), and therefore Area(quadrant $ADCG$) = Area(semicircle ABC). Now subtract the area of the region shared by semicircle ABC and quadrant $ADCG$—that is,

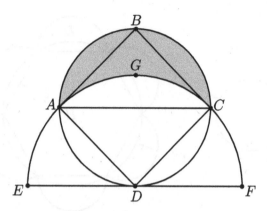

FIGURE 7.9. The area of the lune $ABCG$ equals the area of triangle ACD.

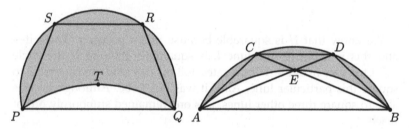

FIGURE 7.10. Two of Hippocrates's squarable lunes.

the segment ACG—to obtain

$$\text{Area(lune } ABCG) = \text{Area(triangle } ACD).$$

We know that any triangle is squarable, thus the lune is squarable.

Hippocrates also discovered the two other squarable lunes in figure 7.10, which are constructed from trapezoids with three congruent sides. The relative lengths of the sides on the left-hand trapezoid are

$$PS:RS:QR:PQ :: 1:1:1:\sqrt{3}.$$

The circular arc PQT is constructed so that it is similar to the arcs on the three congruent sides of the trapezoid. Hippocrates proved that

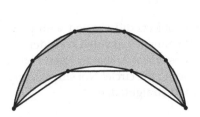

FIGURE 7.11. Two more squarable lunes.

the area of the lune equals the area of the trapezoid. The right-hand lune is even more complicated to describe as it involves the diagonals of the trapezoid. The trapezoid is constructed so that

$$AC:CD:DB:AE:BE :: \sqrt{2}:\sqrt{2}:\sqrt{2}:\sqrt{3}:\sqrt{3}.$$

Hippocrates proved that the area of the lune is equal to the area of the pentagonal region *ACDBE*.

It is natural to ask whether there are other squarable lunes. In 1724 Daniel Bernoulli (1700–1782) proved that there are infinitely many squarable lunes.[9] However, this allows us to count nonconstructible lunes. That is, it may be possible to square a lune with a compass and straightedge, but the lune itself cannot be constructed with these tools. What if we rephrase the question to ask whether there are other constructible lunes (and all of Hippocrates's lunes are constructible) that can be squared?

In 1771 Euler, who was a friend of Daniel Bernoulli, extended Bernoulli's work and in the process discovered two more constructible lunes that are squarable.[10] Euler proved that the lunes shown in figure 7.11 are equal in area to the polygons defined by the chords. Although Euler did not know it, these lunes were not new: they had appeared five years earlier in the graduate dissertation of Daniel Wijnquist, a Finnish mathematician who studied under Martin Wallenius (1730–1772).[11]

Euler conjectured that there are no more than these five.[12] In 1840 Thomas Clausen (1801–1885), who was aware of only one of Hippocrates's squarable lunes, rediscovered the set of five lunes. He made the same conjecture.[13] The Euler–Clausen conjecture was proved following a string of partial results by five mathematicians from 1902 to

$2003.^{14}$ Alas, the circle-squaring lune of Hippocrates is not one of these five.

The Greeks' collection of squarable objects did not end with polygons and lunes. We shall see that Archimedes devoted considerable attention to the problem of quadratures. He was able to square other shapes with curved boundaries, including the circle! Not surprisingly, it was not a quadrature by compass and straightedge.

TANGENT
Leonardo da Vinci's Lunes

From what I hear, Leonardo's life is very irregular and uncertain,
and he seems to live for the day only.... He devotes much of his
time to geometry, and has no fondness at all for the paintbrush.
—Friar Pietro da Novellara, in an April 3, 1501, letter to Isabella d'Este,
who wished to have Leonardo paint her portrait.[1]

EVEN LEONARDO DA VINCI (1452–1519), the great Renaissance
polymath, got in on the action. Leonardo became interested in geometry late in his life; he began studying Euclid and Archimedes some
time between 1496 and 1504, when he was about 50 years old. But once
he started, he became a passionate geometer. And, not surprisingly,
he put his own unique, artistic spin on his work.

Like many who came before him and many who came afterward,
Leonardo was drawn to the famous classical problems of antiquity.
Hundreds of his notebook pages are devoted to doubling the cube and
squaring the circle. Leonardo biographer Martin Kemp wrote that the
quadrature problems[2]

> became an intellectual itch which [Leonardo] found impossi-
> ble to scratch satisfactorily. Each new bout of scratching only
> served to stimulate fresh itches. Even the most devoted admirer
> of Leonardo must wonder if the whole matter had got rather out
> of hand.... There can be little doubt that his motivation in these
> studies was a complex compound of intellectual and aesthetic
> satisfactions. Even the intellectual aspects ultimately assumed
> the air of conundrums or games, rather than belonging to the
> mainstream of mathematical science.

The following note from November 30, 1504, was written vertically
among some of Leonardo's geometrical diagrams:[3]

> The night of St. Andrew's Day I came to the end with the squaring
> of the circle: and it was the end of the light and of the night, and

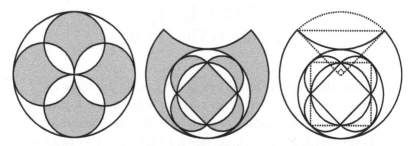

FIGURE T.8. The shaded regions on the left and in the center are squarable.

of the paper on which I was writing; the conclusion came at the end of the hour.

It is not clear what he meant by this—that he thought he'd succeeded or that he'd simply given up for the night.

In his study of quadrature problems, he investigated areas of circles, squares, hexagons, lunes, lenses (the intersection of two circles), and sectors and segments of circles. In one two-page, pen-and-ink composition, Leonardo drew 176 circles and semicircles, each containing lunes, segments, and so on (ca. 1513). Each figure is accompanied by information about the relative areas of the shaded regions, nonshaded regions, and the whole.[4]

For example, Leonardo showed that the shaded regions in the first two parts of figure T.8 are squarable (we've redrawn the second one with more detail to show how it is constructed). If the outer circles have radius 1, then the area of the shaded regions in each example is 2, and it is easy to construct a square with area 2. We leave these fun geometrical exercises as a challenge to the readers.[5]

Although he did not make any lasting mathematical contributions, Leonardo rediscovered a beautiful result about lunes that was first observed by the Muslim mathematician and astronomer Hasan ibn al-Haytham (965–ca. 1040), often known as Alhazen. If we take any right triangle and draw semicircles on all three sides, as shown in figure T.9, then they form two lunes. The lunes may not be squarable individually, but Leonardo and Alhazen proved that in every such case, the sum of the areas of the lunes is equal to the area of the triangle. Thus, together the two lunes are squarable.

The proof is quite short if one knows Euclid's proposition VI.31, which is a little-known gem—a generalization of the Pythagorean

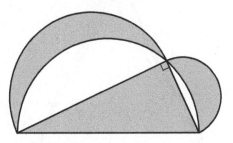

FIGURE T.9. The lunes of Alhazen.

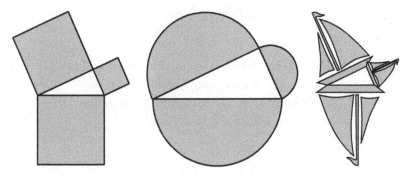

FIGURE T.10. The area of a figure on the hypotenuse of a right triangle equals the sum of the areas of similar figures on the legs.

theorem. It says that if we put similar figures—any similar figures—on the three sides of a right triangle, then the area of the figure on the hypotenuse equals the sum of the areas on the legs. In figure T.10, we see this illustrated with squares (the usual Pythagorean theorem), semicircles, and sailboats.

The lune example is simply the semicircle example, except the semicircle on the hypotenuse is drawn inward instead of outward. Let A, B, C, and T be the areas of the semicircles on the two legs, the semicircle on the hypotenuse, and the triangle, respectively. The generalized Pythagorean theorem gives $A + B = C$. The two lunes can be obtained by taking the region formed from two smaller semicircles and the triangle and removing the large semicircle. Algebraically, this says that the sum of the areas of the lunes is $(A + B + T) - C = C + T - C = T$.

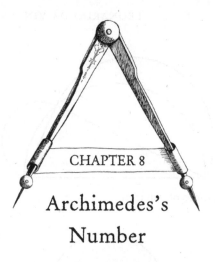

CHAPTER 8

Archimedes's Number

If the Diameter be 7 Inches in length, then the Circumference
is twenty two Inches.
—William Mather, *The Young Man's Companion:
Or, Arithmetick Made Easie*, 1710[1]

IN CHAPTER 6 WE learned that many societies discovered the constant that we now call π. They knew either that for any circle, C/d is a constant value that is approximately 3 or that A/r^2 is a constant of about the same value.

Proving these theorems and obtaining good bounds on π was not easy. The proofs required Eudoxus's method of exhaustion. And to obtain a solid understanding of π—both as the circle constant and the sphere constant—required the clever and careful work of the greatest Greek mathematician, Archimedes.[2] Some even deduced that these constants were equal. But we have no evidence that they tried to prove these facts. Moreover, the approximations were extremely crude and did not come from a careful mathematical investigation of the properties of circles.

The Area of a Circle

In chapter 7 we saw that given any two circles with areas A_1 and A_2 and diameters d_1 and d_2, $A_1/A_2 = d_1^2/d_2^2$. This result was used by Hippocrates, proved by Eudoxus, and published by Euclid in his *Elements*. Eudoxus began his proof with an inscribed and a circumscribed square and repeatedly doubled the number of sides to obtain regular polygons that were as close to the circle as required. Although the details of the proof are subtle and clever, the idea of approximating the circle by polygons was not original to Eudoxus.

The idea may be due to Antiphon the Sophist (480–411 BCE), an orator and statesman, in his attempts to square the circle. Antiphon began with either an inscribed equilateral triangle or a square and repeatedly doubled the sides to obtain polygons that became more and more circular. According to later writers, Antiphon believed that if he doubled the number of sides enough times, the resulting inscribed polygon would *be* a circle. Simplicius wrote,[3]

> Antiphon thought that in this way the area of the circle would be used up, and we should some time have a polygon inscribed in the circle the sides of which, owing to their smallness, coincide with the circumference of the circle. And as we can make a square equal to any polygon ... we shall be in a position to make a square equal to a circle.

Aristotle believed Antiphon was using improper geometric techniques, and thus geometers should not even give his work the benefit of a close look. He wrote, "Thus it is the geometer's business to refute the quadrature by means of segments, but it is not his business to refute that of Antiphon."[4] More recent scholars argue that Antiphon's arguments were misunderstood.[5] Regardless, Antiphon had the essential idea behind the method of exhaustion, which would be used so effectively by Eudoxus and Archimedes.

Later, the sophist Bryson of Heraclea (fl. late fifth century BCE) used the areas of both inscribed and circumscribed polygons to bound the area of the circle. He thought that the true area could be obtained from them, presumably using some sort of averaging technique.[6] Aristotle did not like this approach either, calling the arguments "sophistic" and "eristic."[7] Yet Archimedes would later use a similar technique (using perimeters instead of areas) to obtain bounds for π.

Eudoxus built upon these ideas to prove that $A_1/A_2 = d_1^2/d_2^2$. We will sketch this proof as it appeared in book XII of *Elements*. It is a double proof by contradiction. For the sake of contradiction, assume that $A_1/A_2 \neq d_1^2/d_2^2$. This leads to two possibilities: either $A_1/A_2 < d_1^2/d_2^2$ or $A_1/A_2 > d_1^2/d_2^2$. The first case implies that there is some area $S < A_2$ such that $A_1/S = d_1^2/d_2^2$. Begin with a square inscribed in the second circle, and double the number of sides to obtain an octagon, double again to get a 16-gon, and so on. By proposition XII.16, after each doubling the difference between the areas of the circle and the polygon decreases by more than half. Thus eventually[8] we obtain a polygon P_2 inscribed in the second circle such that $S < \text{Area}(P_2) < A_2$. Next inscribe a polygon P_1 in the first circle that is similar to P_2. By proposition XII.1, $\text{Area}(P_1)/\text{Area}(P_2) = d_1^2/d_2^2$, and by our assumption, this equals A_1/S. This implies that $S/\text{Area}(P_2) = A_1/\text{Area}(P_1)$. But because $A_1 > \text{Area}(P_1)$, it follows that $S > \text{Area}(P_2)$, which is a contradiction. Euclid made quick work of the second case by transforming it into the first case. It follows that the ratios must be equal.

By the start of the third century BCE, the Greeks were able to prove only this one result about circles. In essence, they proved that A/r^2 is a constant. But they did not have a rigorous proof that C/d was also a constant, nor that they were the same one. For this, they had to wait for the supreme genius of Greek mathematics: Archimedes.

Archimedes of Syracuse

We know when Archimedes died—in 212 BCE at the hands of a Roman soldier. But we do not know when he was born. His friend Heracleidese wrote a biography that, alas, is lost. According to a poem written in the twelfth century CE by Johann Tzetzes (ca. 1110–1180), Archimedes was 75 years old when he died. So, by this admittedly flimsy evidence, we give 287 BCE as his birth year.

Archimedes was born in Syracuse, on the eastern side of Sicily. We believe that his father was an astronomer and his grandfather was an artist. Reviel Netz observed that Archimedes's name comes from *arche*, which means "principle, rule, number one," and *medos*, which means "mind, wisdom, wit." So Archimedes likely means "the mind of the principle." This, he argued, is a reference to Archimedes's father's interest in the "new religion of beauty and order in the cosmos."[9]

Although we have no firm evidence, we believe Archimedes studied in Alexandria. Later, when he was living and working in Syracuse, he wrote letters to Eratosthenes (most well known for calculating the circumference of the earth), Conon (an astronomer), and Dositheus (about whom we know little) in Alexandria to spread his work to a wider audience, which was likely still very small.

Like many great historical figures, Archimedes has been the subject of fantastical, often apocryphal, tales. It is difficult to know what is true, what is fiction, and what is embellished truth. Some tales—such as his naked run in the streets screaming "Eureka!" after devising a way to use water submersion to detect gold forgeries[10] and his invention of a weapon of mirrors that could concentrate the power of the sun to set invading ships on fire[11]—are likely false.

However, even if we discard these stories, we are left with a picture of a creative genius who was equally comfortable proving the deepest theorems of pure mathematics as understanding the physical world and inventing useful machines.

His feats of engineering are legendary. He invented a contraption (now known as the Archimedean screw) that gives a mechanical process for pulling water up from the depths. He devised powerful weapons of war. His interest in astronomy and engineering led to the creation of a mechanism that showed the motion of the sun, the moon, and the five planets, and a device that could measure the apparent diameter of the sun. He discovered that by using compound pulleys, a small force can be used to move heavy objects. Plutarch described how Archimedes demonstrated this for King Hiero:[12]

Archimedes therefore fixed upon a three-masted merchantman of the royal fleet, which had been dragged ashore by the great labours of many men, and after putting on board many passengers and the customary freight, he seated himself at a distance from her, and without any great effort, but quietly setting in motion with his hand a system of compound pulleys, drew her towards him smoothly and evenly, as though she were gliding through the water.

This yielded one of Archimedes's famous remarks, as repeated by Pappus: "Give me a place to stand on, and I will move the earth."[13]

Netz referred to Archimedes as both "playful" and "sly." He once gave two false theorems—on purpose—to try to catch others who claimed his results as their own. He wrote that he did this "so that those who claim to discover everything, producing no proof themselves, will be confuted, in their assenting to prove the impossible."[14]

Whether accurate or fanciful, Plutarch, in his "Life of Marcellus," gave a romantic portrait of Archimedes as an absentminded thinker who, despite being an engineering genius, loved pure mathematics above all. He wrote that we should not doubt the stories of[15]

how, under the lasting charm of some familiar and domestic Siren, he forgot even his food and neglected the care of his person; and how, when he was dragged by main force, as he often was, to the place for bathing and anointing his body, he would trace geometrical figures in the ashes, and draw lines with his finger in the oil with which his body was anointed, being possessed by a great delight, and in very truth a captive of the Muses.

Moreover,[16]

Archimedes possessed such a lofty spirit, so profound a soul, and such a wealth of scientific theory, that although his inventions had won for him a name and fame for superhuman sagacity, he would not consent to leave behind him any treatise on this subject, but regarding the work of an engineer and every art that ministers to the needs of life as ignoble and vulgar, he devoted his earnest efforts only to those studies the subtlety and charm of which are not affected by the claims of necessity. These studies, he thought, are not to be compared with any others; in them the subject matter vies with the demonstration, the former supplying grandeur and beauty, the latter precision and surpassing power.

The Circumference Constant

We typically define π as the quotient of the circumference of a circle and its diameter. Stated another way, $C = \pi d$, or perhaps more familiarly, $C = 2\pi r$. But hidden in this definition is a theorem: regardless of what circle we start with, C/d will yield the same number. However, this famous relation—and hence, implicitly, the theorem behind it—has no name attached to it. If one were to ask a handful of mathematicians, "Who first proved that C/d is a constant?," their responses would likely be "Isn't that in Euclid's *Elements*?" or "It is obvious; all circles are similar," or, most frequently, "I don't know."

One expects to see this proposition in Euclid's *Elements*, perhaps stated as "The circumferences of circles are to one another as their diameters." (That is, $C_1/C_2 = d_1/d_2$.) But on this topic, Euclid was silent. Nevertheless, it is likely that Euclid and his predecessors knew it—it was inherited knowledge. On the circular motion of the stars, Aristotle wrote, "It is not at all strange, nay it is inevitable, that the speeds of the circles should be in the proportion of their sizes."[17] When Aristotle referred to the size of a circle, he meant the diameter; so the circumference is proportional to the diameter. Thus, we may suspect that Eudoxus, Euclid, and the rest knew the theorem but were unable to prove it.

The argument that "all circles are similar" is intriguing and has some merit. If we scale a figure by a factor of k, then all lengths increase by this same factor. A circle with circumference C and diameter d becomes a circle with circumference kC and diameter kd, and $(kC)/(kd) = C/d$. This "obvious" argument probably explains why the circumference constant was rediscovered by so many cultures. But this response is mathematically unsatisfying; it is not a simple process to turn it into a rigorous proof. Euclid defined similarity for polygons but not for circles or other curves. It is not clear how he would have done so. He could have said that two circles are similar provided $C_1:d_1 :: C_2:d_2$, but that is what we want to prove! Moreover, we need to know how to measure the lengths of curves—or, in the geometric setting more familiar to Euclid, to compare the lengths of curves. And this was not covered in Euclid's *Elements*.

The theorem on circular areas was proved before the theorem on circumferences because arc length is inherently more complicated than area (just ask any first-year calculus student!).[18] Gottfried Leibniz wrote that "areas are more easily dealt with than curves, because they can be cut up and resolved in more ways."[19] In particular, Euclid could apply his fifth common notion—"The whole is greater than the part"[20]—to areas, but not to curves. (*Elements* began with five common notions. These differed from his postulates in that they were intended to be common to all demonstrative sciences, not just to geometry.) If a triangle is inscribed in a circle, as in figure 8.1, the filled-in triangle is a subset of the circular disk. By Euclid's common notion, the area of the triangle is less than the area of the circle. Although it is intuitively clear that the circumference of the circle is greater than the perimeter

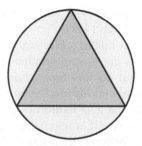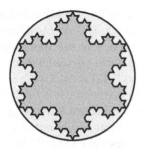

FIGURE 8.1. An equilateral triangle and the beginning of a Koch snowflake, inscribed in a circle.

of the triangle, Euclid's common notion does not guarantee this, and it is not true in general for inscribed polygons. The perimeter of the polygon on the right in figure 8.1 is greater than the circumference of the circle. In fact, this shape is the fourth iteration in the construction of a fractal called the Koch snowflake, which has finite area but infinite perimeter!

The theorem that C/d is a constant is almost certainly due to Archimedes. Clearly he was aware of the result; it is implicit in and can be easily proved from his mathematics. But we have no record that he stated it explicitly. The treatise that would most likely contain the theorem, *Measurement of a Circle*, has not come down to us fully intact. The extant version has only three propositions and is not a faithful copy of Archimedes's original (one of the three results is clearly incorrect as stated). Dijksterhuis referred to it as "scrappy and rather careless."[21] In the version we have, Archimedes proved a result that, when combined with the theorem on circular area in *Elements*, can be used to establish that C/d is constant. Dijksterhuis pointed out that "it is quite possible that the fragment we possess formed part of a longer work, which is quoted by Pappus under the title *On the Circumference of the Circle*, and that the latter also dealt with the more general question as to the ratio between the length of an arc of a circle and that of its chord."

The first proposition of *Measurement of a Circle* is "The area of any circle is equal to a right-angled triangle in which one of the sides about the right angle is equal to the radius, and the other to the circumference, of the circle."[22] That is, $A = \frac{1}{2}Cr$ (see figure 8.2).

An observant reader may gasp at the sight of this proposition and wonder whether Archimedes succeeded in squaring the circle. He was, after all, able to show that the area of a circle equals the area of a

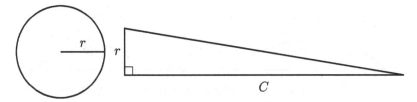

FIGURE 8.2. The area of a circle with radius r and circumference C equals the area of a right triangle with legs r and C.

triangle, and it is easy to use a straightedge and compass to draw a square with the same area as a triangle. Isn't this the circle squaring that the Greeks were seeking? It is not, and Archimedes never claimed it was. The triangle is not constructible: It is easy to draw a segment with length equal to the radius. The problem arises when we have to draw the base of the triangle, which is the length of the circumference. This is the problem of rectifying the circle. Indeed, an implication of this theorem is that the problems of squaring and rectifying the circle are equivalent. If it was possible to rectify the circle, we could construct Archimedes's triangle, and from this, a square with the same area. Conversely, if it was possible to square a circle, we could construct the triangle, and thereby rectify the circle.

With Archimedes's result and Eudoxus's theorem about circular areas, we are but a few algebraic steps away from proving that C/d is a constant and that it is the same as the area constant:

$$\frac{C}{d} = \frac{2A/r}{2r} = \frac{A}{r^2} = \pi.$$

That's it. The numbers C/d and A/r^2 are equal and are the same value for every circle. That is what we now call π.

In his proof, Archimedes used two key inequalities without justification: if a circle has circumference C, and P_{in} and P_{circ} are the perimeters of inscribed and circumscribed regular polygons, then

$$\text{Perimeter}(P_{in}) < C < \text{Perimeter}(P_{circ}).$$

These inequalities would follow if, as in figure 8.3, given a chord AB in a circle, the included arc ADB, and two segments tangent to the circle, AC and BC, we knew that $AB < ADB < AC + BC$.

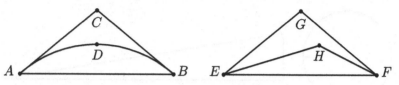

FIGURE 8.3. Archimedes originally assumed $AB < \text{Arc}(ADB) < AC + BC$, whereas Euclid proved $EF < EH + FH < EG + FG$.

One may argue that these inequalities are obvious, especially the first, because a line is the shortest distance between two points. Yet Euclid proved both inequalities for line segments, namely that in figure 8.3, $EF < EH + FH < EG + FG$.[23] According to Proclus, the Epicureans scoffed at the first inequality—called the *triangle inequality*—saying it was so intuitive that even an ass knew it was true: if you put food at one vertex of a triangle and an ass at another, it would walk along the straight line between them and not along the other two.

We believe that a younger, less sophisticated Archimedes took these inequalities for granted, but later in his career realized that they were not trivial. He observed that Euclid's five postulates were not enough to prove the theorem. In his later work *On the Sphere and Cylinder*, he added two new postulates.[24]

He began *On the Sphere and Cylinder* by stating that a curve is *concave in the same direction* "if any two points whatever being taken, the straight lines between the two points either all fall on the same side of the [curve], or some fall on the same side, and some fall on the [curve] itself, but none on the other side."[25] Curves ABC, ADC, and AC in figure 8.4 are concave in the same direction but AEC is not.

Then he stated the following postulates:[26]

1. That among [curves] which have the same limits, the straight line is the smallest.

2. And, among the other [curves] (if, being in a plane, they have the same limits): that such [curves] are unequal, when they are both concave in the same direction and either one of them is wholly contained by the other and by the straight line having the same limits as itself, or some is contained, and some it has as common; and the contained is smaller.

Returning to figure 8.4, (1) implies that AC is shorter than ABC, ADC, and AEC, and (2) implies that ABC is shorter than ADC.[27]

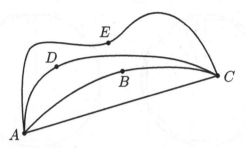

FIGURE 8.4. Archimedes's axioms imply that AC is shorter than ABC, which is shorter than ADC.

With these postulates in place, Archimedes was ready to make rigorous statements about lengths of curves. First, "Assuming these it is manifest that if a polygon is inscribed inside a circle, the perimeter of the inscribed polygon is smaller than the circumference of the circle; for each of the sides of the polygon is smaller than the circumference of the circle which is cut by it."[28] And second, "If a polygon is circumscribed around a circle, the perimeter of the circumscribed polygon is greater than the perimeter of the circle."[29]

We are now ready to sketch Archimedes's proof that $A = \frac{1}{2}Cr$, where the circle has radius r, area A, and circumference C.[30] Archimedes used the method of exhaustion. Let T be a right triangle with legs of length r and C. Suppose the theorem is false; that is, Area$(T) \neq A$. We will show that Area$(T) < A$ and Area$(T) > A$ are both impossible.

Suppose Area$(T) < A$; that is, $A -$ Area$(T) > 0$. Using a result from *Elements*, Archimedes began with an inscribed square and repeatedly doubled the number of sides until he obtained an inscribed regular n-gon P_{in} such that $A -$ Area$(P_{in}) < A -$ Area(T). So Area$(P_{in}) >$ Area(T). An *apothem* of a regular polygon is a segment from the center of the polygon that is perpendicular to one of its sides. Let r' be the length of the apothem of P_{in}. Then $r' < r$ (see figure 8.5). We can divide our n-gon into n triangular pieces, and the area of each of these triangles is $\frac{1}{2} \cdot$ base $\cdot$ height $= \frac{1}{2}sr'$, where s is the length of one side of the polygon. Because Perimeter$(P_{in}) = ns$, Perimeter$(P_{in}) < C$, and $r' < r$, it follows that

$$\text{Area}(P_{in}) = n\left(\tfrac{1}{2}sr'\right)$$
$$= \tfrac{1}{2}r'(ns)$$

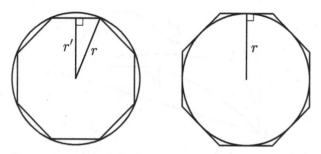

FIGURE 8.5. The lengths of the apothems of an inscribed and a circumscribed polygon are r' and r, respectively.

$$= \tfrac{1}{2}r' \, \text{Perimeter}(P_{\text{in}})$$
$$< \tfrac{1}{2}r'C$$
$$< \tfrac{1}{2}rC$$
$$= \text{Area}(T).$$

But $\text{Area}(P_{\text{in}}) > \text{Area}(T)$. So this is a contradiction.

Next, suppose $\text{Area}(T) > A$. Archimedes began with a circumscribed square, and he repeatedly doubled the numbers of sides to obtain a sequence of circumscribed regular polygons. He proved that each time the number of sides doubles, the difference between the areas of the polygon and the circle is more than cut in half. Thus eventually this process yields a circumscribed regular polygon P_{circ} such that $\text{Area}(P_{\text{circ}}) - A < \text{Area}(T) - A$. So $\text{Area}(P_{\text{circ}}) < \text{Area}(T)$. In this case, the apothem of P_{circ} is the radius of the circle. Arguing as before,

$$\text{Area}(P_{\text{circ}}) = \tfrac{1}{2}r \cdot \text{Perimeter}(P_{\text{circ}}) > \tfrac{1}{2}rC = \text{Area}(T),$$

which is a contradiction. Thus, the theorem is proved.

Archimedes's Bounds on π

Although Archimedes never stated the invariance of C/d explicitly in *Measurement of a Circle*, it is loosely implied by the third proposition, in which he gave his famous bounds:[31] $223/71 < C/d < 22/7$.

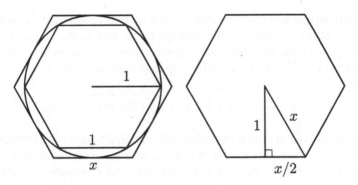

FIGURE 8.6. An inscribed and a circumscribed hexagon.

As with the previous theorem, Archimedes used inscribed and circumscribed polygons to approximate the circle. In the previous proof he used the areas of the polygons, but for this one he used perimeters. He began with regular hexagons and doubled the number of sides four times to obtain a pair of 96-sided polygons! Figure 8.6 shows a circle with an inscribed and a circumscribed hexagon. We do not show inscribed and circumscribed 96-gons because on these book pages they would be so close to circular that they would be indistinguishable from the inked circular curve on the page.[32]

Consider a hexagon inscribed in a circle of radius 1. The side of an inscribed hexagon is precisely the radius of the circle, so the perimeter of the hexagon is 6. This is a crude approximation of the circumference of the circle, which is 2π. So this polygon gives a lower bound of 3 for π. Next let x be the length of a side of a circumscribed hexagon (which is also the distance from the center of the circle to one of the vertices), as in figure 8.6. The apothem of the hexagon is 1, and by the Pythagorean theorem, $x = \frac{2}{3}\sqrt{3}$. Hence, $\pi < \frac{1}{2}(6 \cdot \frac{2}{3}\sqrt{3}) = 2\sqrt{3}$. Thus, the hexagons yield the following bounds: $3 < \pi < 2\sqrt{3} = 3.464\ldots$.

Of course, we can plug $\sqrt{3}$ into a calculator to obtain an accurate rational approximation, but Archimedes did not have this luxury. In fact, his numerical system was much more cumbersome than our Hindu–Arabic numerals, so it was extremely challenging to come up with a good approximation using hand calculations. It is a much-discussed mystery how he did so, but Archimedes gave, without explanation, the remarkable bounds $265/153 < \sqrt{3} < 1351/780$, which are correct to four decimal places.[33]

But this was just the beginning. As Archimedes continued to double the numbers of sides of the polygons, he repeatedly applied the Pythagorean theorem and took square roots of ever larger values. We see, for instance, Archimedes's approximations such as

$$\sqrt{5{,}472{,}132 + \tfrac{1}{16}} > 2339 + \tfrac{1}{4} \quad \text{and} \quad \sqrt{4{,}069{,}284 + \tfrac{1}{36}} < 2017 + \tfrac{1}{4}.$$

In the end he obtained the remarkable rational bounds $25{,}344/8069 < C/d < 29{,}376/9347$ for π.[34] However, as pleased as he must have felt after accomplishing this, the two fractions would be quite difficult to remember. So, at the expense of widening the bounds slightly, he gave the much more elegant bounds[35]

$$\frac{223}{71} < \frac{25{,}344}{8069} < \frac{C}{d} < \frac{29{,}376}{9347} < \frac{22}{7}.$$

Archimedes's bounds were just the beginning. Generations of so-called digit hunters would devote hours, days, weeks, and years to computing more and more digits of π. Until the middle of the seventeenth century and the invention of calculus, all digit hunters used Archimedes's technique of inscribing and circumscribing a circle with polygons.

Spheres

Geometry students know that π is also a sphere constant; it shows up in the formulas for both the volume and the surface area of the sphere. We thank Archimedes for these too.

Elements contained some results on the volumes of solid objects. Euclid's proofs about the volumes of parallelepipeds (box-like figures with parallelograms as faces) and prisms do not differ too much from his proofs about the areas of polygons. That is, these proofs don't require any significantly new ideas beyond those he used in his planar geometry.

But the more complicated solid objects required something different—Eudoxus's method of exhaustion. In book XII Euclid proved several of Eudoxus's theorems about volumes using this technique. He proved that the volume of a triangular pyramid is 1/3 the volume of a prism with the same base and altitude (see figure 8.7). He also

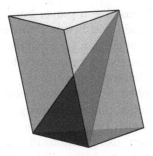 

FIGURE 8.7. The volume of a pyramid is 1/3 the volume of the prism containing it. The volume of a cone is 1/3 the volume of the cylinder containing it.

proved that the volume of a cone is 1/3 the volume of the cylinder that contains it. In modern terminology, if the height of the cone is h and the radius of the base is r, then the volume is $\frac{1}{3}\pi r^2 h$.

The sphere was even trickier than the pyramid and the cone, and it stumped Eudoxus. The best he could do was to prove that the volumes of two spheres are in the same proportion as the cubes of their diameters (that is, $V_1:V_2 :: d_1^3:d_2^3$). This is analogous to his theorem about the areas of circles (see page 109). And just as that theorem implied that there is an area constant for circles, this theorem asserts that there is a volume constant for spheres (V/d^3).

Moreover, just as arc length is trickier than planar area, so too is surface area more subtle than volume. The mathematicians before Archimedes had nothing to say about the surface area of a sphere.

This was the state of mathematics when Archimedes attacked the sphere problems. In some cases, Archimedes may have proved results that were known to mathematicians who preceded him but that they were unable to prove. That may have been the case with the theorem relating the area of a circle to its circumference. However, for the volume and the surface area he had to come up with the statement of the theorem and prove it.

In most mathematical writing we see only the final product—a theorem and its proof—and have no information about how the mathematician discovered the result. We are fortunate that Archimedes left us this information about his sphere results. His treatise *On the Method of Mechanical Theorems* was long thought lost, but in 1906 it was discovered in a palimpsest—a tenth-century Greek copy of Archimedes's work that was erased, rebound, and reused as a prayer book in the thirteenth century. After being lost again and nearly destroyed by mold, it was purchased anonymously at auction for $2 million, and from 1998 to 2008 was painstakingly analyzed by conservators and physicists using advanced imaging technology.

In *The Method*, Archimedes described his ingenious method of discovering these results using, of all things, physics. He imagined cutting the solid objects into thin slices (much as Bonaventura Cavalieri [1598–1647] did many years later). Then, using notions of centers of gravity and the law of the lever, he obtained the volume and surface area of a sphere. However, Archimedes did not recognize this approach as fully rigorous, so he went on to give geometric proofs of the results in *On the Sphere and Cylinder*.

Today we present the volume and surface area of a sphere using formulas, but Archimedes related them to the volumes and areas of known objects. He stated that "every sphere is four times a cone having a base equal to the greatest circle of the circles in the sphere, and, as height, the radius of the sphere."[36] Thanks to Eudoxus (and Euclid) we know that the volume of a cone is $\frac{1}{3}\pi r^2 h$. So, in essence, Archimedes's result gives us the formula $V = \frac{4}{3}\pi r^3$. On the surface area of a sphere, he stated, "The surface of every sphere is four times the greatest circle of the circles in it."[37] That is, the surface area is four times the area of the circular disk bounded by the equator. So $S = 4\pi r^2$.

Archimedes then brought all four of his results—the circumference and area of the circle and the volume and surface area of the sphere— together in one concise, elegant corollary: "And, these being proved, it is obvious that every cylinder having, as base, the greatest of the circles in the sphere, and a height equal to the diameter of the sphere, is half as large again as the sphere, and its surface with the bases is half as large again as the surface of the sphere."[38] In other words, for a sphere inscribed in a cylinder (including its top and bottom disks), as in figure 8.8,

$$\frac{3}{2} = \frac{\text{Volume (cylinder)}}{\text{Volume (sphere)}} = \frac{\text{Surface area (cylinder)}}{\text{Surface area (sphere)}}.$$

Although this looks like it is a statement about two three-dimensional figures—the sphere and the cylinder—the cylinder is a way of representing the circle. The volume of the cylinder is the area

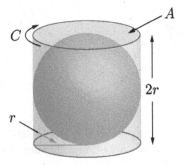

FIGURE 8.8. Both the volumes and the surface areas of a sphere in a cylinder are in 3:2 ratios.

of the circular disk (A) times the height of the cylinder ($2r$). So

$$\frac{\text{Volume (cylinder)}}{\text{Volume (sphere)}} = \frac{2rA}{\frac{4}{3}\pi r^3} = \frac{2r \cdot \pi r^2}{\frac{4}{3}\pi r^3} = \frac{3}{2}.$$

Similarly, the surface area of the cylinder is the area of the top and bottom (both A) plus the area of the side, which is the height ($2r$) times the circumference (C). So

$$\frac{\text{Surface area (cylinder)}}{\text{Surface area (sphere)}} = \frac{A + A + 2rC}{4\pi r^2} = \frac{2\pi r^2 + 2r \cdot 2\pi r}{4\pi r^2} = \frac{3}{2}.$$

Archimedes recognized his work on π for the grand achievement that it was. He began *On the Sphere and Cylinder* by asserting that these results should sit alongside Eudoxus's volume formulas:[39]

> Now these properties were all along naturally inherent in the figures referred to [sphere and cylinder], but remained unknown to those who were before my time engaged in the study of geometry. Having, however, now discovered that the properties are true of these figures, I cannot feel any hesitation in setting them side by side both with my former investigations and with those of the theorems of Eudoxus on solids which are held to be most irrefragably established.

Plutarch wrote, "Although [Archimedes] made many excellent discoveries, he is said to have asked his kinsmen and friends to

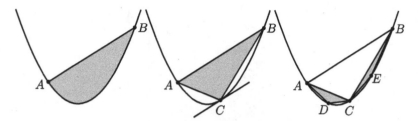

FIGURE 8.9. The area of the parabolic segment (left) is 4/3 the area of triangle *ABC* (center), which is four times the sum of the areas of the two smaller triangles (right).

place over the grave where he should be buried a cylinder enclosing a sphere, with an inscription giving the proportion by which the containing solid exceeds the contained."[40]

Because of these results about the nature of π and his accurate bounds on the value, we should call π *Archimedes's number*.

More Quadratures

Archimedes's work on the areas of curved figures extended beyond circles and the surfaces of spheres. The Greeks had known about conic sections for a century by the time Archimedes started proving theorems. However, nothing was known about areas enclosed by them until Archimedes came along. In his treatise *On the Quadrature of the Parabola*, Archimedes proved that the region bounded by a parabola and a chord of the parabola, *AB* (see figure 8.9), is squarable. In particular, if *C* is the point on the parabola at which the tangent line is parallel to *AB*, then Area(sector) $= \frac{4}{3}$ Area(*ABC*).

Archimedes gave not one, but two, proofs of this fact. The first used a mechanical argument similar to that in *The Method*.[41] The second one was a purely geometric proof that was a model application of Eudoxus's method of exhaustion.

In this second proof, Archimedes observed that the chords *AC* and *BC* cut off two more parabolic segments. In these, he inscribed triangles *ACD* and *BCE* so that the tangent lines to the parabola at *D* and *E* are parallel to *AC* and *BC*, respectively. Archimedes proved that

$$\text{Area}(ACD) + \text{Area}(BCE) = \frac{1}{4}\,\text{Area}(ABC).$$

At this point, we are left with four parabolic segments in which we can inscribe four triangles. Repeating this process indefinitely adds 2^n triangles in the nth step. Moreover, the total area of the triangles added at the $(n+1)$st step is one-quarter the area added in the nth step, and hence has area $\frac{1}{4^n}$ Area(ABC). If we continue this process of adding triangles infinitely many times, they will fill the entire parabolic segment. Thus, using today's notation and terminology, the area of the parabolic segment is the sum of a geometric series:

$$\text{Area(segment)} = \left(1 + \frac{1}{4} + \frac{1}{4^2} + \cdots\right) \cdot \text{Area}(ABC).$$

The well-known geometric series formula[42] yields Area(segment) = $\frac{4}{3}$ Area(ABC).

Of course, Archimedes did not write the area as an infinite sum. He used a double reductio ad absurdum argument to show that the sum could be neither larger nor smaller than $\frac{4}{3}$ Area(ABC).

Archimedes's *Book of Lemmas* contains two more area problems. This work is a collection of 15 theorems that comes to us through an Arabic translation. There is strong evidence that Archimedes did not write the *Book of Lemmas* (the text refers to Archimedes by name in several places, which he would not do if he had written it), but scholars believe the theorems are his. In it he introduced two geometric figures—or more precisely, two families of figures—that he called *arbelos* (which translates as leather cutter) and *salinon* (salt cellar). These figures are geometric regions bounded by semicircles with collinear diameters. The *arbelos* in figure 8.10 is formed from semicircles with diameters AB, AC, and BC (C can be placed anywhere on AB). The segment CD is perpendicular to AB. Archimedes showed that the area of the *arbelos* is the same as the area of the circle with diameter CD.

The *salinon* in figure 8.10 is formed from semicircles with diameters PQ, PR, RS, and QS with $PR = QS$. Then the area of the *salinon* equals the area of the circle with diameter TU.

One imagines that all of Archimedes's work on area-related problems—the parabola, the *arbelos*, the *salinon*, the spiral (which we discuss in chapter 11), and of course the circle—were undertaken with a view to the ultimate prize: the quadrature of the circle. Yet as Knorr

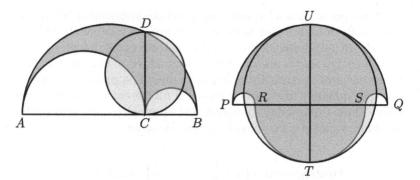

FIGURE 8.10. Archimedes's *arbelos* (left) and *salinon* (right).

wrote, "Despite a career of diligent and masterly effort,... the circle quadrature remained as elusive as ever."[43]

The Death of Archimedes

During the Second Punic War (218–202 BCE), Rome was concerned that the Kingdom of Syracuse might become allied with its adversary, Carthage. In 214 BCE, after failed diplomatic efforts, Rome began a siege of Syracuse by sea and land, which was led by the general Marcus Claudius Marcellus (42–23 BCE). The walled city was well fortified. Its famous resident, Archimedes, helped design the city's defenses. Plutarch gave this colorful description of Archimedes's defenses in action:[44]

> When, therefore, the Romans assaulted them by sea and land, the Syracusans were stricken dumb with terror; they thought that nothing could withstand so furious an onset by such forces. But Archimedes began to ply his engines, and shot against the land forces of the assailants all sorts of missiles and immense masses of stones, which came down with incredible din and speed; nothing whatever could ward off their weight, but they knocked down in heaps those who stood in their way, and threw their ranks into confusion. At the same time huge beams were suddenly projected over the ships from the walls, which sank some of them with great weights plunging down from on high; others were seized at the prow by iron claws, or beaks like the beaks of cranes, drawn

straight up into the air, and then plunged stern foremost into the depths, or were turned round and round by means of enginery within the city, and dashed upon the steep cliffs that jutted out beneath the wall of the city, with great destruction of the fighting men on board, who perished in the wrecks. Frequently, too, a ship would be lifted out of the water into mid-air, whirled hither and thither as it hung there, a dreadful spectacle, until its crew had been thrown out and hurled in all directions, when it would fall empty upon the walls, or slip away from the clutch that had held it.

In 212 BCE, after two years of fighting, the Romans breached the first wall. Eventually, and with the help of someone on the inside, they gained access to the citadel. The Roman soldiers were permitted to plunder the city and murder or enslave its residents. The Roman historian Livy (59 BCE–ca. 17 CE) described Archimedes's death at the hands of a Roman soldier while doing what he loved: mathematics:[45]

The city was turned over to the troops to pillage as they pleased, after guards had been set at the houses of the exiles who had been in the Roman lines. Many brutalities were committed in hot blood and the greed of gain, and it is on record that Archimedes, while intent upon figures which he had traced in the dust, and regardless of the hideous uproar of an army let loose to ravage and despoil a captured city, was killed by a soldier who did not know who he was. Marcellus was distressed by this; he had him properly buried and his relatives inquired for—to whom the name and memory of Archimedes were an honour.

The scene is depicted in figure 8.11.

Thus came to an end the life of one of history's greatest mathematicians, physicists, and engineers. Surprisingly, Archimedes is not so well known as many lesser mathematicians. As Netz wrote, "Archimedes, among the truly great, is relatively neglected. There is a Newton industry and an Einstein industry, but there isn't an Archimedes industry, and there ought to be one."[46]

And what of his tomb with the sphere and the cylinder? According to the Roman lawyer, orator, philosopher, and politician Marcus Tullius Cicero (106–43 BCE), the monument was built. In his *Tusculan*

DEATH OF ARCHIMEDES.

FIGURE 8.11. The death of Archimedes (an engraving based on a painting by Gustave Courtois). (L. Viardot, 1883, *The Masterpieces of French Art Illustrated, Vol. II*, Philadelphia: Gebbie & Co.)

Disputations, which was written in approximately 45 BCE, over a century and a half after Archimedes died, Cicero described his discovery of Archimedes's tomb in Sicily:[47]

> I will call up from the dust and wand a humble and obscure man of that same city, Archimedes, who lived many years after Dionysius. When I was quaestor in Sicily, I found, hedged in and overgrown with briers and brambles, his tomb, unknown by the Syracusans, who did not believe in its existence. I retained in my memory certain verses which I had heard were inscribed on his monument, in which it was said that a sphere with a cylinder was placed on the top of his tomb. After making thorough search (for there are a great many tombs close together near the gate Achradina), I noticed a column very little higher than the surrounding shrubbery, with the figures of a sphere and a cylinder on it. I at once said to the Syracusans, some of their chief men being with me, that I thought that this column was what I had

been looking for. Many laborers with scythes were sent in to clear and open the place. When the entrance was accessible, I stood over against the base of the column, on which was an inscription with the latter parts the several verses almost half obliterated.

By the end of the Greek era, much was known about the number we now call π. However, the problem of squaring the circle was still unsolved, and it had a reputation as an unsolvable problem. Eight centuries after Archimedes, Simplicius—who was widely read in the Middle Ages—pointed out that the problems of squaring and rectifying the circle were unsolved, and moreover, that it may be possible to prove that they are impossible to solve:[48]

The reason why one still investigates the quadrature of the circle and the question as to whether there is a line equal to the circumference, despite their having remained entirely unsolved up to now, is the fact that no one has found out that these are impossible either, in contrast with the incommensurability of the diameter and the side (of the square).

As a final note, although we are describing Archimedes's death, we are not finished discussing Archimedes's contributions to the problems of antiquity—not even close! He will return in chapters 9, 10, and 11.

TANGENT
Computing π at Home

One day I was explaining to [my friend, an actuary,] how it should be ascertained what the chance is of the survivors of a large number of persons now alive lying between given limits of number at the end of a certain time, I came, of course, upon the introduction of π, which I could only describe as the ratio of the circumference of a circle to its diameter.—"Oh, my dear friend! That must be a delusion; what can the circle have to do with the numbers alive at the end of a given time?"—"I cannot demonstrate it to you; but it is demonstrated."—"Oh! Stuff! I think you can prove anything with your differential calculus: figment, depend on it."
—Augustus De Morgan (1806–1871), *Budget of Paradoxes* [1]

SUPPOSE YOU *NEED* to find π, but you don't have a scientific calculator, you don't have access to the internet, you don't even have an old reference book containing the value of π. How can you obtain a good approximation using items on hand?

Polygons. Follow Archimedes's lead and use circumscribed and inscribed polygons. Find or draw a regular n-gon—bigger is better and larger n is better. Find the center: If n is even, draw lines between two pairs of opposite vertices; the diagonals meet at the center. If n is odd, use the perpendicular bisectors of two sides instead. Measure two of the following: the length of one side (s), the distance from the center to a corner (r_c), and the perpendicular distance from the center to an edge (r_i). You can find the third using $r_i^2 + (s/2)^2 = r_c^2$. Compute the perimeter of the polygon, $P = ns$. Because $2\pi r_i < P < 2\pi r_c$, we conclude that $P/(2r_c) < \pi < P/(2r_i)$. For instance, a standard US stop sign (which has a relatively small $n = 8$) measures $s = 327$ mm, $r_i = 395$ mm, and $r_c = 428$ mm. We conclude that $3.06 < \pi < 3.31$.

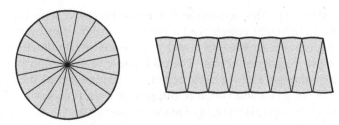

FIGURE T.11. The rearranged sectors of the circle resemble a rectangle.

Circumference. Measure the diameter of a bicycle tire, d. Use chalk to mark the side of the tire where it meets the ground. Mark the ground at that point as well. Roll the bicycle in a straight line for n revolutions, ending chalk mark down. Mark the ground again. Measure the distance between the chalk marks on the ground, l. It follows that $l = n\pi d$, so $\pi \approx l/(nd)$.

Circumference. Find a point in the center of an empty parking lot, and make a mark with a piece of chalk. Tie a string around the chalk. Have a friend hold the other end of the string at the central point, and draw a large circle. Measure the string. Say it has length r. Take a longer piece of string (approximately six times as long), and carefully place it along the circular chalk line until it meets its starting point. Measure this string. Say it has length C. Then $\pi \approx C/(2r)$.

Circular area. Use a compass to draw an accurate circle of radius r on a piece of paper. Cut out the circle. Now cut the paper circle into equally sized wedges as if it were a pizza—the more wedges the better. For instance, cut it in half, then in quarters, then in eighths, and so on. Reassemble the pieces as shown in figure T.11. This is approximately a rectangle with height r. Measure the width of the rectangle. Say it is l. Then the area is rl, but it is also πr^2, so $\pi \approx l/r$.

Circular area. Obtain some graph paper—the finer the better. Open a compass to r graph paper units. Draw a circle. Count the number of graph paper squares that are completely inside the circle. Call this A_l. Now count the number of squares that are not completely outside the circle (that is, the circle may pass through some of the squares). Call this number A_u. So defined, A_l and A_u are lower and upper bounds on the area. So $A_l/r^2 < \pi < A_u/r^2$.

Circular area. This technique requires a precision scale that can measure lightweight objects. Get a piece of cardboard or heavy card stock paper with dimensions x and y, measured in centimeters, say. Suppose it weighs w_0 grams. Then the paper is $w_0/(xy)$ grams per square centimeter. Draw a circle of radius r centimeters on the paper. Carefully cut it out and weigh it. Suppose it weighs w_1 grams. Then the area of the circle is $w_1 xy/w_0$. Hence, $\pi \approx w_1 xy/(w_0 r^2)$.

Spherical volume. Find a ball that is close to a perfect sphere, such as a ball from a pool or billiards set. An American-style pool ball has diameter 2.25 inches, which is 5.715 cm. Partially fill a graduated cylinder with water (well-marked scientific equipment will give more accurate measurements than a kitchen measuring device). Observe how much water it contains. Put the ball in the water so it is completely submerged and measure again. The difference is the volume of the ball. A pool ball will displace slightly less than 100 ml of water. The volume of the sphere is $\frac{4}{3}\pi r^3 = \frac{\pi}{6}d^3$. Solving for π we obtain the formula $\pi = 6V/d^3$. Using our values for the pool ball we find that $\pi \approx 6(100)/5.715^3 \approx 3.2$.

Sums and products. There are many elegant (and not so elegant) formulas for π that involve infinite sums or products. We could add or multiply the first several terms to obtain an approximate value for π. The more terms we use, the better the approximation.

A (*simple*) *continued fraction*[2] is an expression of the form

$$a_0 + \cfrac{1}{a_1 + \cfrac{1}{a_2 + \cfrac{1}{a_3 + \cdots}}},$$

in which the sequence of a_k eventually terminates (implying that the number is rational) or continues forever (it is irrational). Continued fractions have a long and interesting history and there are many beautiful theorems about them. For instance, if we take any continued fraction and cut it off at the nth term, we obtain the nth *convergent* of the continued fraction. The convergents are the best rational approximations to the number. Here the word "best" has a mathematical meaning. Take any convergent for a real number x and write it as a reduced fraction a/b. Then a/b is close to x, and every rational number that is closer to x has a larger denominator.

This brings us to the simple continued fraction for π that begins[3]

$$3 + \cfrac{1}{7 + \cfrac{1}{15 + \cfrac{1}{1 + \cfrac{1}{292 + \cdots}}}}.$$

We can find very good approximations by using the convergents of this continued fraction. For instance, the first convergent is 3/1. The second convergent is one of Archimedes's bounds for π (see page 118), $3 + 1/7 = 22/7$. The next several are 311/99, 355/113, and 99,733/31,746. This last convergent gives the remarkably accurate approximation 3.141592641.... Thus, if we know the first few terms of the continued fraction for π, we can find remarkably good approximations using only a simple calculator.[4]

Needle drop.[5] In 1777 Georges-Louis Leclerc, Comte de Buffon, devised an ingenious probabilistic method for estimating π. Find a surface ruled with equally spaced parallel lines—a hardwood floor is a typical example. Suppose the lines are w units apart. Now take a needle or a toothpick of length l and drop it on the surface (see figure T.12). For simplicity, assume that $l < w$. It is a nice exercise in trigonometry and calculus to show that the probability the needle will lie across a line is $2l/(w\pi)$. If we drop the needle n times and it lies across a line c times, then $c/n \approx 2l/(w\pi)$. In other words, $\pi \approx 2ln/(cw)$.

Italian mathematician Mario Lazzarini, in 1910, allegedly dropped 3408 needles in which $l = 5w/6$ and 1808 crossed a line. Thus he obtained the approximation

$$\pi \approx \frac{2(5w/6)(3408)}{1808w} = \frac{355}{113} = 3.14159292\ldots.$$

Lazzarini almost surely fabricated these numbers to obtain his unrealistically accurate approximation. Notice 355/113 was Zu Chongzhi's approximation of π and the value we obtain from adding the first four terms of the continued fraction for π.

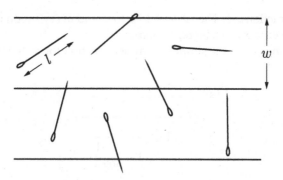

FIGURE T.12. Needles of length l falling on a hardwood floor with boards of width w.

Pendulum. The period of a pendulum, T, is related to its length l by the formula $T = 2\pi\sqrt{l/g}$, where $g = 9.8$ m/s^2 is the acceleration of gravity. This formula holds as long as the amplitude of the pendulum is not too large. Thus we can use a pendulum to estimate π. Take a long piece of string and attach a heavy weight to the end. Measure the length of the pendulum from the point of attachment to the center of mass of the weight. This is l. Start the pendulum swinging with a small amplitude. Time the period (for instance, time how long it takes to swing n times and divide by n). Then $\pi \approx \frac{T}{2}\sqrt{g/l}$.

Scientific calculator. Suppose you have a scientific calculator. Press the π key. Done. OK, now what if your π key has fallen off? The following iterative procedure assumes that you know π approximately and that the sine key is not broken. Begin with any approximation for π. Call it p_0. Plug it into the function $f(x) = x + \sin x$ to obtain a new number $p_1 = p_0 + \sin(p_0)$. If p_0 approximates π to n decimal places of accuracy, then p_1 has $3n$ decimal places of accuracy. For instance, $3.14 + \sin(3.14) = 3.1415926529\ldots$. Note that for this to work, the calculator must be in radian mode, not degree mode.[6]

Probability. Two positive integers are *relatively prime* if their greatest common divisor is 1. For example, 10 and 21 are relatively prime, but 10 and 15 are not (5 divides both numbers). In 1881 Ernesto Cesàro (1859–1906) proved[7] that the probability that two randomly chosen positive integers are relatively prime is $6/\pi^2$. There are several ways to use this theorem to approximate π. We could take a

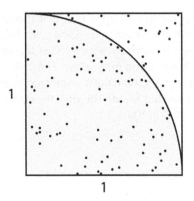

FIGURE T.13. Using darts to estimate π.

nonprobabilistic approach and count the number of relatively prime pairs (a, b) for $a, b \leq n$. If there are m such pairs, then m/n^2 is approximately $6/\pi^2$, so $\pi \approx \sqrt{6n^2/m}$. For instance, there are 100 pairs of numbers less than or equal to 10 (counting, for instance, $(2, 3)$ and $(3, 2)$ as different). Of these, 63 pairs are relatively prime.[8] This gives us $\pi \approx \sqrt{6 \cdot 100/63} = 3.086 \ldots$. There are 6087 relatively prime pairs of numbers between 1 and 100 out of 10,000 total pairs. So $\pi \approx \sqrt{6 \cdot 10{,}000/6087} \approx 3.1395 \ldots$. If we go up to pairs from 1 to 1000 we have 608,383 relatively prime pairs, which gives an approximation for π of $3.1404 \ldots$.

We could use a probabilistic approach to find π. In a room of $2n$ people, have each person choose a partner. Each person chooses a random number—any number of digits is fine—and the partners determine whether their numbers are relatively prime. If there are m relatively prime pairs, then $\pi \approx \sqrt{6n/m}$. This will likely be a very crude approximation both because n is small and because people are not good at choosing numbers at random (or at determining whether two numbers are relatively prime!). Another approach is to use a computer to generate random numbers. We used a spreadsheet to generate 1000 pairs of numbers between 1 and 1000. Of those, 609 pairs were relatively prime. This yields $\pi \approx \sqrt{6 \cdot 1000/609} \approx 3.1388 \ldots$.

Integration by darts. The quarter circle in figure T.13 has area $\pi/4$. We can estimate the area by throwing darts at it. We aren't going to throw actual darts—we will generate random pairs of numbers (x, y), where $0 \leq x \leq 1$ and $0 \leq y \leq 1$. The fraction of darts—or points—that land in the quarter circle will approach $\pi/4$ (because the area of the 1×1 square is 1) as the number of darts goes to infinity. Thus, this fraction can be used to approximate π. This technique of computing area is known as *Monte Carlo integration*. The 100 points in figure T.13 were generated using a spreadsheet's random number generator. Of the 100 points, 79 landed inside the quarter circle. This gives us the approximation $\pi \approx 4 \cdot 79/100 = 3.16$.

CHAPTER 9

The Heptagon, the Nonagon, and the Other Regular Polygons

Imagine a vast sheet of paper on which straight Lines, Triangles, Squares,
Pentagons, Hexagons, and other figures, instead of remaining fixed in their
places, move freely about, on or in the surface, but without the power
of rising above or sinking below it, very much like shadows—only
hard with luminous edges—and you will then have a pretty correct
notion of my country and countrymen.
—Edwin A. Abbott, *Flatland: A Romance of Many Dimensions* [1]

IN THE MATHEMATICAL literature, it is common to see references
to the *three* problems of antiquity. Many scholars omit the problem of
constructing regular *n*-sided polygons from the list. This is a shame. It
is a fascinating problem with a long and interesting history. It certainly
deserves to sit with the other three problems.

Regular polygons are prominently featured in Euclid's *Elements*.
Book IV is dedicated to inscribing and circumscribing geometric
figures. In it Euclid constructed many polygons, but there are infinitely many that neither he nor any other Greek mathematician could

construct. Of the regular polygons with 25 or fewer sides, they were unable to construct more than half, namely those with 7, 9, 11, 13, 14, 17, 18, 19, 21, 22, 23, and 25 sides. The first two of these—the heptagon and nonagon—came to be poster children for this problem of antiquity. We will encounter several mathematicians who tackled the heptagon, and the nonagon ties this problem directly to angle trisection.

The status of this fourth problem is somewhat surprising because regular polygons are ubiquitous in the worlds of art, architecture, and engineering. Perhaps it is because all the most common polygons are constructible. Equilateral triangles, squares, and hexagons can tile the plane and thus form the templates for decorative tessellations.

Despite the appearance of the seven-pointed star in the *Game of Thrones* series[2] and the nonagon in songs by They Might Be Giants and Peter Weatherall,[3] we do not often encounter either polygon "in the wild."

The Hexagon and Equilateral Triangle

It's elementary to inscribe an equilateral triangle and a regular hexagon in a circle. Euclid's first construction is an equilateral triangle. True, it is not inscribed in a circle, but we could use proposition IV.2 to transfer a similar copy into a circle. However, that's needlessly complicated. Euclid gave a quick construction of a regular hexagon in a circle in proposition IV.15, and from this it's easy to obtain an inscribed equilateral triangle—just join every other vertex with lines.

The key fact behind inscribing a regular hexagon in a circle is that the side length of the hexagon is the same as the radius of the circle. We begin with a point A on a circle with center P (see figure 9.1). Draw and extend the line AP to find the point D on the circle. Construct a circle with center A passing through P. The circles meet at points B and F. Then draw and extend lines BP and FP. They meet the original circle at C and E. Then $ABCDEF$ is a regular hexagon, and ACE is an equilateral triangle.

The Square

Euclid constructed a square in book I, but the elementary construction of a square inside a circle appears in book IV. Given a point A on a circle with center P, construct the diameter AC (see figure 9.2). Then

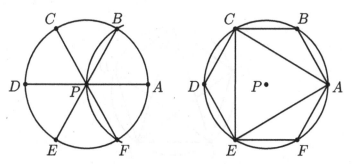

FIGURE 9.1. Constructing a regular hexagon and an equilateral triangle.

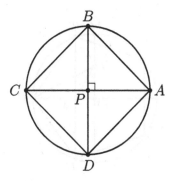

FIGURE 9.2. Inscribing a square in a circle.

construct a line perpendicular to *AC* passing through the center of the circle. It meets the circle at points *B* and *D*. Then *ABCD* is a square.

The Pentagon

Euclid's recipe for inscribing a regular pentagon in a circle is remarkable. He used nearly all the geometry he developed in the first four books. At 35 steps, the construction is somewhat tedious to carry out. But the main idea is to construct an inscribed isosceles triangle, *ABC* in figure 9.3, in which the base angles are twice the vertex angle (a 36°–72°–72° triangle), and then to bisect ∠*BAC* and ∠*ABC* to find *D* and *E* on the circle. Then *ABCDE* is a regular pentagon.

In 1893 Herbert Richmond presented the following easier method of constructing the pentagon. Begin with two perpendicular diameters

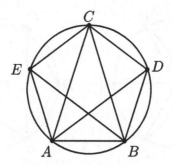

FIGURE 9.3. Euclid's construction of a regular pentagon.

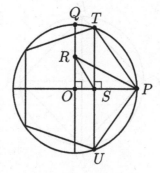

FIGURE 9.4. Richmond's pentagon construction.

OP and OQ (as in figure 9.4). Let R be the midpoint of OQ. Construct the segment PR and bisect $\angle ORP$ to find the point S on OP. Construct a line perpendicular to OP through S. It intersects the circle at T and U. Then P, T, and U are three of the five vertices of the pentagon. Use the given side lengths to find the other two vertices.[4]

The Pentadecagon and Other Constructible Polygons

In the final proposition of book IV, Euclid described how to inscribe a regular 15-gon (a pentadecagon) in a circle. Inscribe an equilateral triangle and a regular pentagon so they share a vertex A (as in figure 9.5). Let AB and AC be sides of the triangle and pentagon, respectively. Then arc AB is 1/3 of the circumference and AC is 1/5

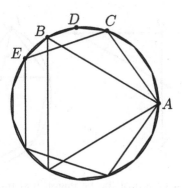

FIGURE 9.5. The pentadecagon.

of the circumference. It follows that BC is $1/3 - 1/5 = 2/15$ of the circumference. Thus, if we bisect arc BC to obtain D, we can construct two sides of the 15-gon: CD and BD. We can use them to construct the rest of the 15-gon. Although Euclid did not point this out, all the vertices of the equilateral triangle and the pentagon are vertices of the 15-gon. In fact, because arc AE is 2/5 of the circle, arc AB is 1/3 of the circle, and $2/5 - 1/3 = 1/15$, then segment BE is a side of the pentadecagon.

At this point, it is natural to ask which regular polygons are constructible. Euclid constructed regular polygons with 3, 4, 5, 6, and 15 sides. He did not say so, but he knew that if he were given a regular n-gon inscribed in a circle, then he could bisect all of the arcs bounded by the vertices of the polygon to obtain a regular $2n$-gon. This process could be repeated. We could begin with a pentagon and double the number of sides to obtain a decagon, a 20-gon, a 40-gon, and so on. In short, Euclid and all later Greeks could construct regular n-gons in which n has the form 2^k, 2^k3, 2^k5, or 2^k15 for some k, leaving open the question of whether this list was complete.

Before we leave this discussion, let's return to the construction of the 15-gon. It was created by first constructing the equilateral triangle and the pentagon. It is not a coincidence that $15 = 3 \cdot 5$. Suppose p and q are relatively prime. Then, by a standard result of number theory, there exist integers a and b, one positive and one negative, such that $1 = ap + bq$. If we divide both sides of this expression by pq we obtain $\frac{1}{pq} = a \cdot \frac{1}{q} + b \cdot \frac{1}{p}$. This relationship tells us that if we inscribe a p-gon and a q-gon so that they share a vertex, then counting vertices in the same

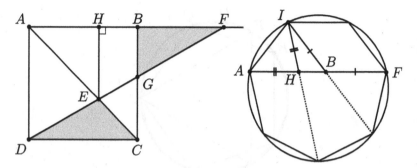

FIGURE 9.6. A construction of the heptagon that may be due to Archimedes.

direction, the $|a|$th vertex of the q-gon and the $|b|$th vertex of the p-gon are neighboring vertices of the regular pq-gon.

For example, 3 and 5 are prime, so they are relatively prime to each other. We are guaranteed the relation $1 = 2 \cdot 3 + (-1) \cdot 5$. In particular, dividing both sides by 15 we obtain $2/5 - 1/3 = 1/15$. So as we saw in figure 9.5, the first vertex of the equilateral triangle and the second vertex of the pentagon are vertices of the 15-gon.

What does this tell us about constructible regular polygons? If we are able to construct regular polygons with prime numbers of sides $p_1, \ldots, p_m$, where the p_i are distinct odd primes, then it is possible to construct a regular $(2^k p_1 \cdots p_m)$-gon. So, just for the sake of argument, if we *were* able to construct a regular heptagon, then we would also be able to construct regular polygons with $21 = 3 \cdot 7$, $35 = 5 \cdot 7$, $105 = 3 \cdot 5 \cdot 7$, $42 = 2 \cdot 3 \cdot 7$, $140 = 2^2 \cdot 5 \cdot 7$, and so on, sides.

Such knowledge does not enable us to construct any new regular polygons—we know only the constructibility status of two odd primes, 3 and 5—but at least we have a framework in place that we can apply if needed. Are the 7-gon, 11-gon, 13-gon, 17-gon, and so on, constructible? If the answer is yes for any of these, then we will obtain a slew of other constructible polygons.

However, even if we knew the status of all odd primes, this would be a far cry from a complete answer to the question of which regular polygons are constructible. For instance, if p is an odd prime and the regular p-gon is constructible, we don't know whether the regular p^2-gon, p^3-gon, and so on, are constructible. The constructibility of an equilateral triangle does not guarantee the constructibility of the nonagon—because the 3^2-gon isn't constructible!

Archimedes's Heptagon Construction

We possess only one example of an ancient Greek construction of a heptagon. Of course, it was *not* a compass-and-straightedge construction. It is attributed to Archimedes, but all we have is a corrupted Arabic translation from the ninth century, which was revised in the eighteenth century. At that time, it was known as *Book of the Construction of the Circle Divided into Seven Equal Parts, by Archimedes.*[5]

This clever construction of a heptagon hinges on an unorthodox construction. We begin with a square $ABCD$, its diagonal AC, and AB extended as in figure 9.6. We then perform the unusual construction: Draw a line through D so it intersects AB at F in such a way that triangles CDE and BFG have the same area. This is the move that cannot be carried out with a compass and straightedge. Once this is accomplished, construct a perpendicular from E meeting AF at H. Now we are ready to construct the heptagon. Take the segment $AHBF$ and construct a point I so that $HI = AH$ and $BI = BF$. Then A, I, and F are vertices of a regular heptagon. Use a compass and straightedge to construct the circle containing the three points, and construct the remaining vertices. (Note that lines HI and BI intersect the circle at two other vertices as well.) We omit the proof that this is a heptagon.[6]

TANGENT
It Takes Time to Trisect an Angle

The strongest of all warriors are these two—time and patience.
—Leo Tolstoy, *War and Peace* [1]

IN THIS TANGENT, we show two unusual trisectors: a clock and a cylinder.

The clock as a construction tool Leo Moser observed that it is possible to trisect any angle θ using an ordinary analog clock (assuming it keeps accurate time). [2] Wait until noon (or simply set the clock to 12:00). Line up the hands of the clock along one of the lines that make up the angle to be trisected, and place the center of the clock on the vertex of the angle. Then wait until the minute hand points along the second line. By this time, the hour hand has moved an angle $\theta/12$ (see figure T.14). Use a compass and straightedge to double this angle twice to obtain the angle $\theta/3$.

Using a cylinder as a construction tool We are able to solve some of the problems of antiquity if we use a three-dimensional tool: a cylinder.

Suppose we have a compass, a straightedge, and a paper-wrapped cylinder, and that we'd like to trisect an angle. [3] We begin with our angle $\angle ABC$ on the bottom of the cylinder, in which A and C are on the edge and B is in the center (see figure T.15). A piece of paper covers the curved surface of the cylinder. Mark A and C on the paper. Then remove the paper and lay it flat on a flat drawing surface. Trisect the segment AC (see endnote 9 on page 370), yielding points D and E. Rewrap the cylinder, lining up the points A and C. Then BD and BE trisect $\angle ABC$.

This same procedure can be used to divide $\angle ABC$ into any number of equal parts. Moreover, if we use the entire side of the paper as our

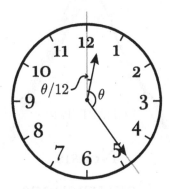

FIGURE T.14. When the minute hand has moved through an angle θ, the hour hand has moved an angle $\theta/12$.

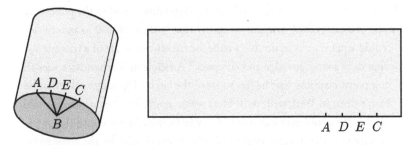

FIGURE T.15. Trisection of an angle using a paper-covered cylinder.

segment AC, we can use this procedure to produce regular polygons with any number of sides.

And of course, the bottom edge of the paper has length $2\pi r$, where r is the radius of the cylinder. It is not difficult (using the techniques developed in chapter 15) to produce from that a square with side length $\sqrt{\pi} r$, and thereby square the circular base of the cylinder.

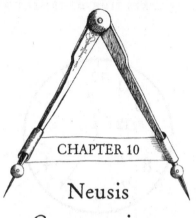

CHAPTER 10

Neusis
Constructions

The keen eye of W. B. Ransom (Tufts College) discovered in the public press a tale of a calculating machine company that "offered \$1000 to anyone who could square a circle, double a cube, or trisect one angle of a triangle by using only a straight-edge and compass." A fellow in Mathematics sues the company, claiming that he has squared the circle. The judge rules that he hasn't done it. Well, well, well! Here we go again! This seems just like old times. We could give one word of advice to the company, and that is to specify that the straight-edge shall be unmarked. Also let the angle to be trisected be an arbitrary angle. Then the company may rest comfortably in the knowledge that it has been proved that no one of the three constructions can be done. And how we wish that would become generally known.
— "Mathematical Miscellany," *Mathematics Magazine*, 1948[1]

ALTHOUGH THE GREEKS used the compass and straightedge as the basis for their geometry, they did occasionally stray from this ideal. We have seen already that they used other curves—such as the conic sections. They also used other tools—such as Eratosthenes's mesolabe. We now consider what seems at first to be a trivial variant on the traditional straightedge—putting two marks on it.[2] The curves we can

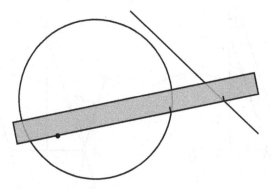

FIGURE 10.1. A neusis construction.

draw with a compass and marked straightedge are still lines and circles, but we have one new construction technique at our disposal, and with it we are able to solve some of the problems of antiquity.

The Marked Straightedge

The new technique is known as *neusis*, which means "verging" or "inclining toward." If we have a point and two curves, which can be either lines or circles, then we can place the straightedge so that the edge passes through the point and the two marks align with the two curves. Then we can trace along the straightedge (see figure 10.1).

Pappus gave the following neusis construction of a regular pentagon. Begin with segment AB, which will be one side of the pentagon (see figure 10.2). Construct a perpendicular bisector to AB and a circle with center B and radius AB. Put marks on the straightedge a distance AB apart. Then perform the neusis construction: Draw a line through A so the distance between the circle and the perpendicular line equals the distance between the marks; say it meets them at C and D, respectively. Then D is a vertex of the pentagon. Constructing the other vertices is straightforward. We omit the proof that this is a regular pentagon, but note that if we were to draw the pentagram inside it, then AD and BE would meet at C.

This construction is a nice initial example because it is easy to carry out and is much more efficient than the complicated approach with

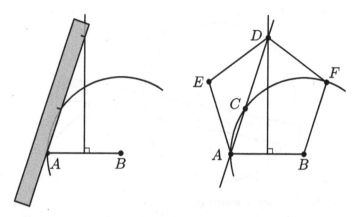

FIGURE 10.2. A neusis construction of a regular pentagon.

an unmarked straightedge. However, the true benefit of using the neusis technique is not that it is more efficient, but that it allows us to construct figures that are not constructible using the Euclidean tools.

In this example, we marked the straightedge during the construction based on a length in the problem. What if we were given a premarked straightedge? If we were given one with marks a units apart, can we make it behave like one with marks b units apart? Surprisingly, yes! All we must do is scale the relevant parts of the figure by a factor of a/b relative to some point O,[3] perform the neusis construction with our marked straightedge, then scale everything back to the original size (by a factor of b/a relative to O). It is tedious and time consuming, but possible. The upshot is that it makes no difference if we begin with a premarked straightedge or if we mark it during the construction.

Neusis is an old Greek technique. The first known use was in Hippocrates's construction of one of his lunes (the right-hand lune in figure 7.10 on page 102). In this case, as for the pentagon, the neusis construction can be avoided—the lune is Euclidean constructible.[4] The most famous use of neusis is Archimedes's elegant angle trisection, which cannot be conducted with the Euclidean tools. We will present Archimedes's construction in the next section. According to Pappus, Apollonius wrote a treatise in two books on the neusis technique. But unfortunately it is lost. In chapter 14 we shall see that the sixteenth-century French mathematician François Viète was inspired

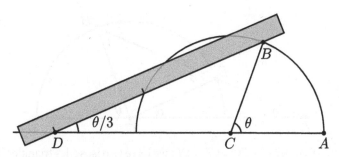

FIGURE 10.3. Archimedes's neusis construction that trisects angle θ.

by Pappus's writings and proposed extending Euclid's postulates by adding the option of neusis constructions.

Archimedes's Trisection

Archimedes's elegant method of trisecting an angle is found in his *Book of Lemmas*. Suppose we want to trisect an acute angle $\theta = \angle ACB$ (see figure 10.3).[5] For simplicity, suppose AC and BC are the same length as the distance between the marks on the straightedge. Draw the circle with center C passing through A and B. Now perform the neusis construction: Maneuver the straightedge so that it passes through B and the marks coincide with the circle and the line AC extended leftward. If the straightedge crosses this extension at D, then we claim that $\angle BDC = \theta/3$.[6]

Let E be the point of intersection of BD and the circle (see figure 10.4). By construction $BC = CE = DE$, and hence the triangles CDE and BCE are isosceles and hence have their base angles equal. Let α and β denote the angles shown in the figure. By elementary geometry, $2\beta = \alpha + \theta$ and $\beta = 2\alpha$. Substituting the second equation into the first and solving for α, we obtain $\alpha = \theta/3$.

Pappus's *Collection* was written over 500 years after Archimedes, and it contains a variation of Archimedes's technique that uses a neusis construction between two lines that likely dates back to the third century BCE.[7] We wish to trisect an acute angle $\theta = \angle CBF$; we may assume BC is the diagonal of a rectangle $BFCG$, as in figure 10.5. Place marks on the straightedge a distance $2BC$ apart, and perform a neusis construction: Draw a line through B so the distance between CF and

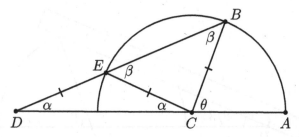

FIGURE 10.4. Because $BC = CE = DE$, we have two isosceles triangles, BCE and CDE.

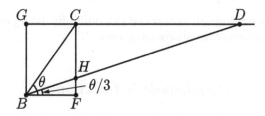

FIGURE 10.5. Pappus's neusis angle trisection.

CG is $2BC$. If the line intersects CG at D and CF at H, then $DH = 2BC$, and $\angle DBF = \theta/3$.

To justify this claim, locate the midpoint E of DH and draw the segment CE (see figure 10.6). We claim that $BC = EH = DE = CE$. It follows immediately from our construction that the first three of these segments have the same length; we must justify only that they are the same length as CE. We know that CDH is a right triangle, so when we inscribe it in a circle, DH is its diameter. But that implies that E is the center of the circle, and hence EH, DE, and CE are all radii of the circle, so CE has the same length as the others.

Next, draw a circle with center C and radius BC, and extend DG so it meets the circle at A, as in figure 10.7. At this point, we may apply Archimedes's result (figure 10.7 is upside down compared to figure 10.6, but we have kept the letters the same) and the fact that AD and BF are parallel. We see that $\angle DBF = \angle BDC = \frac{1}{3}\angle ACB = \frac{1}{3}\angle CBF = \theta/3$.

An added bonus of the ability to trisect any angle is that we can construct the regular nonagon. Begin with a circle with center O, and construct the vertices of an equilateral triangle, A, B, and C (see

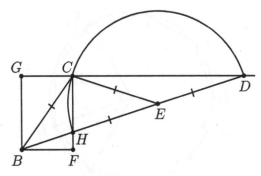

FIGURE 10.6. In Pappus's neusis construction $BC = EH = DE = CE$.

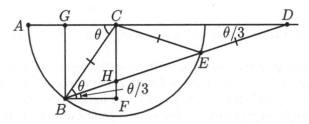

FIGURE 10.7. Pappus's neusis construction is Archimedes's construction in disguise.

figure 10.8). Then construct a point D on the circle so OD trisects the central angle $\angle AOC$. It follows that CD is a side of the nonagon.

Neusis for Cube Doubling

Nicomedes (fl. ca. 250 BCE) discovered the first neusis construction that solved the problem of finding two mean proportionals and hence the problem of doubling the cube. We do not know anything about the life of Nicomedes—even when he lived. However, from the clues available, we know that he was a contemporary of Archimedes.

According to Eutocius, Nicomedes was very proud of his neusis solution and thought it was much better than Eratosthenes's solution using the mesolabe. He also invented a new curve called the conchoid to carry out the neusis construction. We will say more about this curve in chapter 11. Rather than sharing Nicomedes's somewhat

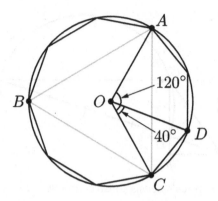

FIGURE 10.8. The nonagon is neusis constructible.

tedious construction, we will give a simplified version of it by Isaac Newton.[8]

Newton gave the following method of finding the two mean proportionals between segments of length a and b where $a < b$: Bisect a segment AB of length b at the point C (see figure 10.9). Draw a circle with center A and radius AC. Then find a point D on the circle so that CD has length a. Extend the lines BD and CD. Now perform a neusis construction: Draw a line through A so that the distance between the lines BD and CD is $b/2$. Suppose this line meets BD and CD at E and F, respectively. Then AE and DF are the two mean proportionals; that is, if they have lengths x and y, then $a/x = x/y = y/b$.

The proof relied upon two classical geometric theorems—ones that today's geometry students may not know. First, Menelaus's theorem: the line BE passes through triangle ACF, so $BC \cdot DF \cdot AE = AB \cdot CD \cdot EF$. Hence, $bxy/2 = ab^2/2$, or equivalently, $a/x = y/b$. To complete the proof we must show that these expressions also equal x/y. Extend AF to produce a secant line intersecting the circle at G and H, as in figure 10.10. Notice that EF has the same length as the radius of the circle, so FG is x units long. By the intersecting secants theorem,[9] $FH \cdot FG = CF \cdot DF$. So $x(x + b) = y(y + a)$. From these two algebraic expressions we obtain our desired conclusion,

$$\frac{x}{y} = \frac{y+a}{x+b} = \frac{y+(xy/b)}{x+b} = \frac{y(b+x)}{b(x+b)} = \frac{y}{b}.$$

Newton's construction is particularly nice for doubling the cube— that is, when $b = 2a$ (for the sake of simplicity, assume $a = 1$). We begin

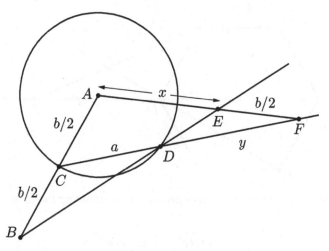

FIGURE 10.9. Newton's method of constructing two mean proportionals using a neusis construction.

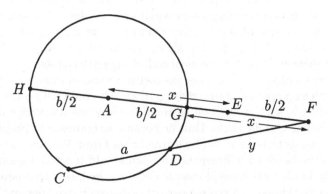

FIGURE 10.10. We can apply the intersecting secants theorem to CF and FH.

with a regular hexagon with side length 1. Let C be the center of the hexagon and A, B, and D be vertices arranged as in figure 10.11. Draw the lines CD and BD. Now perform the neusis construction: Construct a line through A so that the distance between lines BD and CD is 1, and suppose the line meets these two lines at E and F, respectively. Then the length of AE is $\sqrt[3]{2}$.

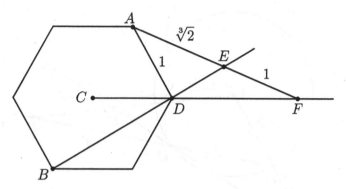

FIGURE 10.11. Newton's neusis procedure for doubling the cube.

Viète's Heptagon

There is no known classical neusis construction of the regular heptagon. We saw Archimedes's neusis-like construction on page 143. However, it wasn't a true neusis construction; it required finding a line that created two regions of equal area. As far as we know, the sixteenth-century mathematician François Viète discovered the first neusis construction of the heptagon. We will have much more to say about Viète and his mathematical contributions in chapter 14.

Viète wanted to solve more geometric problems than were possible with a compass and straightedge alone. He could have introduced more curves, such as the conic sections. But he wanted to stick with the straight line and the circle. Thus he became an enthusiastic proponent of the neusis technique that he learned about from Pappus—so much so that he believed mathematicians should add it to their geometric toolkit. In his 1593 *A Supplement to Geometry*, he introduced new postulates that allowed a neusis construction between two lines and a line and a circle. Indeed, he began *A Supplement to Geometry* with[10]

> To supply the defect of geometry, let it be conceded
> *To draw a straight line from any point to any two given lines,*
> *the intercept between these being any possible predefined distance.*

Some of the constructions in this short treatise were new and some were not. Viète gave Archimedes's method of trisecting an angle (see figure 10.3).[11] It is likely that Viète discovered this proof on his own because Archimedes's *Book of Lemmas* did not appear in Latin until

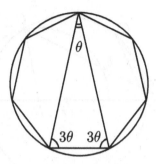

FIGURE 10.12. Three vertices of the regular heptagon form a triangle with angles θ, 3θ, and 3θ.

1659.[12] He also showed how to find two mean proportionals using a neusis construction.[13] Although at a glance his construction looks new, the basic geometry behind it is very similar to Nicomedes's solution. We do not know whether Viète was aware of the original construction. He also gave a neusis construction of the elusive regular heptagon and, as far as we know, it was new, and it was the first such construction.

Just as Euclid's construction of the regular pentagon required finding the isosceles triangle with angles θ, 2θ, and 2θ (see figure 9.3 on page 140), the key to Viète's construction was also an isosceles triangle, one with angles θ, 3θ, and 3θ, implying that $\theta = (360/14)°$ (see figure 10.12).

Here is a sketch of Viète's construction.[14] Suppose we want to inscribe a heptagon in a circle with center O and diameter AB (see figure 10.13). First, find a point C on the circle so that $AC = OA$. Next construct the point D on OA so that $OD = \frac{1}{3}OA$. Draw the circle with center D and radius CD. Now perform the neusis: Construct a line through C meeting this smaller circle and the extended diameter AB at F and E, respectively, so $EF = OA$. (This is Archimedes's angle trisection, so $\angle AEC = \frac{1}{3}\angle ADC$.) Lastly, find a point G on the circle so that $EG = OA$. Then BG is a side of the heptagon.

We will not give Viète's full proof. Using his newly created algebra, he was able to show that OH (where H is the second point of intersection of the line EG and the circle) and AG are parallel (see figure 10.14). We will start our proof there. Denote $\angle OAG$ by θ. Because AG and OH are parallel, $\angle BOH = \theta$. Also, because the inscribed angle $\angle BAG$ and central angle $\angle BOG$ share the same arc, $\angle BOG = 2\angle BAG = 2\theta$. Hence, $\angle GOH = \theta$, and OH bisects $\angle BOG$. Next, because $EG = OG$, triangle

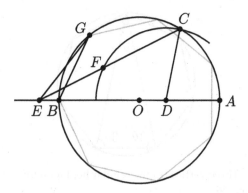

FIGURE 10.13. Viète's neusis construction of a heptagon.

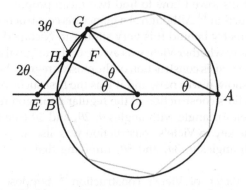

FIGURE 10.14. The key to Viète's construction of a heptagon is that triangle GHO has angles $\theta, 3\theta, 3\theta$.

EOG is isosceles, so $\angle GEO = 2\theta$. Hence, the exterior angle $\angle GHO$ of triangle EHO is 3θ. Lastly, because triangle GHO is isosceles, $\angle HGO = 3\theta$. So triangle GHO has angles $\theta, 3\theta$, and 3θ. By our earlier discussion, we know that $\theta = (360/14)°$. Thus the central angle $\angle BOG = 2\theta = (360/7)°$, and hence BG is the side of a heptagon.

Thus, this slight modification of Euclid's postulates, which gives geometers the freedom to use a marked straightedge, enables us to trisect any angle and to double the cube. It also allows us to construct polygons such as the 7-gon, the 9-gon, and, by the reasoning in chapter 9, the $(2 \cdot 7)$-gon, the $(4 \cdot 9)$-gon, the $(3 \cdot 5 \cdot 7)$-gon, and so on.[15]

TANGENT
Crockett Johnson's Heptagon

Sudden a thought came like a full-blown rose,
Flushing his brow, and in his pained heart
Made purple riot: then doth he propose
A stratagem, that makes the beldame start.
—John Keats, "The Eve of St. Agnes," 1820[1]

ANOTHER CLEVER NEUSIS construction of a heptagon appeared in 1975, and it came from a very unlikely source: a man named David Johnson Leisk (1906–1975), who is better known by his nom de plume Crockett Johnson.[2] Johnson is the author, illustrator, and cartoonist who drew the 1940s comic strip *Barnaby* and wrote a popular series of children's books that began with *Harold and the Purple Crayon* (1955).[3]

Johnson had no formal mathematical training, but late in life he became infatuated with mathematics—geometry in particular. Starting in 1965, Johnson brought mathematics into his art, producing at least 117 mathematically inspired paintings. Each painting is an artistic rendering of some piece of mathematics—often a famous theorem, but occasionally his own mathematical discoveries. The Smithsonian currently has 80 of his pieces.[4]

Johnson especially loved the problems of antiquity. They were frequent subjects of his artwork and his own mathematical musings. He even wrote an article in 1970 giving a method of approximately squaring the circle. It produced an approximation of $\sqrt{\pi}$ accurate to five decimal places. Then he created a painting illustrating his technique.

In the fall of 1973, Johnson made a mathematical discovery while eating at an outdoor café in Syracuse on the island of Sicily—the birthplace of Archimedes. He was playing with a menu, a wine list, and some toothpicks. He realized that if he laid seven of the toothpicks in a zigzag configuration between the menus he could produce an isosceles triangle (see figure T.16).[5] Moreover, it was not just any isosceles

FIGURE T.16. Seven toothpicks between two menus produce a triangle with angles θ, 3θ, 3θ.

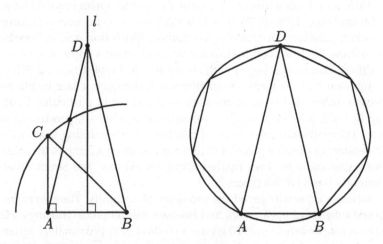

FIGURE T.17. A neusis construction of a heptagon.

triangle—the angles were θ, 3θ, and 3θ. Johnson knew that this triangle was key to constructing a regular heptagon.

It is easy to prove that Johnson's triangle has this form. Suppose $\angle ACB = \theta$. Then, because triangle CDE is isosceles, $\angle CED = \theta$. So the exterior angle of triangle CDE, $\angle BDE$, is 2θ. Because triangle BDE is isosceles, $\angle DBE = 2\theta$. Lastly, observe that triangles ABC and ABE are similar, so $\angle ABE = \theta$. It follows that $\angle ABC = 3\theta$, and by symmetry, $\angle BAC = 3\theta$.

This was the inspiration he needed to devise the following neusis construction of a heptagon. Suppose we begin with a line segment AB (see figure T.17). Construct a segment AC perpendicular to and with the same length as AB. Also, construct the perpendicular bisector to AB; call it l. Then construct a circle with center B and radius BC. Now we perform the neusis construction: Construct a line AD so that D is on l and the distance from D to the circle is the same as the length of AB. Then ABD is the desired triangle. Finally, circumscribe the triangle. These three points, A, B, and D, are vertices of the inscribed heptagon. Use a compass and straightedge to complete the construction.

Johnson's construction is simple and elegant and would have pleased the ancient Greeks, although his proof is decidedly modern: it requires the law of cosines and several trigonometric identities.[6,7]

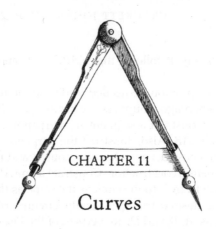

CHAPTER 11

Curves

Euclid alone has looked on Beauty bare.
Let all who prate of Beauty hold their peace,
And lay them prone upon the earth and cease
To ponder on themselves, the while they stare
At nothing, intricately drawn nowhere
In shapes of shifting lineage; let geese
Gabble and hiss, but heroes seek release
From dusty bondage into luminous air.
O blinding hour, O holy, terrible day,
When first the shaft into his vision shone
Of light anatomized! Euclid alone
Has looked on Beauty bare. Fortunate they
Who, though once only and then but far away,
Have heard her massive sandal set on stone.
—Edna St. Vincent Millay[1]

IT IS AMAZING that we can solve such a wide range of problems using only the compass and straightedge. Lines and circles, despite their apparent simplicity, are powerful tools in the pursuit of solutions to thorny geometric conundrums. However, despite this power, ably showcased in Euclid's *Elements*, and despite Plato's admonition against introducing other curves into geometry, Greek mathematicians could not resist tasting the forbidden fruit.

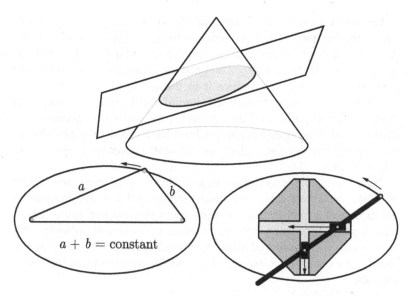

FIGURE 11.1. An ellipse can be the cross section of a cone (top), or it can be drawn using two pins and a loop of string (left) or with a trammel of Archimedes (right).

We can define curves in a variety of ways. We can describe them as a locus of points—it is the set of points that satisfy some conditions. A circle is the locus of points that are a fixed distance from a given point (the center of the circle). All the conic sections can be introduced in this way. For instance, an ellipse is the set of points so that the sum of the distances from two points (the focal points) is a constant. The ellipse and the other conic sections can also be given as the boundary of slices of a solid cone. Curves can be defined by coordinated motion; that is, a point that moves according to some rules, and as it moves it traces out the curve.

These definitions often lead to the invention of specialized devices that can draw the curve—some simple and some complex. The compass, which can draw a circle, is a simple device. So is a string and two pins, which can draw an ellipse—put a loop of string around two pins, which are located at the focal points, pull the loop taut, then move the pencil as if it is a planet in orbit, tracing the ellipse. The so-called trammel of Archimedes is an example of a more complicated drawing device (see figure 11.1). In this contraption, two shuttles slide through tracks and the pencil at the end of an attached arm draws the ellipse.

We will encounter a number of clever drawing instruments in the following pages. It was because these new curves were often described in terms of mechanical devices that Plato rejected them as not geometrical. He wrote, "The good of geometry is set aside and destroyed, for we again reduce it to the world of sense instead of elevating and imbuing it with the eternal and incorporeal images of thought."[2] It was not until the seventeenth century that geometers—starting with Descartes—began pushing back against the prohibition and urging that some so-called "mechanical" curves be adopted fully into geometry.

These new curves could be objects of study in their own right: for instance, as we saw, Archimedes computed the area between a parabola and a line. But as we shall see in this chapter, some curves were invented for one purpose: to solve a problem of antiquity. We have already encountered this idea. In chapter 5 we saw that we could use a compass, a straightedge, and the conic sections to double the cube.

The Quadratrix

In the fifth century BCE, the sophists were a group of non-Athenian, itinerant scholars. They traveled around ancient Greece teaching anyone who would pay how to succeed in business, politics, the law, and so on. The sophists were highly trained in rhetoric and would often teach their pupils how to use argumentative language to convince others, even when arguing a falsehood. Today "sophism" means "specious but fallacious argument, either used deliberately in order to deceive or mislead, or employed as a means of displaying ingenuity in reasoning."[3]

Plato was a particularly harsh critic. In his dialogue *Sophist*, the Elean Stranger said, "The sophist is nothing else, apparently, than the money-making class of the disputatious, argumentative, controversial, pugnacious, combative, acquisitive art."[4] Plato, who came from a wealthy family, was strongly opposed to charging a fee for teaching. He thought education should be for only the aristocracy and not the common man, but the sophists would teach anyone who could pay.

But even among the sophists, Plato held one man, Hippias of Elis (born ca. 460 BCE), in low regard. Hippias appeared in three of

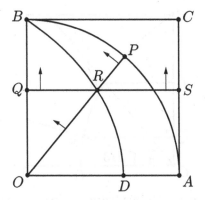

FIGURE 11.2. The curve *BRD* is a quadratix.

Plato's dialogues—*Hippias Major, Hippias Minor,* and *Protagoras*—and is portrayed as a boastful, egotistical money-grubber.

Hippias's self-confidence, while unattractive, may have been well founded. He was a polymath who wrote and spoke extensively on a broad range of subjects: rhetoric, art, music, politics, astronomy, mythology, history, and mathematics. He had a prodigious memory, which he credited to a system of mnemonics. He prided himself on his complete self-sufficiency, even making his own garments.

Apparently Hippias also created new mathematics. Although none of his writings survived, Proclus, in his fifth-century (CE) commentary on book I of Euclid's *Elements*, wrote that Hippias discovered a curve, later called the *quadratix*. If true,[5] this may be the first example of a mathematical curve other than a line or a circle.

Hippias didn't invent the quadratix as a homework problem to torment his students. He introduced the curve because it had one very specific application: it could be used to trisect any angle. Given a straightedge, a compass, and a quadratix, Hippias could trisect any angle given to him. Remarkably, as we shall see, the quadratix can also be used to solve three of the four problems of antiquity! It can be used to construct every regular polygon and to square the circle (although Hippias did not know this last feature).

To construct a quadratix, begin with a square *OABC* and inscribe in it a quarter circle with center *O*, passing through *A* and *B* (see figure 11.2). Have a point *P* on the circle move from *A* to *B* with constant speed. In that same interval of time have the points *Q* and *S* move

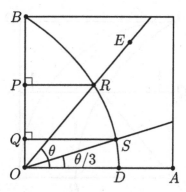

FIGURE 11.3. We can use the quadratrix to trisect an angle.

with constant speed from O to B and A to C, respectively. Draw the segments OP and QS. So at any instant,

$$\frac{OQ}{OB} = \frac{\text{Arc } AP}{\text{Arc } AB} = \frac{\angle AOP}{\angle AOB}.$$

Then R, the point of intersection of OP and QS, traces the quadratrix BD.

It is straightforward to trisect an angle using a quadratrix. Suppose we wish to trisect an acute angle $\theta = \angle AOE$ (see figure 11.3). Construct a square on the side OA and in it draw a quadratrix. The line OE intersects the quadratrix at a point R and the line through R parallel to OA intersects OB at P. Use the compass and straightedge to find the point Q that is $1/3$ the distance from O to P. The line through Q parallel to OA intersects the quadratrix at S. Because QS is $1/3$ as far above OA as is PR, $\angle AOS = \frac{1}{3}\angle AOE$. More formally, by the defining properties of the quadratrix,

$$\frac{\angle AOS}{\angle AOE} = \frac{\angle AOS}{\angle AOB} \cdot \frac{\angle AOB}{\angle AOE} = \frac{OQ}{OB} \cdot \frac{OB}{OP} = \frac{OQ}{OP} = \frac{1}{3}.$$

In fact, there is nothing special about trisecting the angle. We can use this same technique to construct angles $\theta/2$, $\theta/3$, $\theta/4$, or θ/n for any positive integer n. In particular, we could use this technique to inscribe a regular n-gon inside a circle for any n.

Remarkably, the quadratrix enables us to square the circle, as well. Proclus wrote,[6]

For the squaring of the circle, there was used by Dinostratus, Nicomedes, and some other more recent geometers a certain curve which took its name from this property; for it is called by them "square-forming" [quadratrix].

Nothing else is known about the first use of the quadratrix for squaring a circle. Because Dinostratus (fl. ca. 350 BCE) lived before Nicomedes, he is generally credited with the discovery. However, Knorr argued convincingly that using the quadratrix for squaring the circle was due to Nicomedes, who named the curve, and not Dinostratus. The mathematical techniques and arguments involved are clearly influenced by the work of Archimedes, and these were unavailable to Dinostratus, who predated Archimedes.[7]

The crucial fact, which we will not prove,[8] is that if the circular arc AB has radius r, then the segment OD in figure 11.2 has length $2r/\pi$. In particular, because we can construct this segment of length $2/\pi$ (taking $r = 1$), it is possible to construct a segment of length π, and hence, one of length $\sqrt{\pi}$.

The use of the quadratrix for squaring the circle is somewhat controversial and has been since the Greek period.[9] The commentator Sporus of Nicaea (ca. 240–ca. 300 CE) observed that the point D, where the quadratrix apparently intersects the line OA, is not well defined using our construction. The two lines OP and QS in figure 11.2, which define the quadratrix, coincide when they are horizontal, and thus they do not have a single point of intersection. However, the curve does approach D in a limiting fashion.[10]

The Conchoid

We mentioned that Nicomedes used a neusis construction to solve the problem of finding two mean proportionals. In fact, he did something slightly different. In his work *On Conchoid Lines*, he introduced the *conchoid* curve to perform a broad class of neusis constructions.[11] This way, he could use the curve rather than the marked straightedge. He used the conchoid to find two mean proportionals and to trisect an angle.

Suppose we have a point O, a line l, and a line or circle c, and we would like to find a line through O such that the distance between l and c along this line is some distance k. If we had a marked

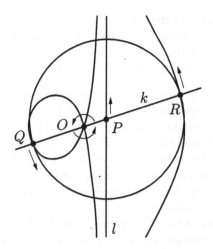

FIGURE 11.4. The conchoid.

straightedge, we could perform a neusis construction. What if we had only an unmarked straightedge? This is where Nicomedes's conchoid comes in.

Let us look at the construction of the conchoid, which is a pair of curves determined by the point O, the line l, and the value k (see figure 11.4).[12] Consider a circle with radius k and center P, where P is on l. The line OP intersects the circle at points Q and R. Now imagine sliding P along l; then Q and R trace out the two branches of the conchoid. Notice that there are infinitely many conchoids, which look quite different depending on whether the distance between l and O is greater than k (as in figure 11.4), less than k, or equal to k.

There is a clever mechanical device, which may have been invented by Nicomedes, that can draw a portion of one branch of a conchoid (figure 11.5). The *trammel of Nicomedes* consists of two pieces. A T-shaped piece has a slot running along the top of the T and a peg on the trunk of the T. Affix the T to the paper with the slot aligned along the line l and the peg over the point O. The second piece has a pencil at one end, a peg in the middle, and a slot at the other end (the distance from the peg to the pencil is the constant k). The peg (which corresponds to the point P) rides in the slot of the T, and the slot fits over the peg at O. As the pegs slide in the slots, the pencil traces out the conchoid.

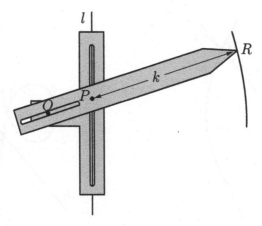

FIGURE 11.5. The trammel of Nicomedes can draw a section of the conchoid.

Here's how to use the conchoid to solve a neusis problem. Suppose we want to draw the line that passes through a point O so that the distance between two lines l_1 and l_2 is k. Then use O, l_1, and the trammel of Nicomedes to construct a conchoid (see figure 11.6). If the conchoid intersects l_2 at R, then OR is the desired neusis line. This is easy to see—if P is the point of intersection of the line OR with l_1 then, by the construction of the conchoid, PR is k units long.

Notice that l_2 may intersect both branches of the conchoid multiple times; each point of intersection gives a new neusis solution. The two dashed lines in figure 11.6 are neusis lines—one from each branch.

Let us see how to use the conchoid to trisect an angle. Essentially, it will be performing Pappus's neusis trisection (on page 149). Suppose we want to trisect $\angle AOB$ in figure 11.7 (for simplicity, assume that $\angle BAO = 90°$). Following Pappus, choose k to be twice the length of BO. Draw the conchoid determined by O, AB, and k. Now draw a line through B parallel to AO. It intersects the conchoid at some point R. Draw the line OR, which intersects AB at P. By the defining property of the conchoid, the length of PR is k. Thus $\angle AOR = \frac{1}{3}\angle AOB$.

Thus, Nicomedes made important progress in solving the problems of antiquity. He showed how the quadratrix could be used to square the circle and how his conchoid could be used to perform neusis constructions that could trisect angles and find two mean proportionals.

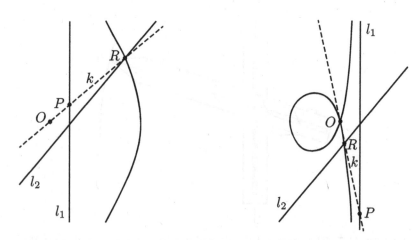

FIGURE 11.6. Using a conchoid to perform a neusis construction: the dashed line passes through O and along this line, the distance between l_1 and l_2 is k.

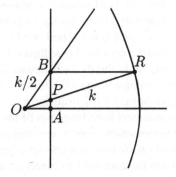

FIGURE 11.7. We can use a conchoid to trisect an angle.

Although the conchoid is not a familiar curve to today's students (or to professional mathematicians), it had a period of popularity after Pappus's *Collection* was published in 1588. Descartes, Fermat, Roberval, Huygens, and Newton all studied the conchoid. Newton in particular liked the conchoid. He believed that simplicity of description was of prime importance in geometry, and he viewed the conchoid as sitting right behind the circle. He wrote, "We ought either to exclude

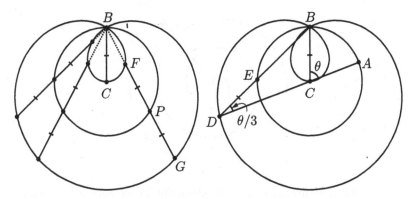

FIGURE 11.8. Points on the limaçon are a distance BC from the circle on the line BP. Archimedes's neusis result shows that $\angle ADB = \frac{1}{3}\angle ACB$.

all Lines, besides the Circle and right Line, out of Geometry, or admit them according to the Simplicity of their Descriptions, in which Case the Conchoid yields to none, except the Circle."[13]

The Limaçon of Pascal

The limaçon is another famous neusis-inspired curve that can be used to trisect an angle. This curve is not of Greek origin. It was drawn in 1525 by Albrecht Dürer (1471–1528) in his *Underweysung der Messung*. But it was first studied mathematically by Étienne Pascal (1588–1651) (father of the more famous Blaise Pascal [1623–1662]) around 1650 and is often called the limaçon of Pascal. The curve's trisecting ability was discovered later.[14]

We create the limaçon as follows. Begin with a circle with center C and radius BC, as in figure 11.8. Every line though B intersects the circle at B and at another point P, say. (When the line is perpendicular to BC, it intersects the circle at one point, in which case $B = P$.) Let F and G be the two points on the line BP that are a distance BC from P. As we let the line vary, these points trace the limaçon.[15]

Suppose we wish to trisect an angle θ. Sketch the angle so that $\theta = \angle ACB$, where A is on the circle (see figure 11.8). Extend AC so that it meets the limaçon at D. Draw the line BD, and suppose it intersects

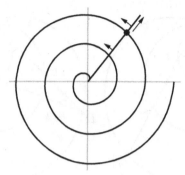

FIGURE 11.9. As the ray rotates and the point moves outward, the point traces an Archimedean spiral.

the circle at E. By the defining property of the limaçon, $DE = BC$. Thus, by Archimedes's neusis construction, $\angle ADB = \frac{1}{3}\angle ACB$.

Spiral of Archimedes

Archimedes gave us our modern understanding of π, he squared geometric shapes with curved boundaries, he discovered a trisection method using a neusis construction, and he constructed a regular heptagon with a neusis-like procedure. He also introduced a spiral that could divide an angle into n equal parts, construct every regular n-gon, and square and rectify the circle.

In *On Spirals*, Archimedes wrote,[16]

> If a straight line drawn in a plane revolve at a uniform rate about one extremity which remains fixed and return to the position from which it started, and if, at the same time as the line revolves, a point move at a uniform rate along the straight line beginning from the extremity which remains fixed, the point will describe a *spiral* in the plane.

Figure 11.9 shows three turns of the Archimedean (linear) spiral.

It is easy to use the spiral to trisect an angle—or more generally to divide the angle into n equal parts. Suppose we begin with an angle $\theta = \angle AOB$, as in figure 11.10. Construct a spiral with the generating ray pointing along OA. Let's say the first intersection of the spiral and

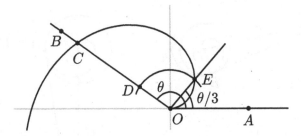

FIGURE 11.10. Archimedes used the spiral to trisect an angle.

the ray OB is C. Then use a compass and straightedge to trisect the line segment OC; say $OD = \frac{1}{3}OC$. Construct a circle with center O and radius OD. It intersects the spiral at E. Draw ray OE. Then $\angle AOE = \frac{1}{3}\angle AOB = \frac{1}{3}\theta$. The explanation is very similar to that for the quadratrix. By the definition of the spiral, because D is 1/3 of OC, the ray has turned 1/3 of the way to the ray OC.

Notice that we could use a similar technique to divide an angle into any number of equal parts. In particular, if we start with a 360° angle, then we can construct the angle $(360/n)°$ for any n. Thus the spiral enables us to construct any regular polygon.

Once we see the angle division properties of the spiral, we can't help but imagine that it was devised for this purpose. However, it is not obvious that the spiral can be used for squaring the circle. This is where Archimedes's brilliance becomes apparent.

Suppose we have a spiral with center O. Let A be the endpoint of the first turn of the spiral (see figure 11.11). Let B be a point on this first turn of the spiral. Construct a line OC perpendicular to OB. Archimedes then showed how to construct a tangent line to the spiral at B. Finding tangent lines to curves became an important topic of conversation 2000 years later when calculus was taking shape. It is likely that this was the first tangent line to a curve other than a circle. Say that the tangent line intersects OC at D. Construct a circle with center O passing through B. Let E be the point of intersection of the circle and OA. Archimedes then proved that segment OD has the same length as the arc of the circle from E to B.

Consider the case in which $A = B$ (which is also E), as in the right-hand diagram in figure 11.11. In this case, OD has the same length as the circumference of the circle with radius OA. Thus, Archimedes used the spiral to rectify the circle! As we know, this is equivalent to being

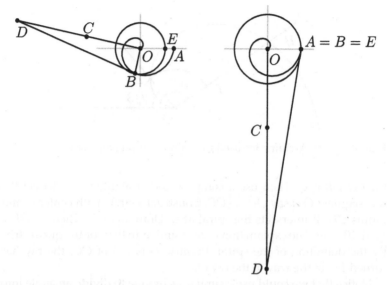

FIGURE 11.11. The length of *OD* equals the length of the circular arc from *E* to *B*. The area of triangle *ADO* (right) equals the area of the circle with radius *OA*.

able to square the circle. Notice that one leg of the triangle *ADO* is the circumference of the circle and one leg is the radius. By Archimedes's work in *Measurement of a Circle*, the triangle and the circle have the same area, and it is straightforward to square this triangle.

One final comment on spirals: When many people think of a spiral, they think of the spiral they see in the cross section of a nautilus shell or in the arrangement of seeds on the head of a sunflower. These are logarithmic spirals, not Archimedean spirals. Logarithmic spirals were introduced by Descartes and were studied by many subsequent mathematicians, including Jacob Bernoulli (1655–1705) who named them *spira mirabilis*, which is Latin for "miraculous spiral."[17] It turns out we can use a logarithmic spiral to divide an angle into any number of equal parts, to construct every regular polygon, and to double the cube. This was shown in a 2012 article by Pietro Milici and Robert Dawson in which they also introduced a compass that can draw a logarithmic spiral.[18]

We cannot give all of the curves that have been invented or repurposed to solve the problems of antiquity. We can't even give all of the Greek curves. For instance, around 100 BCE, Diocles (ca. 240–ca. 180 BCE) introduced the cissoid in his work *On Burning-Glasses* to double the cube.[19] Pappus, in his *Collection*, presented two techniques (without giving the sources) for using hyperbolas to trisect angles.[20] Iamblichus wrote that Apollonius squared the circle using a "sister" curve to the conchoid and that Carpus squared the circle using a curve "of double motion"; but, alas, we do not know what these curves are.[21]

TANGENT
Carpenter's Squares

Then stayed the fervid wheels, and in his hand
He took the golden compasses, prepared
In God's eternal store, to circumscribe
This Universe, and all created things.
One foot he centered, and the other turned
Round through the vast profundity obscure,
And said, "Thus far extend, thus far thy bounds,
This be thy just circumference, O world!"
—John Milton, *Paradise Lost* [1]

IN CHAPTER 5 WE saw the ancient Greek method of finding two mean proportionals using two carpenter's squares (see page 77).[2] In 1928 Henry Scudder showed that he could trisect angles using just one.[3] His carpenter's square needs a mark on one leg so the distance from the corner is twice the width of the other leg—say, one leg is 1 inch wide and there's a mark 2 inches from the corner on the other leg.

To trisect $\angle AOB$ in figure T.18, draw a line l parallel to and 1 inch away from AO; this can be accomplished using a compass and straightedge or the leg of the carpenter's square as a double-edged straightedge. Now perform the step that is impossible using Euclidean tools: Place the carpenter's square so that the inside edge passes through O, the 2-inch mark lies on the line BO, and the corner sits on l (at the point C). The inside edge of the carpenter's square and the line CO trisect the angle. This procedure works for any angle up to 270°, although the larger the angle, the narrower the short leg must be.

In fact, we don't need a carpenter's square to carry out this construction. All we need is a T-shaped device in which the top of the T is 2 inches long (see figure T.18). To see that this device trisects the angle, observe that triangles CFO, CEO, and DEO are congruent. This device was the inspiration for a new compass that can draw an

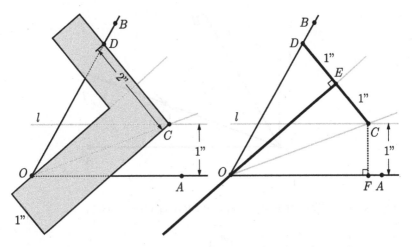

FIGURE T.18. A carpenter's square or a T-shaped tool can be used to trisect an angle.

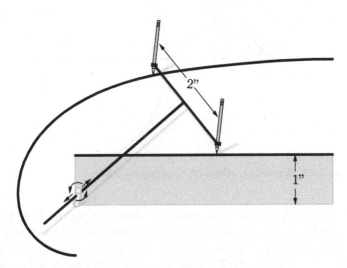

FIGURE T.19. A compass to draw the carpenter's square curve.

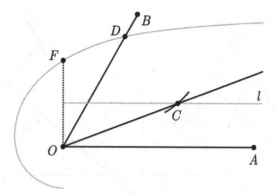

FIGURE T.20. The carpenter's square curve is a trisectrix.

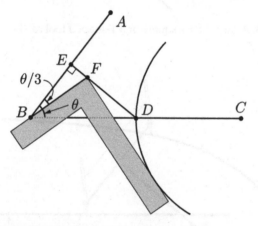

FIGURE T.21. Brooks's method of trisecting an angle.

angle-trisecting curve.[4] The compass, shown in figure T.19, has a straightedge that is 1 inch wide and a T-shaped tool with two pencils. The long arm of the T passes through a device at one corner that can rotate and that allows the T to slide back and forth. One pencil draws a line along the straightedge, the other draws the *carpenter's square curve*.

We use the compass as follows. Suppose we would like to trisect ∠AOB in figure T.20. Place the bottom of the straightedge along OA with the corner (to which the T is attached) at O. Use the compass to draw the straight line *l* and the carpenter's square curve. Say that BO intersects the curve at D. Open an ordinary compass 2 inches. (One way to do this is to draw a line perpendicular to OA at O that intersects the carpenter's square curve at F. It turns out that OF is 2 inches.) Use the compass to draw a circle with center D. It will intersect l at two points. Let C be the rightmost point (viewed from the perspective of figure T.20). Then OC trisects the angle.

In 2007 David Allen Brooks found a way to trisect an angle using an *unmarked* carpenter's square.[5] To trisect $\angle ABC$, bisect BC at D, then draw the segment DE perpendicular to AB (see figure T.21). Draw a circle with center C and radius CD. Next, arrange the carpenter's square so that one edge goes through B, one edge is tangent to the circle, and the vertex F sits on DE. Then $\angle ABF = \frac{1}{3}\angle ABC$.

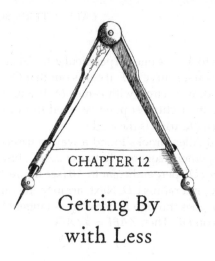

CHAPTER 12

Getting By
with Less

I were better to be eaten to death with a rust than to be scoured
to nothing with perpetual motion.
—William Shakespeare, *King Henry IV, Part 2*[1]

RESTRICTING OUR GEOMETRIC tools to a compass and straight-edge severely limits the number of constructions we can accomplish. In the preceding chapters, we have added tools and curves to our arsenal. In this chapter we take the opposite approach: Can we get by with less? What if we restrict our set of tools further? What if our compass is rusty and won't open? What if we throw out our compass entirely? What if we get rid of our straightedge?

The Rusty Compass

A compass is tedious to use. One must constantly adjust it to the desired opening. If it is an expensive compass that is held fast with a screw, then it is a tiresome and time-consuming process to adjust it. If it is a cheap discount-store compass that stays in place using friction only, then after it has swept out 360°, the jaws may have opened or closed slightly and the pencil doesn't close up the circle. All compasses, high or low quality, are constrained by some maximum

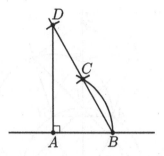

FIGURE 12.1. Constructing a perpendicular line at the point A using a rusty compass.

radius that they can open to (and small circles are a challenge as well). Consequently, there is something to be said for using a compass that is permanently opened to a fixed radius. In the literature, these compasses are known as "fixed" or "rusty" compasses. (In 1694 the British surveyor William Leybourn called the device a "meat fork."[2])

The Greeks were interested in rusty compass constructions, although we did not learn this until the late twentieth century when an Arabic translation of Pappus's *Collection* surfaced containing previously unknown sections. Pappus included a section on the rusty compass because of its practical usefulness.[3] There is no evidence that he wanted to create a new theory of constructibility. He described how to perform constructions such as drawing a perpendicular line from a point on a line, dividing a line segment that is longer than the opening of the compass into n equal parts, doubling the length of a segment, and so on.

Figure 12.1 shows his technique for drawing a line perpendicular to a given line and through a point A on the line. Draw a circle with center A. Let B be a point of intersection of the circle and the line. Draw a circle with center B; it intersects the first circle at C. Draw line BC. Finally, draw a circle with center C; it intersects this second line at D. Because ABD is a 30°–60°–90° triangle, AD is the desired line.

As far as we know, Mohammad Abu'l-Wafa al-Buzjani (940–997/98) was the next to investigate the rusty compass. Abu'l-Wafa was born in Buzjan (present-day Iran) and died in Baghdad (present-day Iraq). He translated and wrote commentaries on the works of Diophantus of Alexandria (fl. 250 CE), Hipparchus of Nicaea (ca. 190–ca. 120 BCE), Euclid, and Al-Khwārizmī (ca. 780–ca. 850). He wrote books on

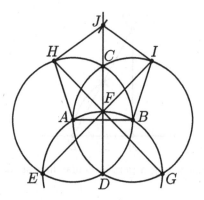

FIGURE 12.2. Mathes Roriczer's approximate construction of a regular pentagon using a rusty compass.

practical arithmetic, geometry for artisans, and astronomy; they contained well-known mathematics and new results. He also created improved trigonometric tables.

Abu'l-Wafa presented a number of rusty compass constructions, such as constructing a regular pentagon, octagon, and decagon on a line segment given a compass open to the length of the segment; and inscribing a square or a regular pentagon in a circle using a compass open to the radius of the circle.[4]

Medieval artisans, artists, architects, craftsmen, masons, and carpenters required a practical understanding of geometry. For instance, they may have had to construct regular polygons for decorative or practical purposes, such as locating spokes on a wheel. Most of these tricks of the trade were transmitted from one generation to the next through apprenticeships and were never put in writing. But one of the rare written accounts—a short, untitled, anonymous technical pamphlet printed at the end of the 1480s—contains an elegant approximation of a regular pentagon using a rusty compass. This pamphlet was later printed with the name *Geometria Deutsch* (*Geometry in German*) and we now believe it was written by the architect Mathes Roriczer.[5]

Begin with a rusty compass opening to the length of segment *AB* (see figure 12.2). Draw circles about *A* and *B*. They meet at *C* and *D*, which we connect with a line. Draw a circle with center *D*. It meets *CD* at *F* and the two circles at *E* and *G*. Lines *EF* and *FG* meet the circles at H and I. Finally, draw a circle with center I. It meets CD at J. Then $ABIJH$ is a good approximation to a regular pentagon.[6]

Until the discovery of the *Geometria Deutsch* it was believed that this construction was discovered by Albrecht Dürer, for it was also in his 1525 work *Underweysung der Messung*.[7] Interestingly, the *Underweysung* also contains some Euclidean constructions, which shows that Dürer had access to Euclidean mathematics even at this early date when they were just becoming available again.

Even Leonardo da Vinci got in on the fun. There are at least 10 examples of rusty compass constructions in his notebooks. But like his contemporaries, he viewed the rusty compass as a tool for artisans and craftsman and not a subject worthy of theoretical investigation.

During this early period of investigation, the emphasis was on practical applications, and in this spirit, the compass was often set to an opening that was useful to the construction. During the sixteenth century, attention shifted to the theoretical question of what constructions can be performed with a rusty compass in which the geometer had no control over the opening—"when the aperture was any invariable one proposed by an adversary."[8]

At this time, a soap-opera-like drama took place in Italy over the solving of cubic equations. It featured broken promises, insulting letters, and mathematical duels between rival sides. We will say more about this story in chapter 13, but in short, Gerolamo Cardano (1501–1576) published a book containing his method for solving cubic equations, but in doing so, he broke a promise not to reveal some secret information given to him by Niccolò Tartaglia (1499 or 1500–1557). Thus began a rivalry between Tartaglia and Cardano—often with Cardano's student (and eventual colleague), the hot-headed Ludovico Ferrari (1522–1565), fighting on his behalf.

In 1547 Ferrari challenged Tartaglia to a mathematical duel. They each proposed 31 mathematical puzzlers for the other person to solve. The first 17 of Tartaglia's problems were geometric constructions requiring a rusty compass.[9] In a bold move, Ferrari did not solve the problems per se, but rather he proved all the propositions of the first six books of Euclid's *Elements* under the rusty compass assumption (in essence, he changed the third postulate).[10] Of course, he omitted the propositions that end with the construction of a circle.[11] Because he did so, all of Tartaglia's problems could be solved.

In 1556 Tartaglia gave his own proof of Euclid's propositions, and he claimed he knew this was possible all along.[12] While it can be expected that Tartaglia knew how to perform the constructions that he sent to Ferrari, we have no evidence, other than his word, that he knew it was possible to prove "all of Euclid." Perhaps this is true, but his work appeared after the published proofs of Ferrari in 1547 (proofs of 12 key propositions were subsequently republished by Cardano in 1550)[13] and Tartaglia's student Giambattista Benedetti in 1553.[14]

The geometry of the fixed compass had a habit of being continually rediscovered. From 1560 to 1700 no fewer than 13 mathematicians wrote about the rusty compass.[15] Many of the appearances consisted of a few constructions—they were viewed more as curious puzzlers than deep mathematical constructions. An exception was an anonymous 24-page pamphlet *Compendium Euclidis curiosi*[16] published in Amsterdam in 1673 and translated into English in 1677, which showed that it was possible to prove all of Euclid using a rusty compass.[17] In the introduction, the author admitted that he had heard that this was possible, but could not find any specific references, so he gave his own proofs. We will return to this mysterious anonymous writer on page 187.

While the Italians proved that it was possible to prove "all of Euclid" by substituting a rusty compass for a Euclidean one, there is no evidence that they made the next, more abstract step of saying that everything constructible by a compass and straightedge is constructible by a rusty compass and a straightedge. It appears that this anonymous author may have made this connection. In the preface, he asserted that he could have included other constructions, but "considering that all flat or plain (plane?) operations may be reduced from these [operations given in this work], these shall suffice."[18]

Of course, as the Italian geometers knew, these construction methods cannot be equivalent. If the opening of the rusty compass is set to 1 inch, then it is impossible to draw a circle with radius 2 inches. But as the Frenchman Jean Victor Poncelet (1788–1867) observed, that is not the right way to think about these constructions.

Poncelet had an interesting life and career. He studied for three years at the École Polytechnique. Then he joined the corps of military engineers, and as a lieutenant in the French army participated in Napoleon's Russian campaign. He was taken prisoner during a retreat and spent more than a year and a half in a camp on the Volga River at Saratov. During his captivity, he worked on mathematics—without any books or collaborators—the notes of which were later published.

During his free time as a military engineer, he worked on geometry (projective geometry in particular). He made important contributions, but was somewhat soured by criticisms by Augustin-Louis Cauchy (1789–1857) and by a priority dispute that lead to tensions with Joseph Gergonne (1771–1859) and Julius Plücker (1801–1868). In 1824 he became a professor of "mechanics applied to machines," and subsequently his research and teaching were focused on applied mechanics. He designed an improved waterwheel and a new type of drawbridge. He is one of the 72 scientists, mathematicians, and engineers whose name is engraved on the Eiffel Tower.

In 1822 Poncelet made the astute observation that to prove that a set of construction tools is equivalent to the compass and straightedge, we should turn our attention away from the drawing of lines and circles and focus instead on *constructible points*—points that can arise as intersections of the lines and circles.[19] In particular, we can boil things down even further and focus on three things. We must show that our construction tools can, like a compass and straightedge, (1) find the point of intersection of two lines, (2) find the points of intersection of a line and a circle, and (3) find the points of intersection of two circles.

If we can accomplish these three things with a rusty compass and a straightedge, then the two sets of drawing tools are equivalent. In the case of the rusty compass we may rephrase (2) as, given a line and two points A and B, can we find the points of intersection of the line and the circle that has center A and passes through B? That is, we would need to find the points of intersection even if the circle itself was never drawn. The third criterion can be similarly rephrased.

When Poncelet made this observation he was considering a slightly different, more general problem that we discuss later in the chapter. A decade later, Jakob Steiner gave a complete and rigorous proof. We can use this Poncelet–Steiner theorem to prove that (1) through (3) are possible using a straightedge and a rusty compass, and thus every construction that can be accomplished using a straightedge and compass can be done with a straightedge and a rusty compass.

Compass-Only Constructions

Lorenzo Mascheroni (1750–1800) taught mathematics at the University of Privia near Milan. He was also a talented poet; there are several Italian editions of his poetry. Nathan Court wrote, "To consider him as 'the greatest poet among mathematicians' is to belittle him, for literary men are just as eager to claim him as one of their own as mathematicians are to consider him as belonging to their clan."[20]

Mascheroni argued that because straightedges were rarely straight and they tended to skid, compasses were inherently more reliable. In his 1797 *Geometria del compasso* (*Geometry of the Compass*),[21] he investigated the geometric constructions that could be accomplished using only a compass.[22] In this approach, we can draw circles of any size—just as in the classical case—but now we have no straightedge to draw lines or line segments. He considered a line drawn if he had constructed two points on it.

Throughout his work, Mascheroni used a locking compass, not a Euclidean collapsing one. And Euclid's proof that anything that can be accomplished with a locking compass can be accomplished with a collapsing compass (see page 50) required the use of a straightedge. However, in 1890 August Adler (1863–1923) proved that Mascheroni did not lose any generality by using a locking compass.[23]

Mascheroni had an unusual admirer: Napoleon Bonaparte. Napoleon liked mathematics generally, and geometry in particular. Mascheroni sympathized with the French Revolution and was a fan of Napoleon. His 1793 book *Problemi per gli agrimensori* (*Problems for Surveyors*) contained a dedication, written in verse, to the military leader. The two met in 1796 when Napoleon invaded northern Italy, and they discussed their shared interest in geometry. Later, Mascheroni wrote a poem at the start of his *Geometria del compasso* for Napoleon called "A Bonaparte l'Italico." Mascheroni called one problem in his book "Napoleon's problem" because Napoleon enjoyed setting it as a challenge to his engineers: given a circle and its center, use only a compass to divide the circumference into four equal parts.[24] In 1797, when Napoleon showed the mathematicians Joseph-Louis Lagrange (1736–1813) and Pierre-Simon Laplace (1749–1827) some of Mascheroni's constructions, Laplace reportedly said, "General, we expected everything of you except lessons in geometry."[25]

Mascheroni's work became well known throughout Europe. *Geometria del compasso* was quickly translated into French and in 1825, German. Court wrote, "By outbidding Plato in 'puritanism' Mascheroni brought the question of the role of construction tools in geometry to the fore at a time that was ripe and ready to deal with it."[26]

As with the rusty compass, to prove that compass-only constructions are equivalent to compass-and-straightedge constructions, it

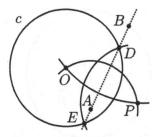

FIGURE 12.3. Using a compass to find the points of intersection of *AB* and *c*.

suffices to show that we can find the intersections of two lines, a line and a circle, and two circles. To get a feeling for such constructions, we will show one special case: we find the points of intersection of a circle *c* with center *O* and the line determined by points *A* and *B* in which *A*, *B*, and *O* are not collinear (see figure 12.3). Draw circles with centers *A* and *B* passing through *O*. They intersect at *O* and at another point, *P*. Draw the circle with center *P* and the same radius as *c*. This is the reflection of *c* across *AB*. Thus, it intersects *c* at the sought-after point or points (*D* and *E* in figure 12.3).[27] The case in which *A*, *B*, and *O* are collinear requires significantly more steps.[28]

Mascheroni's construction of the point of intersection of two non-parallel lines required drawing 11 circles, most of which were drawn with a locking compass. Rather than giving his construction, we describe an approach given by Adler in 1890.[29] Adler's construction requires drawing 36 circles. But it is the idea of his approach that is more important than its efficiency.[30]

Adler used *inversions*—an idea that Steiner introduced in 1824. We are all familiar with reflections across a line. The reflection of a point *A* across a line *l* is a point *B* that is the same distance from *l* as *A* that sits on the line perpendicular to *l* through *A*. An inversion is a reflection across a circle and is defined as follows:[31] Let *c* be a circle with center *O* and radius *r*. The inversion of a point *P* (that is not *O*) across *c*, which is *P'* in figure 12.4, lies on the ray *OP*, and the distance from *O* is $r^2/|OP|$.

It is not difficult to see that if *P'* is the inversion of *P*, then *P* is the inversion of *P'*. If one point is inside the circle, then the other is outside. In fact, when one point is close to *O*, the other is far away; the point *O* has no inversion, or it can be taken to be the "point at infinity." When *P* is on the circle, it is its own inversion.

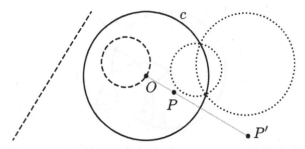

FIGURE 12.4. The inversion of P across c is P'. Also, the inversions of circles and lines are either circles or lines.

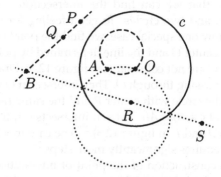

FIGURE 12.5. Using inversions to find the intersection of two lines.

Inversions take circles and lines to either circles or lines. In particular, the inversion of a circle is another circle unless it passes through O, in which case it is a line not passing through O. Conversely, the inversion of a line is a circle through O unless the line passes through O, in which case it is its own inversion. In figure 12.4, the dashed line and the dashed circle are inversions of each other, as are the two dotted circles.

So given points P, Q, R, and S, we find the point of intersection of lines PQ and RS using inversions as follows: Construct any circle c with center O (see figure 12.5). Invert the two lines across c to obtain two circles, both of which go through O and a second point A. Now invert A across c to obtain the point of intersection of the lines, B. The key point for us is that we can do these steps using only a compass.[32]

There are many theorems named after someone other than the discoverer. As the musician David Bowie said, "It doesn't matter who does it first. It matters who does it second."[33] Compass-only constructions are often called Mascheroni constructions. But we now know that Mascheroni was not the first to analyze such constructions.

In 1927 a student browsing a used bookstore came upon an old book on geometry called *Euclides Danicus*[34] by someone named Georg Mohr (1640–1697). He was intrigued. He bought the book and showed it to his professor, Johannes Hjelmslev (1873–1950) of the University of Copenhagen. Fortunately, Hjelmslev recognized the book for what it was—a book on Mascheroni constructions. However, the book was published in 1672, 125 years before Mascheroni published *Geometria del compasso*! Hjelmslev published a German translation the following year.

It is not surprising that *Euclides Danicus* was almost lost to history. There was very little interest in it at the time, partly, no doubt, because it was published only in Danish and Dutch. Also, Mohr did not do a good job of letting the reader know that it contained results that were new and noteworthy. Seidenberg wrote, "It would be easy for an inattentive reader to misjudge the value of the work."[35] Contemporary references to it seem to imply that it was a compilation of Euclid's *Elements*. It is no wonder Mascheroni was unaware of Mohr's work (in the preface, Mascheroni claimed his work was original).

Yet Mohr was not unknown in his time. He was born in Copenhagen but left Denmark to study mathematics and philosophy in the Netherlands. Apparently, he felt enough at home in the Netherlands to get involved in the French–Dutch conflict of 1672–73; he even spent some difficult time imprisoned by French troops. He traveled extensively, spending time in France and England. He met a number of other mathematicians, including Leibniz, who referred to "the Dane Georg Mohr very well versed in geometry and analysis"[36] in a May 12, 1676, letter to Henry Oldenburg (ca. 1619–1677), secretary of the Royal Society of London.

The rediscovery of *Euclides Danicus* is amazing, but it isn't the end of the Georg Mohr story. Earlier in the chapter we discussed the anonymous booklet on the rusty compass called *Compendium Euclidis curiosi*. The authorship of this work was unknown until the 1939 publication of a collection of James Gregory's (1638–1675) correspondences. In a letter to Gregory on September 4, 1675, John Collins (1625–1683) wrote,[37]

There being present with him a Dane George Moohr who lately published in low Dutch, two little Bookes the one named Euclides Danicus where he pretends to perform all of Euclids problems with a paire of Compasses only without Ruler, and another intituled Euclides Curiosus, wherein with a Ruler and a forke (or the Compasses at one opening) he performs the same.

The *Curiosi* attracted more attention than *Euclides Danicus*; it was even translated into English in 1677. We don't know why Mohr published it anonymously. As Kirsti Andersen wrote, "The non-interest in *Euclides Danicus* and the anonymity of *Euclides Curiosus* made Mohr's name quickly forgotten."[38] We are happy that Mohr is getting the recognition he deserves.

Straightedge-Only Constructions

Anything Euclidean constructible is constructible using only a compass. What about straightedge-only constructions? They go back to at least 1759 when they were studied by Johann Lambert (1728–1777), who we will encounter in chapter 21 for his work on π. But a satisfactory answer to the question did not come until the nineteenth century.

We cannot construct everything with only a straightedge that we can with a straightedge and a compass. Here's one way to see it. Euclid's first proposition in book III is to find the center of a circle. This is impossible using only a straightedge. David Hilbert (1862–1943) gave a nice argument, which we sketch here.[39] It is a proof by contradiction. Suppose it *is* possible. Then there is a sequence of straightedge steps that ends with two lines crossing at the center of the circle. He then described a transformation of the plane—think of distorting space so that points move around—that had some very important properties: the circle remains circular, all lines remain lines, and the center of the circle moves to a location that is no longer the center. Then the same sequence of lines (or more precisely, the images of the lines under the transformation) should still find the center of the circle, but it finds the old center, which is no longer the center under Hilbert's transformation. That's a contradiction.

Although straightedge-only constructions are not equivalent to straightedge-and-compass constructions, in 1822 Poncelet conjectured that if we have one circle—*any* circle—and its center, then we can construct every Euclidean constructible point using only a straight-edge.[40] Poncelet was correct, but he didn't prove his conjecture. The Swiss mathematician Jakob Steiner gave the first rigorous proof in 1833.[41]

Steiner was born in Bern, Switzerland, the son of a farmer and tradesman. Because he had to work for his parents, he did not receive much of an education. He learned to write when he was 14 years old. Nevertheless, late in his teens he left home to seek a proper education, eventually becoming a professor at the University of Berlin.

Although Steiner made contributions in a number of areas of mathematics, he is most well known for his brilliant and fundamental work on synthetic geometry—that is, geometry like Euclid's, not the analytic geometry that Descartes and others had made fashionable. Steiner was equally at home finding the overarching themes that tied together seemingly unrelated theorems as he was solving specific problems.

Heinrich Dörrie made the bold claim that Steiner was "the greatest geometer since the days of Apollonius."[42] Steiner's contemporary Carl Jacobi (1804–1851) wrote,[43]

Starting from a few spatial properties Steiner attempted, by means of a simple schema, to attain a comprehensive view of the multitude of geometric theorems that had been rent asunder. He sought to assign each its special position in relation to the others, to bring order to chaos, to interlock all the parts according to nature, and to assemble them into well-defined groups.

According to Johann Burckhardt, "Students and contemporaries wrote of the brilliance of Steiner's geometric research and of the fiery temperament he displayed in leading others into the new territory he had discovered." However, "he often behaved crudely and spoke bluntly, thereby alienating a number of people."[44]

We omit the proof of the Poncelet–Steiner theorem, but we will give an example of a straightedge-only construction.[45] Let's look at Steiner's third problem, which is to construct a line through a point P that is perpendicular to a line AB (see figure 12.6).[46] We may assume that we are able to construct a line parallel to any given line through any given point (this was Steiner's first problem). Recall that we are assuming we are given some circle with center C passing through a point D. First, construct a line parallel to AB through D. This line meets the circle again at E. Now draw the line CD, which meets the circle again at F. Then DF is a diameter of the circle, and hence EF is

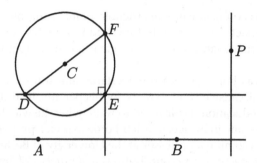

FIGURE 12.6. A straightedge-only construction of a line through P perpendicular to AB.

perpendicular to DE and AB. Finally, construct a line parallel to EF through P.[47]

No good mathematician can leave the hypotheses of a theorem unexamined. Can we do better? Can we shave off some of the hypotheses? Do we really need the center of the circle? We do. As we described earlier, it is impossible to find the center of a circle using only a straightedge. (The first edition of Steiner's book did not explicitly mention that the center of the circle was required, which was an error.)

Hilbert asked whether we could replace a circle and its center with two or three circles without their centers.[48] In other words, if we had two or three circles, could we use a straightedge to find the center of one of them? Detlef Cauer (1889–1918) discovered that in general two nonintersecting circles without centers is not enough. However, if the two circles intersect or are concentric, then we do not need their centers. And any three circles without their centers is also sufficient.[49]

Finally we ask: Do we need an entire circle, or can we get by with a semicircle or a quarter circle? In 1904 Francesco Severi (1879–1961) proved that if we have the center of a circle and an arc of the circle, regardless of how small, the conclusion of the theorem holds.[50] So if we sit down, compass and straightedge in hand, and start to draw our very first circle, and having only drawn a fraction of the circle the compass breaks, we would be fine. We could continue with only a straightedge and perform the needed constructions.

As a final comment, recall that on page 183 we stated that the rusty compass theorem was a corollary of the Poncelet–Steiner theorem.

Indeed, if we had a rusty compass and a straightedge, we could use the compass to draw a circle (or the arc of a circle). Then we could put the rusty compass back in our drawer and by the Poncelet–Steiner theorem, complete the rest of our construction using only the straightedge! Thus, anything constructible using Euclidean tools is constructible with a straightedge and a rusty compass.

TANGENT
Origami

God is a circle himself, and he will make thee one; go not thou about to
square either circle, to bring that which is equal in itself to angles and
corners, into dark and sad suspicions of God, or of thyself, that God can
give, or that thou canst receive, no more mercy than thou hast had already.
—John Donne, Christmas Eve sermon, 1624[1]

CHILDREN LOVE FOLDING colorful squares of origami paper to
make swans, cranes, frogs, flowers, boxes, and boats. Expert origami
artists make stunningly complex objects from this same simple square.
Although origami has historically been an artistic pursuit, it has
recently attracted the attention of mathematicians. Some mathematicians address the practical: Can we create an origami blueprint for an
object? What pattern of folds would yield a particular shape?

Here we approach origami with a more theoretical question. In
Euclidean constructions the basic objects are lines, circles, and their
points of intersection. In origami the basic objects are lines—the fold
lines—and the points of intersection. Thus we ask: What points and
lines are constructible using origami? Can we trisect an angle? Double
the cube? Square the circle? Construct regular polygons?

Just as we have restrictions on how we can use a compass and
straightedge, there is a well-defined set of allowable paper-folding
techniques.[2] Some origami rules have obvious Euclidean counterparts.
The Euclidean assumption that two points determine a line has a clear
origami analogue: it is possible to fold a paper through any two points.
Another folding move allows us to fold a line l_1 onto a line l_2 to form
a new line l_3. Then l_3 is either parallel to and equidistant from l_1 and
l_2 or it bisects one of the angles made by l_1 and l_2. In either case, l_3 is
Euclidean constructible. In fact, every point that is constructible using
a straightedge and compass is constructible using origami.[3]

What makes origami constructions interesting is that there is a
fold that—like the neusis construction—has no Euclidean counterpart.

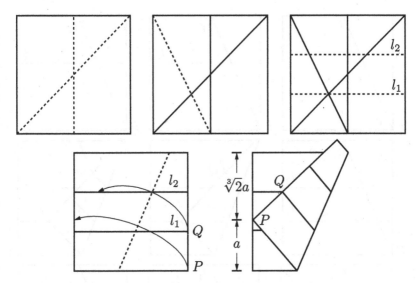

FIGURE T.22. Messer's origami cube doubling.

Suppose we have two distinct points p_1 and p_2 and two nonparallel lines l_1 and l_2. We can fold the paper so that p_1 lies on l_1 and p_2 lies on l_2.[4] This move allows us to construct non-Euclidean points, and hence enables us to solve some of the problems of antiquity.

In 1986 Peter Messer discovered the following origami technique for doubling the cube.[5] The first three steps in figure T.22 yield lines l_1 and l_2, which divide the paper into thirds. It is the last fold that cannot be accomplished using the Euclidean tools. Fold the square so that the point Q lies on l_2 and P lies on the left edge of the square. Then P divides the left edge of the square into segments of length a and $\sqrt[3]{2}a$.

In the 1970s, Hisashi Abe discovered an ingenious method of angle trisection using origami.[6] Assume, as in figure T.23, that we have an acute angle θ formed by the bottom edge of the paper and a line l_1 through the lower-left corner P. Create an arbitrary horizontal fold l_2, then fold the bottom edge up to l_2 to form l_3. Let Q be the left endpoint of l_2. Fold the paper so P and Q meet l_3 and l_1, respectively. With the paper still folded, refold l_3 to create a new fold l_4. Open the paper, and extend l_4 to a full fold extending to P. Finally, fold the lower edge of the paper up to l_4 to create l_5. Lines l_4 and l_5 trisect the angle θ.

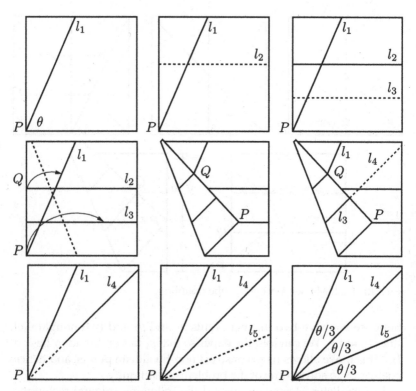

FIGURE T.23. Abe's method of angle trisection using origami.

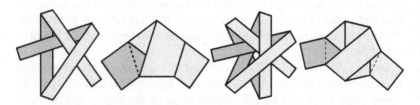

FIGURE T.24. A pentagon and a heptagon by knotting paper.

It turns out that origami gives us precisely the same set of constructible points that we obtain from a compass, straightedge, and conics.[7] In particular, it is possible to fold the elusive heptagon and nonagon.[8] And, perhaps not surprisingly, origami is no help in squaring the circle.

Although it is not origami, we can form regular polygons from paper in another way.[9] Figure T.24 shows that if we take a long strip of paper, knot it in the method shown, and carefully tighten the knots, we can obtain a regular pentagon and a regular heptagon.

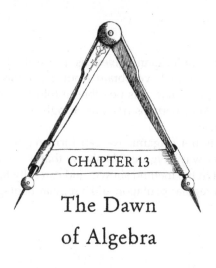

CHAPTER 13

The Dawn
of Algebra

Algebra is generous: she often gives more than is asked for.
—Jean d'Alembert[1]

WE CAN'T FAULT the Greeks—the masters of geometry—for being unable to solve the problems of antiquity, for they are impossible. We also cannot fault them for not *proving* that the problems are impossible to solve. The problems are geometric, but the proofs of impossibility are not geometric; they are algebraic. And algebra did not exist during the Hellenistic period.

Although there have been glimpses of algebra and algebraic thinking throughout the history of mathematics, algebra as we know it—school algebra that teenagers learn and certainly the algebra required to prove the impossibility of these problems—is only a few hundred years old.

In this chapter we take a break from the problems of antiquity and give a whirlwind tour of the history of algebra from the most ancient cultures up to the sixteenth century. It is by necessity a superficial history and hits only some of the major milestones.

What Is Algebra?

Before we discuss the history of algebra, we must be sure we know what algebra is. When we think back to our high-school algebra class, we remember finding the roots of polynomials by symbolic manipulation, factoring, and the quadratic formula; solving systems of equations; puzzling through word problems; and graphing. And indeed, this list captures many of the features of algebra.

Algebra is a field of mathematics in which abstraction plays a large role. As the textbook word problems show, the exact same equation of x's and y's could represent inventories of a clothing store, lengths of the sides of a right triangle, dollars invested in a bank, or positions of cars on a roadway. In an algebra class, the students learn a new algebraic technique and then are asked to apply it to a variety of problems. This universal nature of algebra is one feature that makes it so powerful.

Not only is it abstracted in a way that it can be applied to countless situations, it is abstracted so that it applies to numbers without necessarily having to work with any particular numbers. The algebraic expression $x^2 - y^2 = (x + y)(x - y)$ is true when $x = 1$ and $y = 2$ or when $x = 1,000,001$ and $y = \sqrt{2}$.

Also important is that an algebraic mathematical argument is expressed using a common set of symbols and few—if any—words. We use letters to represent variables and arbitrary constants; $+$, $-$, $\times$, $\div$, and $\sqrt{}$ for operations; $=$, $<$, and $>$ for relations between values; positive, negative, fractional, and real exponents for powers; and so on. This universal language allows mathematicians to ignore the context of the problem and instead focus on the mathematics.

Algebra consists of methods for formally manipulating expressions. If two expressions are equal and we add the same quantity to both, they are again equal. We can apply algebraic rules and procedures to modify an expression, such as $(a + b)^3 = a^3 + 3a^2b + 3ab^2 + b^3$. We can express complicated mathematical procedures succinctly and unambiguously. Given a quadratic equation of the form $ax^2 + bx + c = 0$, we can find the two roots using the quadratic formula

$$x = \frac{-b \pm \sqrt{b^2 - 4ac}}{2a}.$$

Today's algebra students are often asked to graph algebraic expressions. The graph of $y = 2x + 3$ is a line, $x^2 + y^2 = 1$ is a circle, $y = x^2$ is a parabola, and so on. This is not pure algebra; it is a highly useful melding of algebra and geometry, known as analytic geometry. Analytic geometry has become so ingrained in the way we do mathematics that we think of it more as algebra than as geometry. (When we think of geometry, we picture Euclid's coordinate-free synthetic geometry.)

Understanding and using algebra requires a good understanding of numbers. But it took millennia for mathematicians to gain our current understanding of numbers: whole numbers, rational numbers, irrational numbers, zero, negative numbers, complex numbers, and so on.

Algebra in Egypt and Mesopotamia

There was no single inventor of algebra. In fact, the creation of algebra was long and gradual. The earliest cultures had methods of solving certain problems using what today we would consider rudimentary algebraic techniques. The emphasis of their mathematics was practical—agriculture, accounting, business, surveying, monetary transactions, measurements, and so on. And they were often satisfied with approximations rather than exact solutions.

The Egyptian Rhind Papyrus, which we encountered in chapter 6, has a number of problems that required solving linear equations. Problem 32 states, "A quantity, its $\frac{2}{3}$, its $\frac{1}{2}$, and its $\frac{1}{7}$, added together, becomes 37. What is the quantity?"[2] Written in modern symbols, this problem asked for the solution to the linear equation $x + \frac{2}{3}x + \frac{1}{2}x + \frac{1}{7}x = 37$. Although solving a linear equation seems elementary—even trivial—to us today, recall that the Egyptians did not have a positional number system, and they expressed all rational numbers as sums of unit fractions (and $\frac{2}{3}$). In this case, the Egyptian solution is $16 + \frac{1}{56} + \frac{1}{679} + \frac{1}{776}$, and in the process of verifying this solution, the scribe had to compute $\frac{2}{3} \cdot \frac{1}{179} = \frac{1}{1358} + \frac{1}{4074}$, a nontrivial task!

Although this mathematics does not possess all of the criteria of algebra, we certainly see glimpses. The reference to "the quantity" is the unknown quantity we would denote by x. And the Egyptians had definite methods for solving this problem and ones like it.

Unlike the Egyptians, the Babylonians had a positional number system—base 60. This allowed them to solve more difficult problems. They were able to solve systems of linear equations and certain quadratic equations. However, like Egyptian mathematics, Babylonian mathematics that has come down to us is in the form of problems and solutions to those problems. They possessed well-defined procedures for solving the problems, but the algorithms were shown by example. It was through solving the problems that the student learned the procedures needed to solve similar problems. The methods were not presented in abstract form; they were not algebraic.

Moreover, Egyptian and Babylonian problems were written out in words, not in symbols. This early period in the history of algebra is called the *rhetorical stage*. Thus, although there were some early signs of algebra, many of the requirements of algebra were missing in these early cultures.

Greek Geometrical Algebra

Despite the Greeks' extensive advances in geometry and number theory, they did not perform any mathematics that we would recognize as obviously algebraic. Nevertheless, there is a famous scholarly disagreement over whether the Greeks exhibited algebraic thinking and an algebraic approach to mathematics.

In 1885 Hieronymus Zeuthen (1839–1920) coined the term "geometric algebra" to refer to parts of Euclid's *Elements*.[3] Then, for nearly a century, it was an oft-repeated claim that some of Euclid's propositions were algebra disguised as geometry. And in fact this claim was not new—it went all the way back to the days in which algebra was a cutting-edge mathematical discipline. In 1685 John Wallis wrote, "It is to me a thing unquestionable, That the Ancients had somewhat of like nature with our *Algebra*; from whence many of their prolix and intricate Demonstrations were derived. And I find other modern Writers of the same opinion with me therein."[4]

Let's look at some examples. Proposition II.4 from *Elements* states, "If a straight line be cut at random, the square on the whole is equal to the squares on the segments and twice the rectangle contained by the segments."[5] Geometrically, we picture a square sliced horizontally and vertically into two rectangles of the same size and two squares, as in figure 13.1.

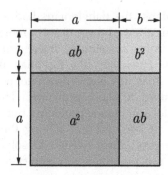

FIGURE 13.1. The geometry behind Euclid's proposition II.4.

The Greeks did not assign measurements to their line segments. Line segments were magnitudes, and the product of two segments was another magnitude—a rectangular area. However, if we were to assign lengths to segments and numerical values to the areas of rectangles, we could view II.4 as an algebraic statement: "If a straight line be cut at random [into lengths a and b], the square on the whole [$(a+b)^2$] is equal to the squares on the segments [a^2 and b^2] and twice the rectangle contained by the segments [$2ab$]." Or, in algebraic shorthand, $(a+b)^2 = a^2 + b^2 + 2ab$.

For a more complicated example, consider proposition II.6. It is equivalent to the algebraic expression $x(x+a) + (a/2)^2 = (x+a/2)^2$. Here is Euclid's version with the x's and a's inserted to make it easier to follow:

> If a straight line [of length a] be bisected and a straight line [of length x] be added to it in a straight line, the rectangle contained by the whole with the added straight line and the added straight line [area $x(a+x)$] together with the square on the half [area $(a/2)^2$] is equal to the square on the straight line made up of the half and the added straight line [area $(a/2 + x)^2$].

Figure 13.2 shows the geometry behind this proposition: the area of an $x \times (x+a)$ rectangle plus the area of an $(a/2) \times (a/2)$ square equals the area of an $(x+a/2) \times (x+a/2)$ square. One could view this proposition as a rigorous geometric justification for algebraic manipulation.

Starting in 1969, a series of articles by a variety of historians of mathematics attacked the idea of Greek geometrical algebra, arguing that Euclid's geometry was just geometry, and our attempt to see algebra

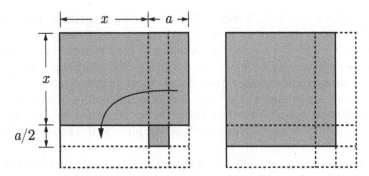

FIGURE 13.2. The geometry behind Euclid's proposition II.6.

in it was "Whig history." We know algebra, and when we look back at what the Greeks did, we can frame it in terms of algebra. That does not mean that the Greeks knew algebra.

It is true that these geometrical theorems have algebraic equivalents. But examine them closely, and we see that many of the hallmarks of algebra are missing. There is no unknown quantity, no algebraic equation, and no symbolic manipulation. In fact, the Greeks did not view magnitudes as numbers at all.

We will not weigh in on this controversy. The belief that Euclid's *Elements* contains geometric algebra is certainly no longer in favor in the math history community, but it is still not settled.[6]

Diophantus of Alexandria

Some time around 250 CE—over five centuries after the heyday of Greek mathematics—Diophantus of Alexandria made important strides in moving algebra away from the ancient rhetorical approach to a more modern one that included symbols for some, but not all, of the ingredients in algebraic equations.

However, Diophantus did not come to algebra from geometry or from problems with real-number solutions. In his most well-known book *Arithmetica*, he wrote about number theory. He presented numerous problems in which the aim was to find positive integer (or positive rational) solutions to polynomial equations with integer coefficients.

He investigated problems such as "Find a square number that can be written as the sum of two squares"; that is, find integers x, y, and z

such that $x^2 + y^2 = z^2$. It was well known before Diophantus that there are infinitely many of these so-called Pythagorean triples: $3^2 + 4^2 = 5^2$, $5^2 + 12^2 = 13^2$, and so on. This equation is easy to solve but, in general, Diophantine problems are notoriously difficult or impossible. We have already encountered (page 34) the most infamous family of Diophantine equations: $x^n + y^n = z^n$. It was in the margin of his copy of Diophantus's *Arithmetica* that Pierre de Fermat stated his "last theorem," that there are no positive integer solutions to this equation when $n \geq 3$.

Diophantus expressed his equations using symbols. He used a symbol like ς for the variable we would call x, and for powers of the variable he used Δ^Υ, K^Υ, $\Delta^\Upsilon\Delta$, ΔK^Υ, and $K^\Upsilon K$ (x^2, x^3, x^4, x^5, and x^6, respectively)—Δ^Υ are the first two letter of *dynamis* (power) and K^Υ are the first two letters of *kybos* (cube). In particular, he did not restrict himself to the third power as later algebraists—who viewed the algebra geometrically—would do. He had a way of writing the reciprocals of these; he described how to manipulate expressions with positive ("forthcoming") and negative ("wanting") terms ("wanting" times "wanting" is "forthcoming"), and he gave many of the basic algebraic rules for manipulating equations that we know well.

Yet Diophantus's algebra is still not modern. He had only one way to express an unknown quantity, but his equations often had more than one. He also did not have a symbolic way of expressing a generic coefficient (for nx he would write "x, however great" or "any x").

India

The lack of an efficient number system was a major roadblock that prevented algebra from developing. The Greeks possessed a clumsy nonpositional base-10 number system, and Roman numerals were a hassle to use in practice. Try computing MMMMDCCCXCVI ÷ XVIII without converting it to decimal notation. Go! The answer? CCLXXII. That's not easy, but today's elementary-school student could show that $4896 \div 18 = 272$ using long division. Algebraists needed a better number system, and that system came from India.

As the French mathematician Laplace wrote,[7]

It is India that gave us the ingenious method of expressing all numbers by means of ten symbols, each symbol receiving a value of position as well as an absolute value; a profound and important idea which appears so simple to us now that we ignore its true merit. But its very simplicity and the great ease which it has lent to all computations put our arithmetic in the first rank of useful inventions; and we shall appreciate the grandeur of this achievement the more when we remember that it escaped the genius of Archimedes and Apollonius, two of the greatest men produced by antiquity.

There is no single inventor of the positional base-10 number system. It developed over a span of a few hundred years. We know that Brahmagupta (ca. 598–ca. 668 CE) was a key player. In his *Brahmasphutasiddhanta*, written in 628, he wrote about methods for performing arithmetic using the base-10 positional number system. He also wrote about 0 and the rules for performing arithmetic with 0 and negative numbers. He wrote,[8]

> The product of a negative and a positive is negative, of two negatives positive, and of positives positive; the product of zero and a negative, of zero and a positive, or of two zeros is zero.

Zero plays two roles in our number system: it is a number signifying nothing, and it is a placeholder when one of the powers of 10 is missing in the expansion of the number—we write 7089 for $7 \cdot 10^3 + 8 \cdot 10^1 + 9 \cdot 10^0$ with the 0 representing the nonappearance of 10^2.

Brahmagupta also gave general instructions for solving quadratic equations. Here is his method for solving $ax^2 + bx = c$:[9]

> Diminish by the middle [number] the square-root of the *rūpas* multiplied by four times the square and increased by the square of the middle [number]; divide [the remainder] by twice the square. [The result is] the unknown.

In today's symbols, we would write $x = (\sqrt{4ac + b^2} - b)/(2a)$. Note that Brahmagupta used *rūpas* to denote the units for the known or constant quantities in the problem. (Interestingly, he used colors to represent the unknown quantities.)

In fact, Brahmagupta's solutions produced two roots for quadratic equations, even when one of the roots was negative. Like Diophantus, Brahmagupta wrote his algebraic expressions in a syncopated style, using a mixture of symbols and abbreviations of words.

The Islamic Empire

While Brahmagupta was writing about mathematics in India, the Islamic religion was forming around the prophet Muhammad in the Arabic peninsula. After Muhammad's death in 632, the Arab leaders extended their control over neighboring countries and beyond. In 750, the Abbasid dynasty gained control of the empire. They moved the capital from Damascus to the new city of Baghdad, which was founded in 762 on the Tigris River, approximately 50 miles from the former city of Babylon. Thereafter, a cultural and intellectual reawakening began, and Baghdad became the center of this movement. Scholars converged on the city, which was to become the Islamic equivalent of Alexandria. And like Alexandria, Baghdad had its own library, the House of Wisdom, which was founded by the caliph Al-Mamūn.

Al-Mamūn's visions of Aristotle in a dream inspired him to have as many Greek works translated into Arabic as he could obtain. Euclid's *Elements* had arrived in Baghdad in the eighth century and was translated during the reign of al-Mamūn's father Hārūn al-Rashīd. But many more were translated under al-Mamūn, including works by Archimedes, Apollonius, Diophantus, and Ptolemy. It is in part due to these translations that we know as much as we do about Greek mathematics. But Islamic mathematicians were more than just caretakers of knowledge. They assimilated, combined, and extended Greek, Mesopotamian, and Indian mathematics. They are known for their contributions to arithmetic, algebra, trigonometry, and geometry.

Muhammad Ibn al-Khwārizmī was a scholar at the House of Wisdom who wrote more than a half dozen books on astronomy and mathematics. His book *Concerning the Hindu Art of Reckoning*[10]—likely based on Brahmagupta's writings—described the Indian numeral system and the methods of computing with it. Al-Khwārizmī's book was so influential that it led to the widespread incorrect belief that our number system had an Arabic origin, not an Indian origin. Even today they are often called "Arabic numerals." In Europe these computational techniques were often called the methods of Al-Khwārizmī, or through mistranslations, algorismi; this is the source of the term "algorithm."

Al-Khwārizmī's most important work was his algebra text *Al-jabr wa'l muqābala* (*The Compendious Book on Calculation by Completion and*

Balancing). The mathematics was purely rhetorical—even the numbers were written as words. (Diophantus's writings containing syncopated algebra had not yet been translated into Arabic.) Nevertheless, there was still a clear notion of an equation, and Al-Khwārizmī did not present algebra through a series of problems, but gave techniques for solving general, abstract, algebraic equations up to the second degree. Inspired by the Greek's rigorous geometric approach to mathematics, he used geometry to prove the algebraic relations. Also, like modern textbook authors, Al-Khwārizmī gave ample problems illustrating how to use the theoretical tools that he described.

He broke down the solution of quadratic equations into five cases. Today we would say that there is one form of a quadratic equation, which we could write as $ax^2 + bx + c = 0$. But Al-Khwārizmī required that all coefficients be positive. So, for instance, $x^2 + 2x = 3$ and $x^2 + 3 = 2x$ were classified separately.

The term *al-jabr* in the book title means "restoration" or "completion" and is the technique of adding the same quantity to both sides of an equality, thereby ridding the equation of subtracted terms. *Al-muqābala*, which means "reduction," or "balancing," is the process of collecting like terms. It is from Al-Khwārizmī's term *al-jabr* that algebra got its name.[11]

Al-jabr was written for nonmathematicians—businessmen, lawyers, engineers, and so on. While Al-Khwārizmī's techniques for solving linear and quadratic equations were not new, his geometric representations of the solutions were. This became the standard for many years.

Al-Khwārizmī was the most famous, but certainly not the only Islamic mathematician to contribute to the history of algebra. For instance, Abū Kāmil (ca. 850–ca. 930) wrote a higher-level algebra text that built upon Al-Khwārizmī's. Again, the geometric justifications were very important. Often a problem would have both an algebraic solution and a geometric justification. His work may be the first one in which irrational numbers were treated as numbers.

The mathematician and astronomer Omar Khayyám (1048–1131)—who may be best known today for the *Rubaiyat*, a collection of poems attributed to him—tackled the difficult problem of solving cubic equations. He did not believe they had algebraic solutions, and instead focused on solving them geometrically by intersecting conic sections.

Europe

The Romans were not interested in abstract mathematics. Instead, they used mathematics as a practical tool for engineering and commerce. But after the fall of the Roman Empire in the fifth century, even this mathematical culture died out. The following five centuries were a relatively dry period for mathematics in Europe.

In the late tenth century, Christian scholars traveling to Islamic cities made Latin translations of science and mathematics texts that they came across and brought them back to Western Europe. The number of translations increased in the eleventh, and especially the twelfth, centuries. These were translations of treatises written by Arab mathematicians such as Al-Khwārizmī and non-Arabs such as Euclid and Ptolemy (ca. 100–ca. 170).

By the turn of the thirteenth century, maritime travel was becoming safer, and commerce was evolving from the bartering of traveling merchants to shipping and economies based on the exchange of money. Merchants lived in cities such as Lucca, Siena, and Florence, Italy. The success of banking and international finance required improvements to the methods of accounting and bookkeeping and an efficient method of performing calculations. Roman numerals were too cumbersome.

The transition to Hindu–Arabic numerals began in earnest in the early thirteenth century, thanks in large part to Leonardo Pisano (1170–1250)—who is better known by his nickname Fibonacci ("son of Bonaccio"). He was born in Italy, but was raised and educated in North Africa where his father was a customs officer. It was there that Leonardo learned about the superior number system, and he recognized how useful it would be not only for theoretical mathematics, but also for practical arithmetic, especially in business and financial accounting.

In 1202, back in Italy, he published the *Liber abaci* (*Book of Calculation*), which described this number system. It was clearly influenced by Abū Kāmil's algebra text; the *Liber abaci* presented 29 of Kāmil's problems almost exactly. Thereafter, schools opened to train merchants to use this new number system. Nevertheless, it still took several centuries before Roman numerals were truly abandoned for the more efficient number system.

In the fourteenth century, Europe suffered the devastating Great Plague, from which one-third or more of Europe's population died. It was also preoccupied with the Hundred Years' War which stretched into the fifteenth century. Consequently, this was another dry period for European mathematics.

Solving the Cubic

Quadratic equations have what is called a "solution by radicals." The quadratic formula gives us a method of expressing the roots of an equation in terms of the coefficients using addition, subtraction, multiplication, division, and the extraction of roots (the square root in this case). A natural question is whether polynomials of higher degrees admit a solution by radicals. In particular, can it be done for the next degree higher: the cubic equations?

The answer for the cubic is yes, and the story of discovering the formula is a Hollywood-worthy tale of intrigue, secrecy, deception, and rivalries. It is also an important episode in the history of mathematics. The tale begins in Italy at the end of the fifteenth century.

Luca Pacioli (1447–1517) was a Franciscan friar who wrote several works on mathematics (including one with a very famous illustrator— his friend Leonardo da Vinci!). In 1494 he wrote *Summa de arithmetica, geometria, proportioni et proportionalita*, an extremely influential textbook written in Italian, not Latin, that gave a comprehensive introduction to the current state of Renaissance mathematics. It included arithmetic, algebra, geometry, trigonometry, and accounting. And because this book was published after Gutenberg's 1450 invention of the printing press, it was one of the first printed mathematics books.

He described how to solve linear and quadratic equations, and he asserted that the cubic equation was as impossible to solve as the problem of squaring the circle. Pacioli was correct about the impossibility of squaring the circle, but on the cubic he was incorrect, as one of his friends, Scipione del Ferro (1465–1526), a professor at the University of Bologna, was about to show.

A cubic equation has the form $ax^3 + bx^2 + cx + d = 0$, but because a is not zero, we can divide through by that value. So we may as well assume $a = 1$. Like al-Khwārizmī before them, the sixteenth-century mathematicians considered only nonnegative coefficients. So, for instance, $x^3 + 6x = 20$, which they called "a cube and things equals a number,"[12] would be seen as a different type of cubic equation from one with the $6x$ on the other side ("a cube equals things and a number"). Thus, when they imagined solving the cubic, they envisioned solving many different cases. Del Ferro may have completely solved every form of the cubic, but we don't know because he kept his discovery a secret.

For much of the history of mathematics and science, an individual making a discovery wrote about it and published it for the world to see. There is both the altruistic desire to disseminate the result as quickly and widely as possible and the more self-serving drive to stake the claim as the person who made the discovery. However, in sixteenth-century Italy the situation was different. Mathematical scholars would challenge others to "mathematical duels"—they would each pose a number of problems to a competitor that the proposer knew how to solve but that the challenger may not. The one who solved the most of the other person's problems was declared the victor. These challenges were a means of obtaining glory and a way to attract more pupils and teaching opportunities. So mathematicians of the day would sometimes refrain from publishing their discoveries; they would hold onto them as secret weapons or "aces up their sleeves" with the hope that they could bring the techniques out as needed during a competition. Knowledge such as this was viewed more like corporate secrets are today

Del Ferro undoubtedly knew that he had discovered something special, so he may have kept it in reserve, thinking it may be useful in a duel some day. We know he was able to solve the "cube and things equals a number" case because he shared that technique with his student Antonio Fiore (ca. 1506–?). This cubic equation is one case of the *depressed cubic*—a cubic that has no squared term.

Unlike quadratic equations that may have no real solutions, every cubic equation must have at least one. In modern notation, del Ferro discovered that a depressed cubic $x^3 + cx + d = 0$ has a root

$$x = \sqrt[3]{-\frac{d}{2} + \sqrt{\left(\frac{d}{2}\right)^2 + \left(\frac{c}{3}\right)^3}} - \sqrt[3]{\frac{d}{2} + \sqrt{\left(\frac{d}{2}\right)^2 + \left(\frac{c}{3}\right)^3}}.$$

This formula holds when c and d are positive or negative, but the case del Ferro shared with Fiore had c positive and d negative.[13] Applying this formula to $x^3 + 6x = 20$ yields the root

$$x = \sqrt[3]{10 + \sqrt{108}} - \sqrt[3]{-10 + \sqrt{108}},$$

which, despite the messy nested roots, is just 2.

At the same time, a much more talented, but self-taught, mathematician Niccolò Fontana (1499–1557) was also working on the cubic. Fontana was known as Tartaglia ("the stammerer"); he had difficulty speaking because of a facial injury dating back to when he was a young teen and a French soldier slashed him across his face with a sword. Tartaglia discovered a method of solving a case of the cubic different from the one Fiore knew—one in which the squared term is present but the linear term is missing.

After del Ferro died, Fiore, who was not a great mathematician, boasted that he could solve cubic equations. In 1535 Tartaglia heard of Fiore's claims and challenged him to a mathematical duel. They each sent 30 problems to the other, and they were given approximately a month and a half to solve them. Fiore placed all of his mathematical eggs in one basket and sent Tartaglia 30 depressed cubics of the type he knew how to solve. Tartaglia, on the other hand, sent a variety of mathematical problems. Tartaglia worked tirelessly and, near the competition deadline, discovered the procedure for solving Fiore's problems. It then took him two hours to solve the 30 problems. Fiore did not fare well with Tartaglia's problems. Tartaglia won the competition, and Fiore faded from view.

Gerolamo Cardano was a brilliant but troubled physician and mathematician. He suffered from many health ailments, he was often preoccupied by family problems, he wasted time and money on a gambling addiction (dice, chess, and cards), and at one point he was arrested and jailed for heresy by the Inquisition (for writing Jesus's horoscope). In his autobiography, he wrote, "I was the victim of . . . many great discouragements and obstacles in my life. The first was my marriage; the second, the bitter death of my son; the third, imprisonment; the fourth, the base character of my youngest son."[14]

Cardano had a difficult personality and repeatedly clashed with others. He wrote, "This I recognize as unique and outstanding among my faults—the habit, which I persist in, of preferring to say above all things what I know to be displeasing to the ears of my hearers. I am aware of this, yet I keep it up willfully, in no way ignorant of how many enemies it makes for me."[15]

Although he may have been unpleasant to be around, he was intelligent. As a child he was a strong student. He eventually obtained a doctorate in medicine and became one of the most sought-after physicians in Europe. He also displayed a talent for mathematics. Although his gambling ruined his fortunes, he developed very advanced ways of thinking about probability. His *Book on Games of Chance* was published after his death. It was the first serious investigation of probability theory. His collected works, encompassing mathematics, medicine, physics, philosophy, religion, and music, fill 7000 pages. Leibniz wrote, "Cardano was a great man with all his faults; without them, he would have been incomparable."[16] One biographer wrote that "even his earliest works show the characteristics of his highly unstable personality: encyclopedic learning, powerful intellect combined with childlike credulity, unconquerable fears, and delusions of grandeur."[17]

Cardano believed Pacioli's assertion that solving the cubic was impossible. But when he learned that Tartaglia had solved 30 cubic equations, he was impressed and fascinated. He begged Tartaglia to show him the method. He offered him access to the Milanese governor as an enticement. He even pledged a vow of secrecy:[18]

> I swear to you by the Sacred Gospel, and on my faith as a gentleman, not only never to publish your discoveries, if you tell them to me, but I also promise and pledge my faith as a true Christian to put them down in cipher so that after my death no one shall be able to understand them.

In 1539 Tartaglia finally relented. He told Cardano how to solve the depressed cubic, although he did not share the proof that it worked.

This information was what Cardano needed to kick-start his work on the problem. Then, thanks to a key insight—a little algebraic trickery—Cardano realized that knowing how to solve the depressed cubic was sufficient to solve any cubic. If we are given a cubic of the form $x^3 + bx^2 + cx + d = 0$, and we substitute $x = y - b/3$, then we obtain a cubic polynomial with variable y that has no squared term. We can solve this depressed cubic for y, then use the linear equation to solve for x. Thus Cardano was able to solve any cubic equation.

Solving a cubic using today's algebraic notation is challenging enough,[19] but in the sixteenth century, with algebra in its infancy, it was a major accomplishment. Cardano proved his results using geometry. To give a flavor of his method, suppose t and u are positive values with $u < t$. Then we can decompose the cube of volume t^3 into cubes of volumes u^3 and $(t - u)^3$, two boxes of volume $tu(t - u)$, one box of

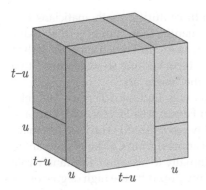 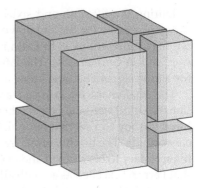

FIGURE 13.3. A geometrical description of a cubic.

volume $u^2(t-u)$, and one box of volume $u(t-u)^2$ (see figure 13.3). This geometric dissection implies that

$$t^3 = u^3 + (t-u)^3 + 2tu(t-u) + u^2(t-u) + u(t-u)^2.$$

Moreover, Cardano presented his mathematics in the rhetorical style. Here is his presentation of the formula for solving a depressed cubic of the form $x^3 + cx = d$:[20]

> Cube the third part of the number of "things," to which you add the square of half the number of the equation, and take the root of the whole, that is, the square root, which you will use, in the one case adding the half of the number which you just multiplied by itself, in the other case subtracting the same half, and you will have a "binomial" and "apotome" respectively; then subtract the cube root of the apotome from the cube root of the binomial, and the remainder from this is the value of the thing.

Although Cardano did not use symbols to perform his algebra, he did use symbols for arithmetic. He used Pacioli's symbols: ṕ for addition, m̃ for subtraction, and Ř for roots. For example, he wrote the root of $x^3 + 6x = 20$, which is $\sqrt[3]{\sqrt{108} + 10} - \sqrt[3]{\sqrt{108} - 10}$, as[21]

$$R.\ v.\ cu.\ R.\ 108.\ \acute{p}.\ 10.$$

$$\acute{m}.\ R.\ v.\ cu.\ R.\ 108.\ \acute{m}.\ 10.$$

The algebraic symbolism that we are familiar with developed slowly over the course of the fifteenth, sixteenth, and seventeenth centuries.

During the fifteenth and sixteenth centuries, mathematicians had different names for the unknown quantity—or "thing"— that today we may represent as x. In Latin it was *res*, in French *chose*, in Italian *cosa*, and in German *coss*. The latter term became so widespread that for a time algebra became know as the "cossic art."

The final character in this drama was Ludovico Ferrari. His relationship with Cardano began when Ferrari was 14 years old. Cardano hired the boy as a servant but quickly recognized Ferrari's mathematical aptitude. The bright Ferrari transitioned from Cardano's servant to his secretary to his student and eventually to his equal. Cardano wrote that Ferrari "excelled as a youth all my pupils by the high degree of his learning."[22] Cardano enlisted Ferrari as an assistant in his task to solve the cubic equation; of course, doing so required breaking his oath to Tartaglia that he would not divulge Tartaglia's method of solving the depressed cubic to anyone.

Ferrari became so well versed in the cubic that when he was 18 years old he discovered a method for reducing the general quartic—a 4th-degree polynomial—to a cubic. Thus, Cardano and Ferrari now possessed the techniques to solve any cubic or quartic equation. But this was all knowledge obtained from eating the forbidden fruit. Because of Cardano's oath to Tartaglia, he and Ferrari were unable to publish their work. They were at a loss as to how to proceed.

In 1543 Cardano discovered that Tartaglia was not the first to discover the method for solving the depressed cubic, a fact he confirmed by reading del Ferro's private papers. Cardano believed that this discovery gave him a green light to publish. Sure, he promised Tartaglia he would keep the solution a secret, but now that he learned about it from a different source, he was freed from that obligation. So Cardano published his and Ferrari's work on solving polynomials up to degree 4 in his 1545 *Ars magna* (*Great Art*).

At the start of the ninth chapter, Cardano gave credit to Tartaglia and acknowledged that he learned the method from him:[23]

Scipio Ferro of Bologna well-nigh thirty years ago discovered this rule and handed it on to Antonio Maria Fiore of Venice, whose contest with Nicolò Tartaglia of Brescia gave Nicolò occasion to discover it. He [Tartaglia] gave it to me in response to my entreaties, though withholding the demonstration. Armed with this assistance, I sought out its demonstration in [various] forms. This was very difficult.

Despite this acknowledgment, Tartaglia was furious. He lashed out at Cardano. Rather than replying to Tartaglia himself, Cardano let the hot-headed and loyal Ferrari carry on the fight on his behalf. There followed an extended back-and-forth—including the exchange of 31 problems apiece that we mentioned on page 181 (17 of Tartaglia's being rusty compass problems). Eventually, Tartaglia and Ferrari decided to hold a debate at the Church of Santa Maria del Giardino in Milan on August 10, 1548. Tartaglia was disappointed not to be debating Cardano, but he was confident going in. However, Ferrari had the home crowd and the first day went terribly for Tartaglia. Tartaglia chose to leave town that evening, and thus Ferrari was declared the victor. Cardano was not at the debate.

Afterward, Tartaglia lost his teaching position at Brescia and Ferrari was invited to lecture in Tartaglia's home, Venice. Despite Tartaglia's groundbreaking work on solving the cubic and his other mathematical contributions—the first translations of Euclid and Archimedes into a modern language (Italian), his formula for the volume of a triangular pyramid given the lengths of its sides, his work with the rusty compass, the physics of projectiles, and so on—he died penniless and largely unknown.

After the publication of *Ars magna*, mathematicians were able to solve linear, quadratic, cubic, and quartic equations. One math historian wrote, "Such a striking and unanticipated development made so strong an impact on algebraists that the year 1545 is frequently taken to mark the beginning of the modern period of mathematics."

The next natural question is how high can we go? In a fascinating twist to this mathematical story, the answer is no further. For the next three centuries, mathematicians tried to crack the general quintic. But it turned out to be impossible to solve. (We mentioned this fact on page 35.) Paolo Ruffini (1765–1822) gave the first proof of the result in 1799, but his work contained a flaw. Abel, unaware of Ruffini's proof, gave his own (correct) proof in 1826. Some higher-degree polynomials are solvable by radicals and some are not, and these were characterized exactly by Évariste Galois (1811–1832).

As one final note on the solution to the cubic, there is one pesky problem with the formula for the depressed cubic: the expression under the square root could be negative. At first that may not seem to be much of a problem. This happens all the time when solving quadratic equations, and when it does, it means the equation has no real roots. But something different happens with the cubic. For example, applying the formula for the depressed cubic to $x^3 = 15x + 4$ yields

$$x = \sqrt[3]{\frac{4}{2} + \sqrt{\left(\frac{4}{2}\right)^2 + \left(\frac{-15}{3}\right)^3}} - \sqrt[3]{-\frac{4}{2} + \sqrt{\left(\frac{4}{2}\right)^2 + \left(\frac{-15}{3}\right)^3}}$$

$$= \sqrt[3]{2 + \sqrt{-121}} + \sqrt[3]{2 - \sqrt{-121}},$$

which looks decidedly nonreal. However, this cubic equation has three real roots: 4 and $-2 \pm \sqrt{3}$. And in fact, this complex-looking root is 4!

As it turns out, any time that a cubic equation has three distinct real roots, the value under the square root will be negative. Historically, these cubic equations were called the *casus irreducibilis* (irreducible case). These cubic equations provided the main impetus to introduce the complex numbers into the realm of numbers. We will discuss the history of complex numbers in chapter 18.

TANGENT
Nicholas of Cusa

Said the man about town, "I have a flair
For squaring the circle, I swear."
But he found that the strain
Was too great for his brain,
So he's gone back to circling the square.
—Unknown[1]

THE FIFTEENTH-CENTURY German theologian, philosopher, and cleric Cardinal Nicholas of Cusa (1401–1464) was not a great mathematician, but he helped revive interest in mathematics at the end of the Middle Ages. He was particularly interested in the nature of infinity, astronomy, and the problems of squaring and rectifying the circle, but his contributions are best described as inspirational rather than rigorously mathematical. As one scholar wrote, "The true strength of Cusanus's mathematics may, therefore, lie in his 'art of discovery' (*ars inveniendi*) rather than in his 'art of demonstration' (*ars demonstrandi*)."[2] Kepler was inspired by Nicholas of Cusa's theory of infinitesimals and referred to him as *divinus mihi Cusanus* (my divine Cusa).

On the classical geometric problems, he held seemingly contradictory beliefs. He asserted that "the area of a circle is incommensurable with that of any non-circle,"[3] yet he gave "proofs" of the quadrature and rectification of the circle.

For Nicholas of Cusa, these problems performed double duty as a metaphor for his philosophical and religious beliefs and as a mathematical challenge. In his most important and well-known philosophical work, *De docta ignorantia* (*On Learned Ignorance*, 1440), he argued that all knowledge of God, truth, and reality is approximate—we may only ever have a finite understanding of the infinite. He gave a geometric analogy:[4]

What is itself not true can no more measure the truth than what is not a circle can measure a circle, whose being is indivisible. Hence reason, which is not the truth, can never grasp the truth so exactly that it could not be grasped infinitely more exactly. Reason stands in the same relation to the truth as the polygon to the circle: the more vertices a polygon has, the more it resembles a circle; yet even when the number of vertices grows infinite, the polygon never becomes equal to a circle, unless it becomes a circle in its true nature.

The problems of squaring or rectifying the circle are not equivalent to saying that a circle *is* a polygon, but he insisted that these too are impossible. The key, in his view, is the meaning of equality. He argued that it is possible to find a polygon whose circumference or area is not greater than, less than, or equal to that of a circle, but that it is possible that they differ by an infinitesimal quantity. In *De circuli quadratura* (1450) he explained,[5]

> Those who hold firmly to the first view [that the circle can be squared] seem to be satisfied with the fact that given a circle, there exists a square which is neither larger nor smaller than the circle. . . . If, however, this square is neither smaller nor larger than the circle, by even the smallest assignable fraction, they call it equal. For this is how they understand equality—one thing is equal to another if it neither exceeds it nor falls short of it by any rational fraction, even the smallest. If one understands the notion of equality in this way, then, I believe, one can correctly say that, given the circumference of a certain polygon, there exists a circle with the same circumference. If, however, one interprets the idea of equality, insofar as it applies to a quantity, absolutely and without regard to rational fractions, then the statement of the others is right: there is no non-circular area which is precisely equal to a circular area.

He elaborated by asserting that the now-well-established intermediate value theorem from calculus fails:[6]

> There can never in any respect be something equal to another; even if at one time one thing is less than another and at another [time] is greater than this other, it makes this transition with a certain singularity, so that it never attains precise equality [with

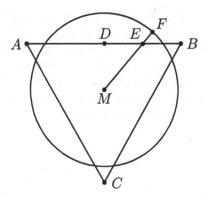

FIGURE T.25. Nicholas of Cusa claimed that the circle and triangle have the same perimeters.

the other]. Similarly, a square inscribed in a circle passes—with respect to its size—from being a square that is smaller than the circle to being a square larger than the circle, without ever arriving at its equal.

However, despite his repeated assertions that these geometric problems are impossible, he also presented several constructions that he claimed squared or rectified the circle (presumably producing polygonal lengths or areas that differed by an infinitesimal amount). He gave the following technique that begins with an equilateral triangle ABC, as in figure T.25. Let D denote the midpoint of AB, E be the midpoint of BD, and M be the center of the triangle. He asserted that the circumference of the circle with a radius equal to 5/4 the length of EM is the same as the perimeter of the triangle.

He wrote, "So you will not think that this is mere conjecture, that one is not led to this claim by any other line of thought, you can draw a rigorous conclusion, which in this case is completely accurate and is dependable to within the smallest rational fraction."[7] He explained that if we used a similar procedure choosing, not E, but a point near D, the resulting circle's radius would be too small, and if we chose a point near B, the radius would be too large. He then argued that there must be a point between D and B for which the radius is correct. Without justification, he chose the midpoint E.

Had this rectification been correct, it would have implied that $\pi = 24\sqrt{21}/35 \approx 3.142337\ldots$, which is within Archimedes's bounds (it is slightly less than 22/7), so it did not fail this basic test of correctness. But his other attempts were not so good. In 1464 the much stronger mathematician Regiomontanus (Johannes Müller, 1436–1476)[8] showed that all of Nicholas of Cusa's other attempts gave a value for π outside Archimedes's bounds.

CHAPTER 14

Viète's
Analytic Art

Algebra is to the geometer what you might call the "Faustian offer"....Algebra is the offer made by the devil to the mathematician. The devil says: "I will give you this powerful machine, it will answer any question you like. All you need to do is give me your soul: give up geometry and you will have this marvelous machine."....Of course we like to have things both ways; we would probably cheat on the devil, pretend we are selling our soul, and not give it away. Nevertheless, the danger to our soul is there, because when you pass over into algebraic calculation, essentially you stop thinking; you stop thinking geometrically, you stop thinking about the meaning.

—Sir Michael Atiyah, 2001[1]

WE ENCOUNTERED FRANÇOIS Viète in chapter 10. He was the geometer who believed so strongly in the power of neusis constructions that he argued it should be added to Euclid's postulates. With this new postulate he was able to solve some of the problems of antiquity and other previously unsolvable problems. Viète is known for his work in geometry and algebra and the interplay between them. He also investigated astronomy, planar and spherical trigonometry, and cryptography.

Viète was born in Fontenay-le-Comte, France, in 1540. He studied to be a lawyer as an undergraduate, but after four years in a practice, he moved on to become a private tutor in mathematics and science. Later, and for much of his life, he worked for the French King Henry III and his successor, and distant cousin, Henry IV.

The late sixteenth century was a time of religious unrest in France. For instance, Viète, a protestant Huguenot, was living in Paris in 1572 when Charles IX authorized the St. Bartholomew's Day massacre of the Huguenots. Viète moved around France throughout his life—between Fontenay-le-Comte, La Rochelle, Paris, Rennes, Beauvoir-sur-Mer, and Tours—sometimes by choice, but other times to serve the king or because he was banished for political reasons.

Viète was not a professional mathematician. In the introduction to one article, he wrote, "I, who do not profess to be a mathematician, but who, whenever there is leisure, delight in mathematical studies...."[2] During one four-year period of banishment, Viète devoted his time to mathematics, and produced much of the mathematics we will discuss.

He also spent time working on cryptography. In 1589 Henry of Navarre, the future King Henry IV, obtained a letter sent to Spain's Philip II. It was written in a code. Henry gave it to Viète to decode, which he was eventually able to do. Because of this work, all intercepted coded messages from Spain could be read.

Viète is not as famous as some of the other founders of modern mathematics, but he plays an important role in our story. In addition to using a compass and marked straightedge to attack the problems of antiquity, he developed algebraic techniques that could be used to solve geometric problems, he helped develop algebraic tools and methods, he introduced new and useful algebraic notation, and he made important contributions to the study of the cubic and quartic equations. His work beautifully tied together the problems of antiquity, the solving of algebraic equations, the solutions to geometric problems, and the geometry of the marked straightedge.

The Analytic Art

Viète was a geometer. All his work in algebra was intended to provide a means for solving geometric problems. He envisioned taking a geometric problem, turning it into an algebraic one, solving it using algebra, and then using that algebraic solution to help solve the geometry problem. But to carry this out, he had to develop the algebra.

In 1591 Viète wrote *Introduction to the Analytic Art* (*In artem analyticem isagoge*). As the title suggests, he called his algebraic techniques the "analytic art," and for much of the seventeenth century algebra became known under that name, or as "analysis." The origin of this term was from the Greeks. To the Greeks, analysis referred to the technique of assuming a solution, then working backward to something known, such as an axiom or a previously proved theorem. This process was followed by "synthesis," in which one begins with knowns and uses a logical mathematical argument to arrive at the conclusion.

In his *Collection*, Pappus wrote,[3]

> In analysis we assume that which is sought as if it were (already) done, and we inquire what it is from which this results, and again what is the antecedent cause of the latter, and so on, until by so retracing our steps we come upon something already known or belonging to the class of first principles, and such a method we call analysis as being solution backwards.
>
> But in synthesis, reversing the process, we take as already done that which was last arrived at in the analysis and, by arranging in their natural order as consequences what were before antecedents, and successively connecting them one with another, we arrive finally at the construction of what was sought; and this we call synthesis.

Viète saw algebra as a superior method of conducting analysis. To solve a geometrical problem he assumed it was already solved and labeled some key unknown quantity as x (or A in his notation). Then he used geometry to produce an algebraic equation involving x. If possible he solved the equation for x. This was the end of the analysis step. Next he took this value of x and turned it into a geometric construction, thus solving the problem. The final step was to prove that the construction was correct.

This description may seem abstract and confusing, so let's look at an example. Suppose the diameter AB of a circle intersects a line l perpendicularly at C, as in figure 14.1.[4] Also, suppose c is some fixed magnitude greater than BC. The geometrical problem is to construct a line AE with E on l that intersects the circle at D and $DE = c$. That is, we must perform a neusis construction between the circle and the line l, passing through A.

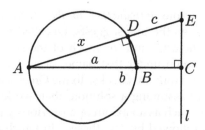

FIGURE 14.1. We must find the line AE so that $DE = c$.

We now perform the analysis: Assume there is a solution. Let x be the length of segment AD. Because ABD and ACE are similar triangles, $AC/AE = AD/AB$ or, letting $a = AB$ and $b = AC$, $b/(x+c) = x/a$. Thus we obtain the quadratic equation $x^2 + cx - ab = 0$, the positive root of which is $x = (-c + \sqrt{c^2 + 4ab})/2$.

At this point, we have completed the analysis of the problem. To solve the problem we must construct a circle with center A and radius equal to this x-value. It will intersect the given circle at the desired point D. It is possible (although we omit the details) to construct such a circle. It is interesting that, in general, neusis constructions are not possible using a compass and unmarked straightedge, but this particular one is.

One of the great benefits of algebra is that it is possible to use mechanical rules to easily manipulate expressions that would be tedious, confusing, and often impossible to express geometrically. But that is also one of its downsides. Traditional geometric analysis would give a clue as to how the synthesis would go. In the best cases, the synthesis is the analysis in reverse. We see here, however, that because the algebraic analysis is far removed from the geometric construction, it can be difficult to turn an algebraic solution into a geometric one.

Viète ends *The Analytic Art* by stating his true aim: "Finally, the analytic art, . . . claims for itself the greatest problem of all, which is TO SOLVE EVERY PROBLEM (*Nullum non problema solvere*)."[5]

Viète pushed algebra further from the syncopated style and closer to the symbolic style we know today. Rather than using *cosa* or its abbreviation *co* for the unknown, he used A or E or some other vowel. For constant values, he used consonants. He wrote,[6]

In order to assist this work by another device, given terms are distinguished from unknown by constant, general and easily recognized symbols, as (say) by designating unknown magnitudes by the letter A and the other vowels E, I, O, U and Y and given terms by the letters B, G, D and the other consonants.

His algebra was still not purely symbolic; he had no symbol for equality and the modern exponential notation had not yet been introduced, so he wrote A *quadtratum* (or A *quad.* or $Aq.$) for A^2, A *cubus* for A^3, A *quadrato–quadrato* for A^4, and so on. But it is not too difficult to transform the cubic equation[7] (in the variable A) "A cubus $- B$ quad. 3 in A, aequetur B quad. in D" into the modern form $A^3 - 3B^2A = B^2D$. Elsewhere he wrote,[8] "If $\frac{A \text{ plano}}{B}$ is to be added to $\frac{Z \text{ quadratum}}{G}$, the sum will be $\frac{G \text{ in } A \text{ planum} + B \text{ in } Z \text{ quadras}}{B \text{ in } G}$," which is

$$\frac{A}{B} + \frac{Z^2}{G} = \frac{GA + BZ^2}{BG}.$$

Notice that all three terms in the expression $A^3 - 3B^2A = B^2D$ have the same degree, in this case 3. The second example, which shows the sum of fractions, also has this property. The letters B, G, and Z are one-dimensional constants and A is a variable representing an unknown area, which is two-dimensional. Thus each fraction has degree 1—a degree-2 term divided by one of degree 1. For Viète, this was essential for any equation, and he called it the "law of homogeneous terms." He wrote, "Homogeneous terms must be compared with homogeneous terms (*Homogenea homogeneis comparari*)."[9] Viète held firmly to the ancient belief that it was possible to add or subtract like terms only; it was impossible to add areas to volumes, for instance. Because of his insistence on homogeneity and its underlying geometric grounding of his algebra, Viète never wrote an expression such as $A^2 + A$, which does not look odd to us at all.

The fact that Viète's algebra included arbitrary constants was groundbreaking. In the past, algebraic methods were presented using examples that were representative of the class of problems. His use of letters to represent arbitrary constants allowed him to write general equations or identities. For example, in 1631 Viète showed that[10] (and here we are using modern notation) $(a + b)^3 = a^3 + 3a^2b + 3ab^2 + b^3$. This short expression says that for *any* values a and b, the left-hand side equals the right-hand side. This identity is much more descriptive, more compact, and infinitely more clear than an example or two that exhibit this same relationship. Moreover, even though we talk about the "Tartaglia–Cardano formula," neither Tartaglia nor Cardano nor anyone else wrote a formula prior to Viète. One needs the idea of arbitrary coefficients to give general formulas. It is difficult to overstate the importance of this new method of writing equations.

Despite his ultimate focus on the underlying geometry of his algebra, Viète began the shift to equations becoming the focus of algebra. Mathematicians became increasingly interested in the relationship between the coefficients of an equation and the roots of the equation. Equations would still be used to solve geometric problems, but they also became interesting objects to study on their own. Mahoney wrote, "The shift in focus is reflected in the different subtitles algebra bears in the sixteenth and seventeenth centuries; as a result of Viète's reformulation of its goals and procedures, the 'art of the coss' becomes the 'doctrine of equations.'"[11]

Angle Trisection and Cubic Equations

At the end of our discussion of the solution to the cubic, we noted the curious fact that the Tartaglia–Cardano formula had difficulties when all three roots are real numbers—this is the so-called casus irreducibilis. When we apply the formula to such polynomials, we obtain an expression for the real roots that contains complex numbers—a situation that would perplex mathematicians for many years.

Viète exposed some surprising connections between angle trisection, the neusis technique, and the casus irreducibilis. His proposition 16 stated the following (see figure 14.2):[12]

> If two isosceles triangles have their sides equal, and the angle at the base of the second is triple the angle at the base of the first, then the cube of the base of the first minus triple the solid made by the base of the first and the square of the common side, is equal to the solid on the base of the second and the same square of the common side.

The statement of Viète's proposition simply refers to two isosceles triangles with legs of equal length. But in the accompanying figure, they are drawn so they share a vertex and the bases are collinear (triangles CDE and BCF in figure 14.2).[13] His figure is simply the diagram accompanying Archimedes's neusis trisection (figure 10.4 on page 150) with the added segment BF, which creates the second isosceles triangle. Using the labels in this diagram Viète's proposition states

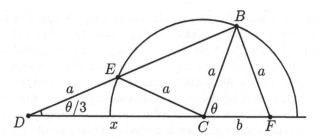

FIGURE 14.2. Viète related angle trisection and cubic equations.

that $(CD)^3 - 3(DE)^2(CD) = (DE)^2(CF)$. If we take the length of the legs of the isosceles triangles to be a, and we let $x = CD$ and $b = CF$, then his proposition states that $x^3 - 3a^2x = a^2b$.

In particular, if we want to trisect an angle θ, then we construct an isosceles triangle BCF with $\angle BCF = \angle BFC = \theta$ and legs BC and BF of length $a = 1$, say. If the length of the base CF is b, then we solve the cubic equation $x^3 - 3x = b$ for x. Then the base angle of the isosceles triangle with sides of length 1, 1, and x is $\theta/3$.

Conversely, suppose we are given a casus irreducibilis cubic equation of the form $x^3 = px + q$. By performing a linear substitution and renaming the constant term, we obtain a cubic equation of the form $y^3 - 3y = b$ with $0 < b < 2$.[14] Then we create an isosceles triangle BCF with sides of length 1, 1, and b. Let the base angle be θ. Now, if we are able to trisect any angle, in particular if we are able to perform a neusis construction, then we can construct the triangle CDE in figure 14.2 with base angles $\theta/3$ and legs of length 1. Then the length of the base of the triangle is a root of the equation.

In short, Viète showed that we can solve the casus irreducibilis geometrically if and only if we can trisect an arbitrary angle. In fact, something stronger is true: a cubic equation can be solved with a compass, straightedge, and angle trisector (like the tomahawk on page 55) if and only if it has three real roots.[15] That is, it is precisely the casus irreducibilis. This fact implies that we can't use a trisector to solve the problem of doubling the cube because $x^3 - 2 = 0$ has only one real root.

In 1591 Viète gave another technique for solving the casus irreducibilis using angle trisection—one that admits a trigonometric solution. Here is the general idea of Viète's work given in modern trigonometric notation.[16] Suppose we begin with a casus irreducibilis cubic $x^3 = px + q$. By performing a substitution, we can transform it into a cubic of the form $4y^3 - 3y = c$ for some constant $-1 \le c \le 1$.[17] Now we bring out a trigonometric identity that you probably did not memorize in your trigonometry class:[18] $4\cos^3(\theta/3) - 3\cos(\theta/3) = \cos\theta$. It is easy to see the resemblance of this identity to the cubic equation. Next find an angle θ such that $\cos\theta = c$. Then $y = \cos(\theta/3)$ is a root of the cubic equation. Moreover, we can use the fact that the cosine function is periodic to find the other two roots. In particular, $\cos(\theta + 360°) = \cos(\theta + 720°) = c$, so $\cos(\theta/3 + 120°)$ and $\cos(\theta/3 + 240°)$ are the other two roots. Finally, use these three y-values to find the three roots of the original equation.[19] Conversely, if we must trisect an angle θ, it suffices to find the root of the cubic equation given above, with $c = \cos\theta$.

This technique works for only certain values of p and q—those that yield a value of c between -1 and 1. But these are precisely the values that correspond to the casus irreducibilis. The bottom line is that we can always avoid complex numbers when solving cubic equations—by using Viète's trigonometric technique for the casus irreducibilis and the Tartaglia–Cardano formula for all other cubics.

So Viète showed that an angle trisector can be used to solve the casus irreducibilis, but he did not stop there. Using his algebra, he showed that every cubic equation could be reduced to one of several standard forms. Then he showed that they could be solved either using angle trisection, as shown earlier, or by finding two mean proportionals. For instance, solving $x^3 = a^2b$ is equivalent to finding the first of two mean proportionals between a and b, $a/x = x/y = y/b$. Moreover, he knew from Ferrari's work that every quartic equation could be reduced to a cubic and thus could also be solved in this way. He stated this in his proposition XXV:[20]

> It is generally true that the problems otherwise insolvable in which cubes are equated to solids or fourth powers to plano-planes, with or without affections, can all be solved by constructing two mean continued proportionals between given terms or by sectioning angles into three equal parts.

Because angle trisection and finding two mean proportionals can both be solved using the neusis technique, Viète concluded that every geometric problem that produces an equation of degree less than 5 can be solved by a compass and marked straightedge!

There is an interesting consequence of Viète's work that he did not mention. As we have seen, angle trisection and finding two mean proportionals can be solved using a compass, a straightedge, and conic sections. So geometric problems that yield an equation of degree less than 5 could be solved using conics. In particular, using Pappus's classification, they are plane or solid problems. Conversely, the intersection of two conics can be found by solving an equation of degree 4 or less. Thus, all plane and solid problems can be reduced to solving equations of degree less than 5.

Viète's Formula for π

Viète gave us further insight into the construction of regular polygons, the trisection of angles, and the construction of two mean proportionals. It turns out that he even contributed, in a way, to the problem of squaring the circle. Before Viète, mathematicians were able to compute π to as high a degree of precision as desired by carefully and tediously applying Archimedes's technique of polygonal approximations. But there were no formulas for π.

In 1593 Viète gave the first numerically precise expression for π. It was a product of infinitely many nested square roots:[21]

$$\frac{1}{\pi} = \frac{1}{2}\sqrt{\frac{1}{2}}\sqrt{\frac{1}{2} + \frac{1}{2}\sqrt{\frac{1}{2}}}\sqrt{\frac{1}{2} + \frac{1}{2}\sqrt{\frac{1}{2} + \frac{1}{2}\sqrt{\frac{1}{2}}}} \cdots ,$$

or, if we take the reciprocal of both sides,

$$\pi = 2\frac{2}{\sqrt{2}}\frac{2}{\sqrt{2 + \sqrt{2}}}\frac{2}{\sqrt{2 + \sqrt{2 + \sqrt{2}}}} \cdots .$$

It is the first known example of an infinite product, and was thus an early step in the study of infinite processes that would become so important to mathematics in the coming centuries.[22]

TANGENT
Galileo's Compass

Wilfrid, you came to request me to resolve equations, to fly on a rain-cloud,
to plunge into the fiord and reappear as a swan. If science or miracle were the
end of humanity, Moses would have left you a calculus of fluxions; Jesus
Christ would have cleared up the dark places of science; ... In our own day,
the greatest miracle would be to square the circle, a problem which you
pronounce impossible, but which has no doubt been solved in the progress
of worlds by the intersection of some mathematical line, whose curves are
apparent to the eye of spirits elevated to the highest spheres.
—Honoré de Balzac, *Seraphina*, 1834[1]

BECAUSE COMPASSES WERE indispensable tools for artisans,
craftsmen, draftsmen, engineers, military commanders, navigators,
astronomers, and surveyors, they got more and more elaborate.
By the Renaissance, the term "compass" (or "compasses") referred
to a variety of tools that could draw circles, ellipses, and other
conic sections (drawing compasses), transfer distances (dividers),
scale figures by some fixed ratio (reduction compasses), measure
spherical or cylindrical objects (calipers), transfer distances on maps
(three-legged compasses), and perform calculations involving ratios
(proportional compasses).

In 1597 Galileo Galilei (1564–1642) had instrument maker Marcan-
tonio Mazzoleni make a multifunction compass that he had designed
(see figure T.26).[2] It is now known as *Galileo's geometric and military
compass* and is known generically as a *sector*. It is a proportional compass made from brass and is about 25.6 cm long. The arms are wide
and flat and both sides are engraved with ruled lines radiating out
from the hinge. It came with a detachable arc (called a quadrant), a
plumb bob, and a foot that enabled it to stand; these add-ons were for
sighting angles and were often omitted in later versions.[3]

The key mathematics behind Galileo's compass—and all proportional compasses—is similar triangles. The engraved lines and their

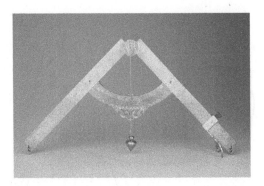

FIGURE T.26. Galileo's compass. (Collection of Historical Scientific Instruments, Harvard University)

marking on both arms are identical, so they can be used to create families of similar isosceles triangles. From a practical standpoint, the user can use a pair of dividers to go back and forth between a drawing and the compass, either placing a divider between the same two marks on the arms to measure the foot of a triangle (or between a mark and the hinge to measure a leg) and then transferring this measurement to the drawing, or the user can measure a quantity on the figure using the dividers, then open the arms of the compass so that the distance between two identical marks equals the span of the dividers.

Although the device was purely a practical one to be used in the field, it also could be used to solve some of the problems of antiquity. It could be used to square regular polygons—a task that is possible with a compass and straightedge—and to square circles—a task that is not. To do so, we use the *tetragonic lines* on the device. These lines are marked with integers 3 (at the far end) through 13 (closest to the hinge), and there is a circle situated between the 6 and 7 (see figure T.27).[4] They are placed on the line so the area of the circle with radius equal to the distance between the two circles is equal to the area of the regular n-gon whose side length equals the distance between the two n's.

To square a circle, open the dividers to the radius of the circle. Then open the compass so the distance between the circle markings on the tetragonic lines match the opening of the dividers. Then the distance between the 4s on the tetragonic scale is the side of the desired square. We could square any regular polygon in a similar fashion.[5]

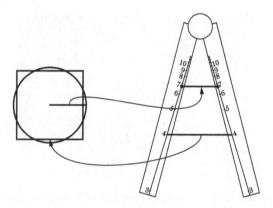

FIGURE T.27. We can use the tetragonic line to square the circle.

We can also use the device to construct regular polygons. Suppose we are given a circle with center C and radius AC, and we wish to inscribe a regular nonagon inside it, as in figure T.28 (a task that is impossible with a compass and straightedge). It suffices to find a point B on the circle so that AB is a side of the polygon. For this task we use the *polygraphic lines*.[6] First, use an ordinary pair of dividers to measure the radius of the circle. Then open Galileo's compass so that the dividers line up with the 9s on both arms. The polygraphic line is ruled so that the distance between points m and m on the two arms is the radius of the circle for which AB is the side of an inscribed regular m-gon. In particular, because the side of a regular hexagon is equal to the radius of the circle that circumscribes it, the distance between 6 and 6 is precisely the length of AB. Thus, we open a drawing compass so that the two tips are on 6 and 6, and we can use this opening to find the point B on the circle.

We can use the *stereometric lines* to add the volumes of any number of similar solids (see figure T.29).[7] In particular, we pick one common length on all of the solids such as the sides of cubes, the heights of similar cylinders, the diameters of spheres, and so on. Open Galileo's compass an arbitrary amount. Use dividers to measure the distances on the solids and find the numbers that correspond to those numbers on the compass. Suppose we have three spheres. One diameter is the same as the distance between the 10 and the 10 on the stereometric line, the diameter of the second one is the same as the distance between the 25 and the 25, and the third is the same as

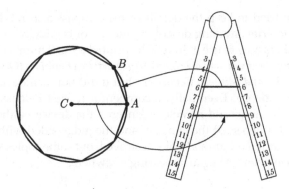

FIGURE T.28. We can use the polygraphic line to draw a nonagon.

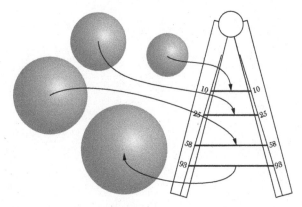

FIGURE T.29. We can use the stereographic line to add volumes.

the distance between the 58s. Then, because $10 + 25 + 58 = 93$, the sum of the volumes of these spheres is the same as the sphere with diameter between 93 and 93. These stereometric lines allow us to double a cube. If the side of a cube spans the distance between values x and x on the compass, then the cube with side length from $2x$ to $2x$ has twice the volume.[8]

These solutions to the problems of antiquity are not ideal. They are only as good as the quality of the device maker. Constructing the device and making accurate markings on the arms is not easy. In particular, they cannot be accomplished with a compass and straightedge!

Galileo tried to keep the details of his compass a secret. For years he did not write anything describing the use of his device. Instead, he taught others how to use it. In 1606 he printed a small number of copies of his notes, *Le operazioni del compasso geometrico et militare* (*Operations of the Geometric and Military Compass*),[9] but it did not include any drawings of the device. Eventually, a competitor named Baldassar Capra obtained the notes and reverse engineered the device—although not perfectly. Galileo sued the copycat, and the judge sided with Galileo. The judge ordered Capra's books destroyed, but some copies survived, and Galileo's device began appearing elsewhere.[10]

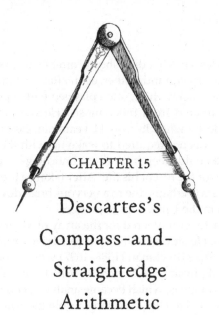

CHAPTER 15

Descartes's Compass-and-Straightedge Arithmetic

This doctrine of Descartes, of which no germ has been found
in the writings of the ancient geometers, . . . [is] *prolem sine
matre creatam* [a child without a mother].
—Michel Chasles, 1875[1]

"*Cogito ergo sum.*"—"I think, therefore I am."

ASK WHO RENÉ Descartes was, and people on the street will likely
produce this familiar philosophical proposition—one that has been
pondered and discussed by generations of undergraduates. More
mathematically inclined individuals may namedrop the Cartesian
plane that they learned about in an algebra class. Probably not many
would say that Descartes wrote an important text that improved mathematicians' understanding of which geometrical problems could and
could not be solved using lines, circles, and other curves.

The philosopher, mathematician, and scientist René Descartes was
born in La Haye, France, in 1596 to a bourgeois family of doctors and
lawyers. His father was a lawyer and a member of the Parlement of

Brittany. Descartes's mother died shortly after he was born, and he was raised by his maternal grandmother, even after his father remarried.

Descartes was a sickly child, often plagued with a persistent cough. He may have suffered from tuberculosis. He was sent to the Jesuit College at La Fleche when he was 11 years old. Because of his weak constitution, he was not required to wake up with the other students at 5:00 a.m., but could sleep in until midmorning—a practice that he would continue for the rest of his life. After taking a break from schooling for a few years (perhaps due to a nervous breakdown), he received a law degree from the University of Poitiers.

In 1618 Descartes volunteered for the army of Maurice of Nassau in the Netherlands. He studied military engineering, but he also met the Dutch scientist Isaac Beeckman (1588–1637) with whom he discussed mathematics and physics. On March 26, 1619, he told Beeckman of his new mathematical ideas, which became analytic geometry.

In 1619 he traveled to Ulm, Germany, to join the army of Maximilian of Bavaria. While there, he had a series of three dreams that he cited as the inspiration for his future inquiries: he must not rely on others—to find true knowledge he must carry out all investigations himself; and he must doubt everything and look for the most basic principles upon which to build his philosophy. At the core: I think, therefore I am.

From 1620 to 1628, Descartes traveled around Europe. It was during this time that he met the French polymath Father Marin Mersenne (1588–1648), who later helped Descartes stay in contact with the scientific world. He eventually landed back in the Netherlands in 1628, where he remained for the next two decades. The Netherlands suited Descartes, who was famous in Europe at the time and wanted to find an urban environment in which he could live in solitude and peace.

In 1649 Sweden's Queen Christina convinced Descartes to travel to Sweden to tutor her in philosophy. It was not a good match—personally or geographically. The two did not get along, the weather was cold, and Descartes had to wake early every morning—something he was loathe to do. He contracted pneumonia and died in Sweden in 1650.

We will not discuss Descartes's works in philosophy, such as *The World*, which he wrote but did not publish after hearing about Galileo's conviction of heresy in 1633, *Meditations on First Philosophy* (1639), *Principles of Philosophy* (1644), and *The Passions of the Soul* (1649). We will also ignore many of his contributions to mathematics and science.[2] Our focus is on his *La géométrie* (*Geometry*), which was one of three appendices to his 1637 *Discours de la méthode* (*Discourse on Method*)—the other two being *La dioptrique* (which discusses optics) and *Les météores* (meteorology).

Descartes's *Geometry*

The nineteenth-century mathematician Michel Chasles (1793–1880) famously described Descartes's *Geometry* as a child without a mother (see this chapter's epigraph). This remark certainly overstates the situation: *Geometry* was built upon the foundation constructed by previous scholars—such as Viète. But, as Judith Grabiner wrote,[3]

> Although he had all these predecessors, Descartes combined, extended, and then exploited these earlier ideas in an unprecedented way... At the very least we may say of the *Geometry* what Thomas Kuhn once said about Copernicus's *On the Revolution of the Celestial Orbs*; though it may not have been revolutionary, it was a "revolution-making text."

Today, when we think of analytic geometry, or Cartesian geometry, we think of perpendicular coordinate systems—x- and y-axes, ordered pairs, and curves represented by equations in x and y. One can't be faulted for assuming that this view of mathematics was proposed by Descartes. A modern reader, flipping through Descartes's *Geometry*, would not recognize much of today's analytic geometry.

As the title suggests, Descartes's aim was to solve geometry problems, but he intended to solve them using a compass and straightedge or more complicated curves that could be drawn using his new compasses. To do so, he introduced variables and algebraic equations. This is where analytic geometry enters the story. The algebra was a means to an end, not an end in itself.

For example, he used algebra to help solve a famous geometric problem stated by Pappus. The problem, which we won't state exactly here, begins with a collection of lines and the objective is to find all points that satisfy a certain condition related to the distances from the lines. The Greeks were able to solve the cases in which there are three or four lines—the set of points is a conic section. Descartes used his new techniques to arrive at this same conclusion and to extend the result to more lines. For instance, he found that the five-line solution is a cubic curve. Today we would be interested in plotting the curve, but Descartes did not sketch it. Instead he was interested in whether it could be constructed using one of his mechanical devices and whether it could help solve other geometric problems.

The mathematics in *Geometry* is difficult to follow—it was then, and it is now. The Dutch mathematician Francis van Schooten (1615–1660) helped clear up some confusion when he published a Latin translation complete with commentary. The first edition appeared in 1649, and an expanded second edition appeared a decade later. His work helped spread the use of Cartesian analytic geometry. As Carl Boyer wrote, Descartes's work represents the "early youth" of analytic geometry; the transition through its "gawky teen-age development" to its "adult stage" took 160 years.[4]

Although Descartes's mathematical arguments are difficult to unravel, his algebra looks very familiar. One historian cautions the modern reader that "the ease that Descartes displays in handling symbolic notation and the familiarity we have with it today must not obscure its profound novelty at the time.... Contemporary mathematicians were not mistaken in using the *Geometry* as a 'Rosetta stone' for deciphering symbolism... The new symbolism turned out to be an essential specific element of the Scientific Revolution."[5]

Arithmetic with Line Segments

One of Viète's biggest handicaps was his homogeneity assumption. Every term of his algebraic equations had to have the same degree. He could not add x and x^2 because to him, that would be adding a length and an area.

One of Descartes's most important contributions to analytic geometry was discovering how to perform arithmetic using line segments—how to add, subtract, multiply, divide, and take nth roots of line segments and yield another line segment—not an area or a volume.[6] This made the homogeneity assumption irrelevant, and it allowed Descartes to turn geometric problems into algebraic problems.

A naive approach is to perform line-segment arithmetic using the lengths of the line segments. This works fine for sums and differences but not for the other arithmetic operations. Suppose AB is 2 cm long and CD is 3 cm long. How long should $AB \cdot CD$ be? Perhaps $2 \cdot 3 = 6$ cm? But what if we measured the same segments in meters? Then $AB \cdot CD$ should be $0.02 \cdot 0.03 = 0.0006$ m $= 0.06$ cm long! If we switched to inches, the length would be $(0.784\ldots)(1.1811\ldots) = 0.9300\ldots$ in $= 2.3622\ldots$ cm (see figure 15.1). In short, the length of

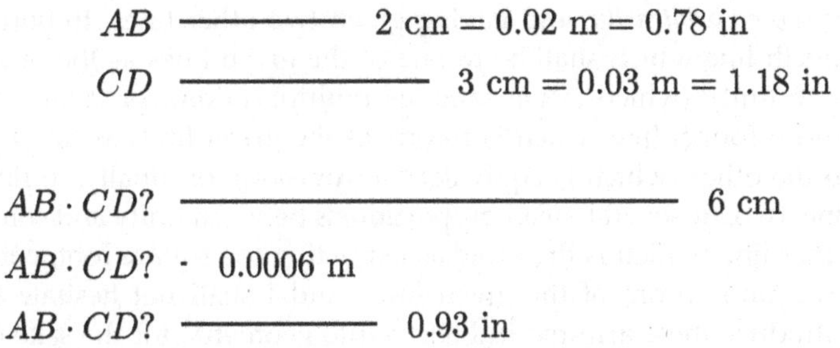

AB ——————— 2 cm = 0.02 m = 0.78 in

CD ————————— 3 cm = 0.03 m = 1.18 in

$AB \cdot CD$? ————————————— 6 cm

$AB \cdot CD$? · 0.0006 m

$AB \cdot CD$? ——————— 0.93 in

FIGURE 15.1. How long is the segment $AB \cdot CD$?

$AB \cdot CD$ will be different depending on which units we choose to measure the segments, be they centimeters, inches, miles, light-years, parsecs, furlongs, angstroms, or cubits.

While this seems like an insurmountable problem, it isn't. To enable the arithmetic of line segments we need to choose our units at the start of a problem—it doesn't matter what they are—and keep them throughout. This was Descartes's key insight. He did not state this assumption in terms of centimeters, inches, or any of the other familiar units; he stated it geometrically.

At the start of a problem we choose a line segment and proclaim that it has unit length. We use this segment throughout the problem as the measuring stick—it is the standard against which all segments are measured. In arithmetic terms, we decide how long "1" is. Once we do that, all the mathematics falls into place.

Descartes wasted no time in initiating this program. The first sentence of *Geometry* boldly stated his claim:[7]

> Any problem in geometry can easily be reduced to such terms that a knowledge of the lengths of certain straight lines is sufficient for its construction.

He continued (we italicized his introduction of the unit length segment),

> Just as arithmetic consists of only four or five operations, namely, addition, subtraction, multiplication, division and the extraction of roots, which may be considered a kind of division, so in geometry to find required lines it is merely necessary to add or subtract other lines; or else, *taking one line which I shall call unity in order to relate it as closely as possible to numbers, and which can in general*

be chosen arbitrarily, and having given two other lines, to find a
fourth line which shall be to one of the given lines as the other
is to unity (which is the same as multiplication); or, again, to
find a fourth line which is to one of the given lines as unity is
to the other (which is equivalent to division); or, finally, to find
one, two, or several mean proportionals between unity and some
other line (which is the same as extracting the square root, cube
root, and so on, of the given line. And I shall not hesitate to
introduce these arithmetical terms into geometry, for the sake of
greater clearness.

In other words, Descartes used ratios to define multiplication,
division, and nth roots. If PQ has unit length, then, according to
Descartes, the product of AB and CD is EF provided $EF{:}AB :: CD{:}PQ$.
When we replace the ratios with quotients and PQ with 1, this becomes
$EF/AB = CD/1$, or equivalently $AB \cdot CD = EF$, making Descartes's definition much more clear. Division is handled similarly. Notice that
the product of two segments is not a unique well-defined segment.
We could replace EF by any segment with the same length—it could
be located in a different part of the Euclidean plane and could be
rotated in any direction. All that matters is its length.

The nth roots are also defined using mean proportionals. For example, EF is the cube root of AB if there is a segment CD such that $AB{:}CD ::$
$CD{:}EF :: EF{:}PQ$. In fraction form this becomes $AB/CD = CD/EF = EF/1$.
A little algebra shows that $EF = \sqrt[3]{AB}$. This idea can be generalized in
a straightforward way to find nth roots.

Descartes made it crystal clear to the reader that he viewed these
geometric operations as a new form of arithmetic. He wrote,[8]

Often it is not necessary thus to draw the lines on paper, but it
is sufficient to designate each by a single letter. Thus, to add the
lines BD and GH, I call one a and the other b, and write $a + b$.
Then $a - b$ will indicate that b is subtracted from a; ab that a is
multiplied by b; $\frac{a}{b}$ that a is divided by b; aa or a^2 that a is multiplied by itself; a^3 that this result is multiplied by a, and so on,
indefinitely. Again, if I wish to extract the square root of $a^2 + b^2$,
I write $\sqrt{a^2 + b^2}$; if I wish to extract the cube root of $a^3 - b^3 + ab^2$,
I write[9] $\sqrt{C.a^3 - b^3 + abb}$, and similarly for other roots. Here it
must be observed that by a^2, b^3, and similar expressions, I ordi-
narily mean only simple lines, which, however, I name squares,

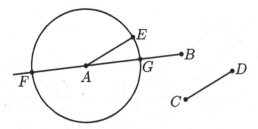

FIGURE 15.2. Addition and subtraction of line segments.

cubes, and so on, so that I may make use of the terms employed in algebra.

Descartes's symbolic notation is very modern. He introduced notation and conventions that are familiar to today's algebra students. He used letters at the end of the alphabet, such as x, y, and z, for unknown quantities and letters earlier in the alphabet for known quantities. He introduced the exponential notation that we use today: a^3 for aaa, a^4 for $aaaa$, and so on. However, his exponents were only positive integers and he used aa and a^2 interchangeably.[10] The modern equal sign had already been introduced by Robert Recorde[11] (ca. 1512–1558), but it was not in wide use. Instead, Descartes used ∞ to represent equality.

Compass-and-Straightedge Arithmetic

Once Descartes established that he could make sense of performing arithmetic with line segments, he turned to the geometrical problem of constructing line segments of specified lengths. He showed that the four arithmetic operations and square roots could be accomplished using a compass and straightedge. Indeed, the title of the first book (or, chapter) of *Geometry*—"Problems the construction of which requires only straight lines and circles"—screams *compass and straightedge*.

   Addition and subtraction of segments is so easy with the Euclidean tools that Descartes did not describe them. Suppose we are given segments AB and CD. Open the compass to length CD, then draw a circle about the point A. With points F and G as in figure 15.2, $BF = AB + CD$ and $BG = AB - CD$.

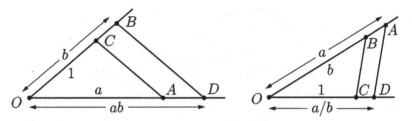

FIGURE 15.3. Multiplication (left) and division of line segments using similar triangles.

Multiplication and division of line segments and the extractions of square roots are accomplished using similar triangles. Suppose we wish to multiply two line segments OA and OB, which we may assume share an endpoint and are not collinear (see figure 15.3). Recall that we have designated a line segment as having unit length. Find a point C on OB so that OC has unit length. Draw AC, and then construct a segment BD parallel to AC so that D lies on OA. By construction, triangles ACO and BDO are similar. Thus $OD{:}OB{::}OA{:}OC$, which, as we discussed earlier, implies that $OD = OA \cdot OB$. Expressed another way, if OA, OB, and OC have lengths a, b, and 1, respectively, then OD has length ab. The procedure for dividing two segments is similar. We simply begin with a different configuration of segments OA, OB, and OC (see figure 15.3).

Lastly, we show Descartes's procedure for taking the square root of a line segment AB of length a. Find a point C on AB, but on the opposite side of A from B, so that $AC = 1$ (see figure 15.4). Construct a circle with diameter BC and a line perpendicular to BC passing through A. Let D be a point of intersection of this line with the circle. The triangles ACD and ABD are similar. Thus $AB/AD = AD/AC$, or, replacing AC with 1, $AD^2 = AB$, or $AD = \sqrt{AB}$.

Descartes gave the following example to illustrate his technique.[12] Suppose that to solve a certain geometric problem he must construct the solution to the equation $z^2 = az + b^2$. For instance, he may have, or he knows he can construct, segments of length a and b, and he wishes to construct another with length z. Because z represents a line segment, he must construct only the positive root of the equation, which he states is $z = a/2 + \sqrt{a^2/4 + b^2}$.

To do so, he uses a compass and straightedge to construct a right triangle LMN with $LN = a/2$ and $LM = b$ (see figure 15.5). Then he draws a circle with radius LN and center N and extends

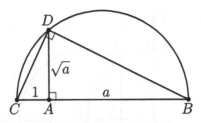

FIGURE 15.4. Descartes's method of constructing square roots.

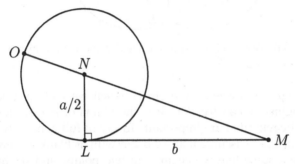

FIGURE 15.5. The segment OM is the solution to $z^2 = az + b^2$.

NM to O. By the Pythagorean theorem, $MN^2 = LN^2 + LM^2$. Moreover, $MN = MO - NO = MO - LN$. Combining these equations we have $(MO - LN)^2 = LN^2 + LM^2$. Rearranging terms gives $MO^2 = 2\, MO \cdot LN + LM^2 = a\, MO + b^2$, which implies that $z = MO$ is the solution to the equation.

Constructible Points and Constructible Numbers

So that we can understand what came next in the investigation of these problems, it is important to be very clear about what we can conclude from Descartes's work on constructibility. Unfortunately, his *Geometry* isn't set up in a modern way, with clearly articulated definitions, carefully stated theorems, and rock-solid proofs. Because we want to present the conclusions in a way that our path forward is more clear, in this section we will transport ourselves to the present day and examine some of the conclusions we can draw from Descartes's work. We will do it in a clear, rigorous, and modern fashion—not in the way Descartes wrote about them.

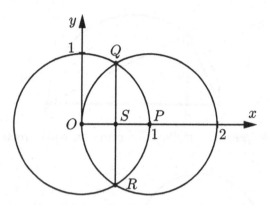

FIGURE 15.6. Two circles and a line yield constructible numbers 0, ±1, ±1/2, ±√3/2, and ±√3.

Although we see lines and circles when we look at a compass-and-straightedge construction, it is the points of intersection of the lines and circles that are important. To say in full generality what is constructible, we assume that we begin with as blank a slate as possible. We assume that we begin with two points, and we make the assumption that they are one unit apart (or equivalently, we begin with a unit line segment). At times, it is convenient to assume we have a coordinate system; in this case, we assume the two points are $(0,0)$ and $(1,0)$. (We could even draw the x- and y-axes using the compass and straightedge if we wanted to.) Now, from these we can define two related notions: constructible points and constructible numbers.

A point P is a *constructible point* if, starting with our two points, there is a sequence of legal compass-and-straightedge moves (recall we gave the "rules of the game" on page 49) such that P is the point of intersection of two constructed curves (lines or circles). A real number a is a *constructible number* if there exist constructible points P and Q (with $P = Q$ a possibility) such that $|a|$ is the length of segment PQ.

Because we begin with the points $O(0,0)$ and $P(1,0)$ (see figure 15.6), then without touching our compass and straightedge, we know that O and P are constructible points and 0, 1, and -1 are constructible numbers. (Note that -1 is constructible because $|-1| = 1$ is the distance between O and P.) Now use the compass to draw two circles of radius OP, one with center O and one with center P. They intersect at $Q(1/2, \sqrt{3}/2)$ and $R(1/2, -\sqrt{3}/2)$. Next draw QR, which intersects OP at $S(1/2, 0)$. Consequently, O, P, Q, R, and S are constructible points.

And, by computing the distances between all of them, we conclude that 0, ± 1, $\pm 1/2$, $\pm\sqrt{3}/2$, and $\pm\sqrt{3}$ are constructible numbers.

Constructible numbers and constructible points are related in another way. If we begin with the points $(0,0)$ and $(1,0)$, then (x,y) is a constructible point if and only if x and y are constructible numbers. For instance, we know that 1 and $-\sqrt{3}/2$ are constructible numbers, so the point $(-\sqrt{3}/2, 1)$ is constructible.

Although constructible points are more natural than constructible numbers, the latter are more useful. One of the major objectives of this book is to determine which numbers are constructible and which are not. Descartes gave one good answer to this question.

In mathematics, a *field* is a collection of numbers that satisfy the basic operations of arithmetic: addition, subtraction, multiplication, and division (by any nonzero number).[13] When we say "satisfy," we mean that it is closed under that operation; that is, the sum, difference, product, and quotient of any two numbers (excluding division by zero) in the field is another number in the field. The real numbers are a field. We can't perform some arithmetic operation on the real numbers and end up with a number that is not real. The rational numbers and the complex numbers are also fields. But the integers are not a field. They are closed under addition, subtraction, and multiplication, but not under division; 1 and 2 are integers, but $1/2$ is not. The natural numbers are even worse—they have the division problem, but they also have a subtraction problem: 1 and 2 are natural numbers, but $1 - 2 = -1$ is not.

Descartes showed that it is possible to use a compass and straightedge to add, subtract, multiply, and divide line segments. Essentially, then, Descartes proved that the set of constructible numbers form a field! So because 1 and $\sqrt{3}/2$ are constructible numbers, so are $1 - \sqrt{3}/2$ and $(\sqrt{3}/2)^3 = 3\sqrt{3}/8$.

In fact, the constructible numbers have more structure than just that of a field. Descartes proved that we can take square roots of segments as well. The square root of any nonnegative constructible number is also a constructible number. So, using our small starter set of constructible numbers, we can conclude that

$$1 + \frac{3\sqrt{3}}{8} + \sqrt{1 - \frac{\sqrt{3}}{2}}$$

is a constructible number.

Carrying this reasoning to its logical conclusion, every real number that can be expressed using the integers, addition, subtraction, multiplication, division, and square roots is a constructible number. So

$$3+\sqrt{\frac{11}{3}-\sqrt{2+\sqrt{3}}}+\sqrt{\frac{15}{34}}+\frac{8-\sqrt{7}}{5+\sqrt{1+\sqrt{3}}}$$

is a constructible number. It may take a lot of compass-and-straightedge moves, but Descartes gave us all the techniques we need to construct a line segment having this length.

Thus we have found a lot of constructible numbers. Every integer, every rational number, and irrational numbers such as the one above are constructible. But are *all* real numbers constructible? No.

Although Descartes stated it differently, and did not prove it, he knew that only numbers of this form are constructible.[14] The proof is elementary, but somewhat messy. In short, if a line is determined by two constructible points, then it has the form $ax + by + c = 0$ where a, b, and c are constructible numbers. Likewise, if a circle is determined by two constructible points, then it will have the form $(x - d)^2 + (y - e)^2 = r^2$ for some constructible numbers d, e, and r. Then, if we solve for the points of intersection of two such lines, a line and a circle, or two circles, the coordinates can be written in terms of the coefficients in the lines or circles, the four arithmetic operations, and square roots. So they have the desired form. Thus, we have a notable statement:

Descartes's constructible number theorem. A real number is constructible if and only if it can be obtained from the integers using addition, subtraction, multiplication, division, and the extraction of square roots.

This is an amazingly powerful result. To show that a number is not constructible, all we must do is show that it can't be written using the four arithmetic operations and square roots. If we can show that $\sqrt[3]{2}$ is not a constructible number, then it follows that it is impossible to double the cube. The number $\sqrt[3]{2}$ is a *cube root* of an integer, not the square root of an integer, so we're done, right?

Not so fast! We have to be careful. A number can be expressed multiple ways. For example, $\sqrt[3]{8}$ does not have the required form, but it is just 2, so it is constructible. Likewise, $\sqrt[4]{1+\sqrt{2}}$ is constructible because a 4th root is a square root applied twice:

$$\sqrt[4]{1+\sqrt{2}} = \sqrt{\sqrt{1+\sqrt{2}}}.$$

Things can get even worse: Surely $\sqrt[3]{7+5\sqrt{2}}$ is not constructible, right? The cube root is a dead giveaway. Or is it? It is not difficult to show that $\sqrt[3]{7+5\sqrt{2}} = 1 + \sqrt{2}$. Simply cube both sides and see.[15] So it is constructible.

We return to the four numbers that we introduced in chapter 1: $\sqrt[3]{2}$, $\cos(\theta/3)$, $\cos(2\pi/n)$, and π. If we can prove that these numbers are not constructible, then we have proved that the cube can't be doubled, the angle θ can't be trisected, the regular n-gon is not constructible, and the circle can't be squared. But, as our cautionary examples show, we need to be careful and not draw conclusions prematurely.

As a footnote to this conversation, recall that in trigonometry class we learned that certain angles yield values with square roots of integers when plugged into the trigonometric functions, such as $\cos(30°) = \sqrt{3}/2$ and $\tan(45°) = 1/\sqrt{2}$. In chapter 1, we saw that if we can construct an angle θ, then we can construct $\cos\theta$, and vice versa; and the same is true for any trigonometric function. Thus, the constructible number theorem tells us that θ is a constructible angle if and only if $\cos\theta$ (or equivalently, $\sin\theta$ or $\tan\theta$) can be expressed using the four arithmetic operations and square roots!

New Curves and New Compasses

One of Descartes's contributions to mathematics was his inclusion of new curves into the field of geometry. For twenty centuries, the line and the circle were the only fully recognized geometric curves. Conic sections were heavily studied, but were second-class citizens. And curves such as the spiral and the quadratrix were certainly not geometric. Descartes was not ready to invite all of these curves to the party, and he still believed that the simpler the curve, the better. But he went a long way to forming a more inclusive household.

Today we assert that Descartes included into geometry all algebraic curves—that is, sets of points satisfying a polynomial equation of two variables. For example, $y = x^2$ yields a parabola and $x^2 - y^2 = 1$ a hyperbola. But this statement is somewhat misleading. Descartes was a geometer, so he would not—and did not—define curves using algebra.

At the heart of this issue is a question that geometers struggled with for centuries: What is an acceptable way to define a curve? Before Descartes, algebraic representations of curves were unavailable. Euclid defined the line and the circle through his postulates. The conic sections were defined as the boundaries of cross-sectional slices of cones. Other mathematicians described curves through the use of simple or complicated drawing devices—a straightedge for drawing a line, a compass for drawing a circle, a trammel of Nicomedes for drawing a conchoid, and so on. The simultaneous motion of two intersecting lines or curves could also define a curve—think of the two moving lines that trace the quadratrix. Such constructions could be literally carried out or they could be a mental exercise in which the mathematician is convinced that the curve is well defined.

Scholars from Plato to Pappus to Descartes believed that some of these procedures were acceptable ways to define a geometric curve and some were not. Unlike his predecessors, Descartes believed that curves constructed using certain drawing devices or by two moving intersecting curves should be considered geometric.[16] As it turned out, curves defined in this way are precisely the algebraic curves.

Almost two decades before writing *Geometry*, Descartes expressed his vision for geometry in a letter to Beeckman:[17]

> So I hope I shall be able to demonstrate that certain problems involving continuous quantities can be solved by means of straight lines or circles only, while others can be solved only by means of curves produced by a single motion, such as the curves that can be drawn with the new compasses (in my view these are just as exact and geometrical as those drawn with ordinary compasses), and others still can be solved only by means of curves generated by distinct independent motions, which are surely only imaginary such as the notorious quadratic curve [quadratrix]. There is, I think, no imaginable problem which cannot be solved, at any rate by such lines as these. I am hoping to demonstrate what sorts of problems can be solved exclusively in this or that way, so that almost nothing in geometry will remain to be discovered.

Greek geometers looked down upon using "mechanical" devices to solve geometric problems. But Descartes pushed back. In

Geometry he argued that circles and lines are drawn using mechanical devices:[18]

> If we say that [the curves] are called mechanical because some sort of instrument has to be used to describe them, then we must, to be consistent, reject circles and straight lines, since these cannot be described on paper without the use of compasses and a ruler, which may also be termed instruments.

He suggested adding a postulate to Euclid's postulates:[19]

> Now to treat all the curves which I mean to introduce here, only one additional assumption is necessary, namely, two or more lines can be moved, one upon the other, determining by their intersection other curves. This seems to me in no way more difficult.

His criteria for the admittance of a curve into the realm of geometry was that it could be drawn by the intersection of moving curves, in which the movements are all dependent on a single motion.[20] For instance, he designed compasses so that moving one arm caused all the other arms to move. He believed that drawing curves like the quadratrix and the spiral, which required coordinating linear and circular movement, required the ability to rectify the circle and that that was a geometric impossibility. He retained the term "mechanical" to describe such curves and considered them to be outside of geometry.

Descartes argued that every curve he could construct using his rules was algebraic. He did not explicitly state that the converse was true—that every algebraic curve can be drawn by one of his instruments—but his work implied that he believed this was so. Perhaps he was silent on this topic because he was unable to prove it. The proof of the converse had to wait almost a century and a half until Alfred Kempe (1849–1922) proved it in 1876.[21] Nevertheless, it is for this reason that Descartes is attributed with introducing all algebraic curves into geometry.

Moreover, whereas Pappus classified geometric problems as plane, solid, or linear depending on whether they could be solved using lines and circles, conic sections, or more complicated curves, Descartes's algebraic representations allowed a more fine-grained classification: the degree of the polynomial.

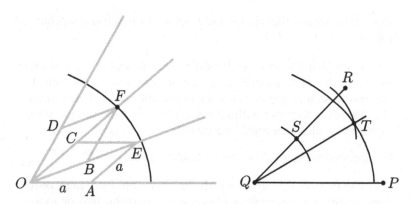

FIGURE 15.7. Descartes's angle trisection compass.

Like Viète, Descartes proved that any solid problem can be solved if we are able to trisect angles and construct two mean proportionals. Thus, he constructed compasses that could perform these tasks.

His four-armed compass shown in figure 15.7 can be used to trisect any angle.[22] It consists of four short rods (AE, BF, CE, and DF) of the same length a. There are hinges at A, B, C, and D, all of which are a units from the hinge at O. The rods are free to slide up and down the two central arms at the points E and F. There is a pencil at F that traces a curve as the mechanism opens and closes.

Suppose we wish to trisect $\angle PQR$. Place the compass along PQ with the hinge O located atop Q. Open the compass and draw the curve. Use an ordinary compass to draw a circle of radius a and center Q. It intersects QR at S. Then draw another circle with center S and radius a. It intersects the first curve at T. Then $\angle RQT = \frac{1}{3}\angle PQR$. By adding more arms to this compass, it is possible to divide an angle into any number of equal angles.

Some time around 1619, Descartes designed the generalization of Eratosthenes's mesolabe shown in figure 15.8.[23] It enabled him to find any number of mean proportionals and to solve certain algebraic equations.[24] This device has a hinge at the point Y. The bar BC is attached to, and forms a right angle with, XY; all of the other bars slide along either XY or YZ maintaining right angles with them. When the hinged device is opened, BC pushes CD which pushes DE, which pushes EF, and so on. The device can be made with as many such arms as desired. The points D, F, H, and so on trace out the dotted curves.

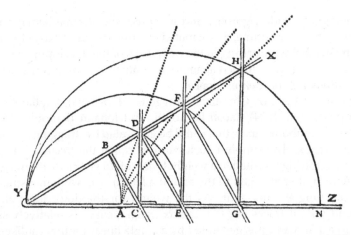

FIGURE 15.8. Descartes's mesolabe compass. (R. Descartes, 1637, *La géométrie*, Leiden: De l'Imprimerie de Ian Maire)

This device has an abundance of similar right triangles, which are the key to solving the geometric problems. For instance, from the similar triangles BCY, CDY, and DEY we obtain $BY{:}CY :: CY{:}DY :: DY{:}EY$. Thus CY and DY are the two mean proportionals between $AY = BY$ and EY, and if we open the compass so that EY is twice AY, then $CY = \sqrt[3]{2} \cdot AY$. Equivalently, if we use our compass to draw the dashed curve AD, then we can find the point D by finding the intersection of the curve with the circle with diameter AY.

Descartes illustrated how this device can be used to solve algebraic equations, such as $x^3 = x + 2$. Take $BY = 1$, open the compass so that $CE = 2$, and let $CY = x$. The geometry implies that $x^3 = x + 2$. So the length of CY is the positive real solution to the cubic equation.

Descartes introduced a huge new class of curves into geometry. But his criteria excluded other important newly discovered curves, such as the cycloid, the logarithmic spiral, and the catenary. Today we use Leibniz's term for nonalgebraic curves: transcendental. The problem—yet again—was that mathematicians were unable to describe these curves exactly. By the end of the eighteenth century, mathematicians had accepted Descartes's algebraic representation of curves, but by definition, there was no such description of transcendental curves.

Today we can represent such curves analytically, using the exponential functions, trigonometric functions, logarithmic functions, infinite series, and so on. But it took years for these functions to be introduced, understood, made rigorous, and accepted into the mathematician's toolkit. In the meantime, mathematicians used the curves to help solve problems that were arising in the early days of the development of the calculus. Thus, they turned again to mechanical means of construction as justification for their existence.

Gottfried Leibniz, Christiaan Huygens (1629–1695), Guillaume de l'Hôpital (1661–1704), Jacob Bernoulli, and Johann Bernoulli (1667–1748) wrote about curves that could be generated by dragging a weight with a string. The simplest such curve is called the *tractrix*. Imagine holding a string attached to a weight that is off to your side (or imagine a stubborn dog attached to the end of a leash). You walk in a straight line keeping the line taut. As the weight (or the dog) drags along behind you, it traces out the tractrix. Such a curve is relatively simple to explain and it is generated by a single linear motion (unlike the quadratrix and the spiral). Thus they argued that it should be included in the class of geometrically acceptable curves. But it is not algebraic.

In 1692 Huygens wrote,[25]

Descartes was wrong when he dismissed from his geometry those curves whose nature he could not express by an equation. It would have been better if he had acknowledged that his geometry was defective in so far as it did not extend to the treatment of these curves; for he was well aware that the properties and uses of such curves could also be investigated by geometrical methods.

Leibniz took this to an extreme. In 1693 he wrote that geometry concerned "anything that can be exactly constructed by continuous motion."[26] He described (but did not make) some very complicated drawing devices, which, in essence, could integrate functions.[27]

As it turned out, no consensus ever developed about which mechanical techniques could be used to produce geometric curves. By the second half of the eighteenth century, the mathematical community had moved on and no longer required constructions for their curves. The analytic view of mathematics pushed by mathematicians such as Euler became the standard. This view of curves and geometry was so successful that the modern student of mathematics has a difficult time recognizing the arguments at the time.

TANGENT
Legislating π

A *little Learning* is a dang'rous Thing;
Drink deep, or taste not the *Pierian* Spring:
There *shallow Draughts* intoxicate the Brain,
And drinking *largely* sobers us again.
—Alexander Pope[1]

ON JANUARY 18, 1897, Taylor Record introduced a piece of legislation to the Indiana House of Representatives. It was authored by a small-town doctor, Edwin J. Goodwin (1825–1902). It passed unanimously.

Goodwin, apparently, made some mathematical discoveries, published his results in the *American Mathematical Monthly*, and copyrighted the results in the United States, England, Germany, Belgium, France, Austria, and Spain. With this bill, he wanted to give Indiana schools the right to use his discoveries for free. Today the infamous House Bill No. 246 is known as the "Indiana π Bill."[2]

Dr. Goodwin was a circle squarer. The bill claimed that "the ratio of the diameter and circumference is as five-fourths to four" (that is, $\pi = 16/5 = 3.2$) and that the value "in present use ... should be discarded as wholly wanting and misleading in practical applications."

Like most cranks, Goodwin had little mathematical training. His inspiration was spiritual, not mathematical; he wrote that he was "supernaturally taught the exact measure of the circle."[3] Also, like most cranks, he was utterly convinced he was correct. He corresponded with scholars at the National Observatory in Washington, DC, and was sure that he'd convinced them of his correctness. He even attempted to secure a speaker's slot at the 1893 World's Fair in Chicago.

When we look closer at his *Monthly* piece, "Quadrature of the circle," we find that it was not a scholarly article but a letter in their "Queries and Information" section, complete with the disclaimer

"Published by request of the author."[4] One wonders why the editors published this nonsense. Perhaps they were looking to fill space in their new journal (it was the first volume of the now-prestigious publication). In the first sentence Goodwin wrote, "The area of a square is equal to the area of the circle whose circumference is equal to the perimeter of the square," which is clearly ridiculous. The second paragraph began, "To quadrate the circle is to find the side of a square whose perimeter equals that of the given circle." Obviously, it is the areas that must be equal. The mathematics does not improve following these errors.

Goodwin's arguments are so bad that it is not clear what value he is proposing for π. In addition to 3.2, mathematicians Arthur Hallerberg and David Singmaster found that Goodwin's convoluted arguments imply no less than eight other values for π ranging from 2.56 to 4.

At first the press was fooled, calling Goodwin "a mathematician of note."[5] But eventually they caught on to the reality of the situation, calling it "the strangest bill that has ever passed an Indiana Assembly."[6] Finally, Purdue University mathematician C. A. Waldo got wind of the legislation. He set the Senate straight.[7] The *Indianapolis News* described the reception the bill received in the Senate:[8]

> The Senators made bad puns about it, ridiculed it and laughed over it. The fun lasted half an hour. Senator Hubbell said that it was not meet for the Senate, which was costing the State $250 a day, to waste its time in such frivolity. He said that in reading the leading newspapers of Chicago and the East, he found that the Indiana State Legislature had laid itself open to ridicule by the action already taken on the bill. He thought consideration of such a proposition was not dignified or worthy of the Senate.

Thus, because of the professor's words, or more likely because of the fear of continued ridicule, the Indiana State Senate let the bill die.

Goodwin was not only a circle squarer, the π bill also asserted that he could trisect angles and double cubes. In 1895 he wrote,[9]

> (A) The trisection of an angle: The trisection of a right line taken as the chord of any arc of a circle trisects the angle of the arc; (B) Duplication of the Cube: Doubling the dimensions of a cube octuples its contents, and doubling its contents increases its dimensions twenty-five plus one per cent.

His trisection argument was the classic blunder shown on page 21, and his cube doubling argument used $\sqrt[3]{2} = 1.26$ rather than $1.25992\ldots.$

Thus, generations of schoolchildren can thank the Indiana legislature for listening to a college professor, thereby preventing them from learning something blatantly false or becoming an international embarrassment. Perhaps today's legislators could learn a lesson from them!

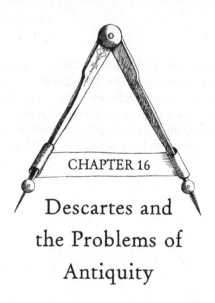

CHAPTER 16

Descartes and the Problems of Antiquity

As the geometrician, who endeavors
To square the circle, and discovers not,
By taking thought, the principle he wants,

Even such was I at that new apparition;
I wished to see how the image to the circle
Conformed itself, and how it there finds place;

But my own wings were not enough for this,
Had it not been that then my mind there smote
A flash of lightning, wherein came its wish.

Here vigor failed the lofty fantasy:
But now was turning my desire and will,
Even as a wheel that equally is moved,

The Love which moves the sun and the other stars.
—The final lines of Dante's *Divine Comedy*[1]

DESCARTES MADE MANY algebraic advancements. But at heart, he was a geometer. He was interested in solving geometric problems, and algebra was a tool to help him accomplish this. He wrote,

> If, then, we wish to solve any problem, we first suppose the solution already effected, and give names to all the lines that seem needful for its construction,—to those that are unknown as well as to those that are known. Then, making no distinction between known and unknown lines, we must unravel the difficulty in any way that shows most naturally the relations between these lines, until we find it possible to express a single quantity in two ways. This will constitute an equation, since the terms of one of these two expressions are together equal to the terms of the other.

Descartes was certain that this procedure was the *right way* to do geometry. After presenting a few examples, he confidently asserted,[2]

> I have given these very simple ones to show that it is possible to construct all the problems of ordinary geometry by doing no more than the little covered in the four figures that I have explained. This is one thing which I believe the ancient mathematicians did not observe, for otherwise they would not have put so much labor into writing so many books in which the very sequence of the propositions shows that they did not have a sure method of finding all, but rather gathered together those propositions on which they had happened by accident.

Gone were the days of clever ad hoc solutions to geometric problems, he claimed. Now all of geometry can be solved using algebra.

Angle Trisection and Finding Two Mean Proportionals

Let us see how Descartes's method works by examining one of his own examples—one that is of particular interest to us. In book 3 of *Geometry* he showed how to trisect an angle using a parabola.[3]

Suppose we want to trisect the central angle $\angle NOP$ in a circle with center O (see figure 16.1). To obtain the segments OQ and OT that trisect the angle, it suffices to construct NQ, which has unknown length x.[4] In order for him to use his algebraic techniques, he needed a

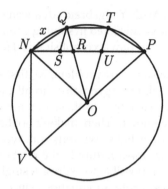

FIGURE 16.1. To trisect $\angle NOP$ we must find $x = NQ$.

unit-length line segment, and he decided it should be the radius of the circle.

Draw NP, which Descartes denoted q, and which intersects OQ at R and OT at U. Draw QS parallel to OT, with S on NP. Triangles NOQ, NQR, and QRS are similar, so $NO{:}NQ{::}NQ{:}QR{::}QR{:}RS$, or rewritten as fractions, $1/x = x/QR = QR/RS$. It follows that $QR = x^2$ and $RS = x^3$—equalities that would have appeared nonsensical to the dimension-conscious geometers before Descartes. Because NQR and PTU are isosceles and $QSUT$ is a parallelogram, $3NQ = NQ + QT + TP = NR + SU + UP = NP + RS$. So $RS = 3NQ - NP$. Replacing the segments with their lengths, we obtain the cubic $x^3 = 3x - q$, one root of which is the length NQ.

Once he got to this point—having an equation to solve—he brought out his algebraic toolkit. He reduced the polynomial as far as possible to obtain a polynomial of minimal degree. This polynomial is fully reduced (at least for a generic q) and the equation is a cubic, so he knew that lines and circles alone would not suffice to construct the root.

Around 1625 Descartes discovered a method to solve any cubic or quartic equation using lines, circles, and a parabola.[5] This was an improvement over Viète and Fermat, whose geometric solutions to such equations were more theoretical and did not exhibit the technique for doing so. His friend and teacher Isaac Beeckman wrote,[6]

Mr Descartes values this invention so much that he avows never to have found anything more outstanding, indeed that nothing more outstanding has been found by anybody.

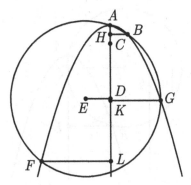

FIGURE 16.2. Descartes used a parabola to solve the cubic equation
$x^3 = 3x - q$.

Here's Descartes's method of solving $x^3 = 3x - q$: Begin with a
parabola opening downward so the distance between the vertex A and
focal point C is 1/2 (see figure 16.2). Draw the line of symmetry; today
this would be the y-axis. Descartes did not draw the equivalent of the
horizontal x-axis, but he measured distances perpendicularly off the
vertical line, as if there were an x-axis. Let D be on the line so that
$CD = 3/2$ and E be to the left of D so that the horizontal line segment
$DE = q/2$. Draw the circle with center E and radius EA. It intersects the
parabola at three points (other than A): B, G, and F. These are the roots
of the equation; that is, if we draw perpendiculars from the points to
H, K, and L on AD, then the roots are the lengths of BH and GK and
the negative of the length of FL. (Descartes pointed out that F represents a "false" root—a negative root—of the equation.)[7] The x-value
we desire—the one that will trisect the angle—is BH. This completes
the construction, and the trisection problem is solved.

Descartes also used a parabola to find the two mean proportionals
between segments of length a and q (see figure 16.3).[8] Recall that finding segments of length z and u so that $a/z = z/u = u/q$ is equivalent to
solving $z^3 = a^2q$. Begin with points A, C, and E so that AC and EC are
perpendicular and have lengths $a/2$ and $q/2$, respectively. Construct a
parabola with vertex A and axis AC so that C is half the distance to the
focal point. Construct a circle with center E passing through A. The
circle and the parabola intersect at A and another point F. Drop a perpendicular from F to a point L on AC. Descartes concluded that FL and
AL are the two mean proportionals (z and u, respectively).[9]

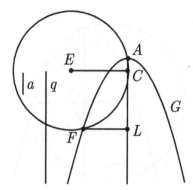

FIGURE 16.3. Descartes used a parabola to find the two mean proportionals between segments of length a and q.

Descartes's Impossibility Claims

Descartes believed, as did generations of mathematicians before him, that it was impossible to trisect an angle or find two mean proportionals using a compass and straightedge. In fact, he would show that a problem was impossible by reducing it to one of these. He wrote,[10]

> It is true that I have not yet stated my grounds for daring to declare a thing possible or impossible.... Solid problems in particular cannot, as I have already said, be constructed without the use of a curve more complex than the circle. This follows at once from the fact that they all reduce to two constructions, namely, to one in which two mean proportionals are to be found between two given lines, and one in which two points are to be found which divide a given arc into three equal parts.

As far as we know, Descartes was the first to attempt to prove that the problems were impossible. In *Geometry* he built up the algebraic framework to discuss the problems—including introducing the notion of an irreducible equation (although his definition is unclear and inconsistent).[11] He zeroed in on the key fact, which became so important in the final proof two centuries later, that solving the problems relied on constructing the roots of irreducible cubic equations. Thus we expect his proof to build from this foundation. Yet when it comes time to make his case that they are impossible to solve, he gives a confusing, imprecise, and geometric—not algebraic—explanation:[12]

> Inasmuch as the curvature of a circle depends only upon a simple relation between the center and all points on the circumference, the circle can only be used to determine a single point between two extremes, as, for example, to find one mean proportional between two given lines or to bisect a given arc; while, on the other hand, since the curvature of the conic sections always depends upon two different things, it can be used to determine two different points.

In other words, he argues that a circle depends on only one thing: the radius. Thus, it can be used to divide something in two—like to bisect an angle. But it is impossible to use a circle to divide something into thirds. For that we need a curve that is determined by more than one parameter, like a conic section. Jesper Lützen, who carefully analyzed Descartes's claims of impossibility, wrote that "Descartes' geometric impossibility proof has a lot of baffling features" and it is "unclear, weird and incorrect."[13] After a close reading of *Geometry*, we conclude that he did not prove the impossibility of these problems—algebraically or geometrically. Lützen wrote,[14]

> When trying to pinpoint Descartes' argument for this impossibility one realizes that, as is often the case in *Geometry*, the line of thought is highly convoluted and involves several different sections of the text. At first reading one may get the idea that Descartes tried to give an algebraic "proof" of the impossibility of the problems, but on second reading one realizes that he actually did not claim to provide such an argument but rather ends up giving a strange geometric argument instead.

Despite this failed proof, Descartes made important advances that got mathematicians significantly closer to proofs of impossibility. First, he recognized that this was a theorem that could be proved, not a vague, imprecise statement about a very difficult or seemingly impossible task. Moreover, he translated the geometric problems into algebraic equations, he introduced many useful algebraic techniques, and he presented the important notion of irreducible polynomials.[15]

Unfortunately, because Descartes made these assertions about the impossibility of the classical problems and because his *Geometry* was so respected and influential, many later mathematicians mistakenly believed that Descartes had proved the impossibility of the trisection of an angle and the doubling of the cube. Recall that in 1775 the Académie des Sciences decided not to allow submissions of articles on these topics. Perhaps because of this common misconception, it took another 200 years to obtain a fully rigorous algebraic proof.

Squaring the Circle and the Rectification of Curves

Descartes also believed that the problem of squaring the circle was impossible. In a 1638 letter to Mersenne he wrote,[16]

> For, in the first place, it is against the geometers' style to put forward problems that they cannot solve themselves. Moreover, some problems are impossible, like the quadrature of the circle, etc.

He recognized that it was a more difficult problem than the others. He could solve those problems using conic sections, but he believed that the problem of squaring the circle was outside geometry—it could be solved only using a nonalgebraic curve like the spiral or the quadratrix.

The problem of squaring the circle is equivalent to the problem of rectifying the circle, and Descartes thought it was impossible to rectify *any* curve. That is, it is impossible to express the length of a curve as the root of an algebraic equation. He wrote,[17]

> On the other hand, geometry should not include lines that are like strings, in that they are sometimes straight and sometimes curved, since the ratios between straight and curved lines are not known, and I believe cannot be discovered by human minds, and therefore no conclusions based upon such ratios can be accepted as rigorous and exact.

This belief—which sounds ridiculously counterintuitive to us—that curves are not like strings that can be straightened into lines was not new to Descartes. There was a long-held belief that lines and curves are fundamentally different and cannot be compared. This belief is often attributed to Aristotle. In his *Physics*, Aristotle wrote about whether it is possible to compare the motion of a body that travels on a circular path with one that moves in a straight line. He concluded it is not:[18]

The fact remains that if the motions are comparable, there will be a straight line equal to a circle. But the lines are not comparable; so neither are the motions.

Archimedes's theorem that the area of a circle with radius r and circumference C equals that of a right triangle with legs r and C reinforced this belief. It showed that rectifying the circle is equivalent to squaring the circle, and because the problem of squaring the circle was believed impossible, so must be the problem of rectifying the circle.

To get around this problem, some scholars in the Middle Ages emended Archimedes's *Measurement of a Circle*. An early version, perhaps from the thirteenth century, added three postulates, one of which was that "there is some curved line equal to any straight line and some straight line equal to any curved line."[19] Another version (which can only be placed between the middles of the thirteenth and fifteenth centuries) said,[20]

> The second of the postulates is that a curved line is equal to a straight line. We postulate this although it is a principle known *per se* and recognized by anybody with a sound head. For if a hair or silk thread is bent around circumference-wise in a plane surface and then afterward is extended in a straight line in the same plane, who will doubt—unless he is hare-brained—that the hair or thread is the same whether it is bent circumference-wise or extended in a straight line and is just as long the one time as the other.

It turned out Descartes was able to rectify the circle—sort of—and this reinforced his belief in its impossibility. Some time before 1628 he gave a method of taking a segment of length C and constructing a segment of length d so that C is the length of the circumference of a circle with diameter d. Thus, it enabled him to construct a segment of length π. However, the procedure required infinitely many steps with a compass and straightedge, so Descartes knew it was not a legitimate Euclidean construction.[21] Consequently, he likely believed that any rectification must require such an infinite process.

We saw some ancient examples of curve rectification: Archimedes rectified the circle using a spiral, and Nicomedes rectified the circle using the quadratrix. But not only are these curves not lines or circles, they are not algebraic, and hence to Descartes were not geometric.

However, Descartes saw curves rectified in his lifetime. In fact, *he* was able to show that a section of the logarithmic spiral (not to be confused with Archimedes's linear spiral) is Euclidean rectifiable![22] This result was discovered independently by Evangelista Torricelli (1608–1647) and Thomas Harriot (1560–1621). Christopher Wren (1632–1723) and Gilles de Roberval (1602–1675) independently proved that an arc of the cycloid is Euclidean rectifiable.[23] The cycloid is formed by a rolling wheel. If we placed a dot on the tread of a bicycle tire, stood off to the side, and watched a cyclist ride by, then the dot would trace out a cycloid.

These discoveries were not enough to change Descartes's mind. He argued that the logarithmic spiral and the cycloid are transcendental curves and are thus not geometric curves—rectifications of mechanical curves do not count. Other mathematicians agreed. Blaise Pascal and René de Sluse (1622–1685) believed the arguments were circular: the curves are rectifiable because drawing them requires the coordination of circular and straight-line motion. In a 1659 letter to Huygens, Pascal shared Sluse's "beautiful remark":[24]

> One ought still admire . . . the order of nature . . . which does not permit one to find a straight line equal to a curve, except after one has already assumed the equality of a straight line and a curve.

Finally, in 1659 and 1660—a decade after Descartes's death— William Neil (1637–1670), Hendrik van Heuraet (1634–ca. 1660), and Fermat each showed that a section of the semicubical parabola $ay^2 = x^3$ could be rectified using Euclidean techniques.[25] It finally disproved Descartes's assertion about straight lines and curves. Bos wrote,[26]

> The central role of the incomparability of straight and curved in Descartes' geometry explains why the first rectifications of algebraic . . . curves in the late 1650's were so revolutionary: they undermined a cornerstone of the edifice of Descartes' geometry.

Yet Descartes—and Aristotle—were correct that it's impossible to rectify the circle. But the proof was still over two centuries away.

TANGENT
Hobbes, Wallis, and the New Algebra

The unhappy Calculator, robbed of all Geometrical defenses, held fast
in the thorny thicket of Numbers, looks in vain to his algebra.
— Johannes Kepler[1]

THOMAS HOBBES (1588–1679) is one of the most well-known and
celebrated philosophers. He was strenuously opposed to the new algebraic methods in mathematics. He was also a circle squarer, an angle
trisector, and a cube doubler: he was a mathematical crank.

Hobbes discovered mathematics relatively late in his life:[2]

> He was ... 40 yeares old before he looked on geometry; which
> happened accidentally. Being in a gentleman's library in ...,
> Euclid's Elements lay open, and 'twas [the Pythagorean the-
> orem]. He read the proposition. "By G—," sayd he "this is
> impossible!" So he reads the demonstration of it, which referred
> him back to such a proposition; which proposition he read. That
> referred him back to another, which he also read. Et sic deinceps,
> that at last he was demonstratively convinced of that trueth. This
> made him in love with geometry.

This discovery inspired Hobbes to base his philosophy on geometry's
deductive structure. But it also inspired him to investigate the problems of antiquity. Unfortunately, his mathematical skills did not match
his enthusiasm. Tackling these famous geometric problems does not
make someone a mathematical crank. What has tagged Hobbes with
this moniker was his undying belief in his (clearly) flawed proofs.

Best practices say, leave mathematical cranks alone. Confronting
them is pointless. But given Hobbes's stature and his stance on politics
and religion, this was not meant to be. The mathematical establishment did not let Hobbes's publications go unchallenged. John Wallis
was his most dogged critic. Hobbes's and Wallis's intense and very
public clash began after Hobbes published his 1655 book *De corpore*,

which contained multiple chapters on mathematics, and continued until his death nearly a quarter of a century later.

Wallis was also late in discovering mathematics. He was a theologian with little mathematics training as a youth. He was inspired to study mathematics in his early 30s after reading William Oughtred's (1574–1660) *Clavis mathematicae* and learning about Cardano's work on the cubic. Unlike Hobbes, Wallis became a creative mathematician with a keen eye for the future direction of the subject. He was appointed Savilian Professor of Geometry at the University of Oxford when he was 32 years old—not because of any mathematical accomplishments but because the previous chairholder was a Royalist. It helped that Wallis aided the Parliamentarians by deciphering captured code during the English Civil War, a war that the Parliamentarians had won.

During his mathematical career, Wallis contributed to algebra, analytic geometry, trigonometry, continued fractions, integration, and infinite series. He was a fan of the new movement toward using algebra in mathematics. For instance, although the conic sections have a three-dimensional origin, in his 1655 *De sectionibus conicis* Wallis worked with them as plane curves described in an analytic fashion. In his work on integration, Wallis came up with the elegant infinite product representation of π:[3]

$$\frac{\pi}{2} = \frac{2 \cdot 2 \cdot 4 \cdot 4 \cdot 6 \cdot 6 \cdot \ldots}{1 \cdot 3 \cdot 3 \cdot 5 \cdot 5 \cdot 7 \cdot \ldots},$$

which was an improvement on Viète's product formula because it does not have square roots. But the convergence was still too slow to be helpful for computing the digits of π. We can also thank Wallis for introducing the now-iconic symbol for infinity: ∞. Perhaps more influential than the theorems he proved was his work promoting and writing about the history, importance, and notation of algebra.[4]

He also wrote about theology, he taught deaf persons to speak, he studied grammar and phonetics, he translated Greek writings, he was one of the founding members of the Royal Society of London, he was the keeper of the archives for his university, and he assisted the university with their legal work. Wallis was self-aggrandizing, competitive, easy to anger, and nationalistic. He was one of the few British mathematicians to attempt the mathematical challenges posed by Fermat (even though he had no training or interest in number theory) and Pascal. For half a century, Wallis helped promote mathematics in Britain. This was at a time—before Newton—that the mathematical center of Europe clearly resided on the continent.

Wallis quarreled with others, but his battle with Hobbes was remarkable for its duration and nastiness. It was about mathematics, but it may also have been due to Hobbes's criticisms of Wallis's book *Arithmetica infinitorum* and for Hobbes's writings about Christianity.

Hobbes was distrustful of the new algebraic symbolism of Viète, Descartes, Oughtred, Harriot, Wallis, and others. He grudgingly admitted that algebra might be useful in discovering theorems, but he asserted that it had no place in the final product. He wrote that "symbols are poor unhandsome, though necessary, scaffolds of demonstration; and ought no more to appear in public, than the most deformed necessary business which you do in your chambers." Of one of Wallis's books he wrote, "It is so covered over with the scab of symbols, that I had not the patience to examine whether it be well or ill demonstrated." Then later, "having examined your pannier of Mathematics, and finding in it no knowledge, neither of quantity, nor of measure, nor of proportion, nor of time, nor of motion, nor of any thing, but only of certain characters, as if a hen had been scraping there."[5]

Hobbes prized geometric justifications. But he ran into trouble when he invented his own geometric definitions and rules. He rejected Euclid's definitions of point and line,[6] arguing that everything has a magnitude, but that the magnitudes—the breadth of a line, for instance—are ignored by mathematicians. Moreover, he defined lines in terms of the motion of points:[7]

> Though there be no body which has not some magnitude, yet if, when any body is moved, the magnitude of it be not all considered, the way it makes is called a *line*, or one single dimension; and the space, through which it passeth, is called *length*; and the body itself, a *point*; in which sense the earth is called a *point*, and the way of its yearly revolution, the *ecliptic line*.

Occasionally Hobbes stated a correct mathematical proposition. Not wanting to give Hobbes credit for these results, Wallis wrote that "it is to be suspected that they are not yours, because they are true, while things of yours are wont to be false."[8]

Hobbes occasionally admitted that his work contained errors, but he was generally unconvinced by Wallis's counterarguments. In his 80s, after years of fighting with Wallis, Hobbes still believed he had solved the problems of antiquity. Of his own accomplishments, he wrote,[9]

In mathematics, he corrected some principles of geometry. He solved some most difficult problems, which had been sought in vain by the diligent scrutiny of the greatest geometers since the very beginnings of geometry; namely these:

1. To exhibit a right line equal to the arc of a circle, and a square equal to the area of a circle, and this by various methods....
2. To divide an angle in a given ratio.
3. To find the ratio of a cube to a sphere....
4. To find any number of continual mean proportionals between two given lines....
5. To describe a regular polygon with any number of sides....

Despite his mathematical shortcomings, Hobbes spoke for a large population of scholars who distrusted the new algebra. Helena Pycior wrote, "In a striking way, then, Hobbes and Wallis stood as prototypes of two kinds of British thinkers who would debate the direction of mathematics for more than a century: Hobbes speaking for the geometric traditionalist and the general scholar; Wallis representing the algebraic or analytic mathematician and, secondarily, the emerging professional mathematician."[10] Euclidean geometry was still dominant in the sixteenth century, and by the eighteenth century calculus was in full force. The seventeenth century was a time of great change in mathematics.

Isaac Newton was an extremely influential voice in mathematics, especially in Britain, but his views of geometry and algebra were inconsistent. Sometimes he embraced Descartes's methods, and other times he spoke out against them. His opinion varied throughout his career and could depend on what and to whom he was writing. And his stated views did not always match up with his usage.

To get a glimpse of Newton's conflicted feelings, consider this excerpt from the appendix to *Universal Arithmetick*, a book that contains his lecture notes from 1673 to 1683:[11]

Equations are expressions belonging to arithmetical computation and in geometry properly have no place except in so far as certain truly geometrical quantities (lines, surfaces and solids, that is, and their ratios) are stated to be equal to others. Multiplications, divisions and computations of that sort have recently been introduced into geometry, but the step is ill-considered and contrary to the original intentions of this science: for anyone who examines the constructions of problems by the straight line and circle devised by the first geometers will readily perceive that geometry was contrived as a means of escaping the tediousness of calculation by the ready drawing of lines. Consequently these two sciences [arithmetical computation and geometry] ought not to be confused. The Ancients so assiduously distinguished them one from the other that they never introduced arithmetical terms into geometry; while recent people, by confusing both, have lost the simplicity in which all elegance of geometry consists.

But Newton also played a large part in the process of confounding the disciplines—using geometry to solve algebraic problems, and vice versa. This was the main purpose of *Universal Arithmetick*. And algebra was an essential ingredient in his newly developed calculus.

There were a number of historical factors that slowed the merging of algebra and geometry. There was an unease about the nature of negative, irrational, and complex numbers and more basically the relationship between discrete numbers and continuous magnitudes. The dimension problem—that the multiplication of two line segments gave area—stymied mathematicians for a long time. Algebra did not always make problems easier; it was often difficult to go back and forth between geometry and algebra, and often a problem was easier to solve geometrically than algebraically. The casus irreducibilis was an instance in which the algebra made the problem worse; the algebraic approach required using complex numbers, but geometrically it could be solved via an angle trisection. There were concerns about exactness: geometric operations are exact, whereas algebraic operations may lead to square roots, which can't be computed exactly. Finally, algebra was perceived as less rigorous than geometry.

Bos wrote,[12]

To us [the merging of algebra and geometry] may also seem a natural and obvious one. But such an impression would be based too much on the hindsight of successful algebraic geometry and on a historiography that commonly treats the emergence of analytic geometry in the seventeenth century as merely laudable, rather than enigmatic. Viewed from the perspective of the late sixteenth century, it was not at all obvious that algebra should be of use as a tool for geometrical analysis.

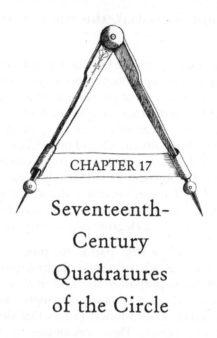

CHAPTER 17

Seventeenth-
Century
Quadratures
of the Circle

Algebra is the analysis of the bunglers in mathematics.
—Isaac Newton[1]

WHILE THE MATHEMATICAL community was still coming to grips with the new field of algebra, there was another revolution happening. Mathematicians were beginning to turn their attention from finite processes to infinite ones. The study of the infinite and the infinitesimal became calculus. This investigation is best known for the many-year controversy between Gottfried Leibniz and Isaac Newton and their followers over who should receive credit for its invention. With the benefit of hindsight we recognize that the groundwork for calculus was laid by many mathematicians, and this set the stage for the simultaneous independent discoveries by these two giants.

With calculus, mathematicians could tackle traditional geometrical problems—tangents, areas, volumes, arc lengths, and so on—with a new toolkit. At this time, "quadrature" no longer meant squaring a region using Euclidean techniques, but rather computing the area of the figure by any means, such as the integral. In particular, in the seventeenth century "the quadrature of the circle" did not always mean what it meant historically.

Christiaan Huygens

Christiaan Huygens was perhaps the greatest mathematician between Descartes and Newton and the greatest physicist between Galileo and Newton. Among his many accomplishments was determining the shape of a hanging chain (it is a catenary, not a parabola), conducting a mathematical investigation of light including putting forward the wave theory, and being a founder of the field of probability. He was also an astronomer (he studied Saturn's rings and discovered the first of Saturn's moons) and an inventor (he improved the design of telescopes and invented the pendulum clock). As a mathematician, Huygens was—at least for most of his life—conservative, preferring Greek geometry and some of Viète's and Descartes's algebra to the new ideas of calculus.

Archimedes showed the simple relationship between the area of a circle and its circumference, thereby proving that the quadrature and the rectification problems are equivalent. In 1657 Huygens extended this work to other conic sections. He discovered that the arc length of the parabola was the same as the area under the hyperbola.[2] He also discovered a relationship between the surface area of a paraboloid—the bowl-shaped surface obtained by revolving a parabola about its axis of symmetry—and the area of the circle.[3] This was the first new result about surface area since Archimedes's work on the sphere. Later, Huygens related the surface areas of the oblate and prolate spheroids and the hyperboloid—surfaces obtained by revolving ellipses and hyperbolas—to the area of a circle.

Huygens was skeptical that it was possible to square the circle, but his results gave others hope that it could somehow be accomplished. Joella Yoder wrote, "This was the promise contained in Huygens's transformation, which substituted the paraboloid, whose progenitor had always been the most amenable of curves, for the unruly circle."[4] Alas, none of these discoveries helped solve the problem of squaring the circle.

Gregory of Saint Vincent

Gregory of Saint Vincent (1584–1667) was a Belgian priest who studied mathematics and astronomy under Christopher Clavius (1538–1612) in Rome. Gregory published one book in his lifetime, and it was a whopper—a 1250-page rambling, disorganized, at-times brilliant work called *Opus geometricum quadraturae circuli et sectionum coni* (*Geometric Work of Squaring the Circle and Conic Sections*).[5] In this 1647 book, which was likely written two decades earlier, he tackled geometry, conic sections, doubling the cube, angle trisection, infinite series, and ideas that would later form the basis of calculus.

For instance, in his work on infinite series he observed that[6]

$$\frac{1}{2} - \frac{1}{4} + \frac{1}{8} - \frac{1}{16} + \cdots = \frac{1}{3}.$$

From this, he produced a method of repeatedly bisecting an angle that would—if performed infinitely many times—trisect an angle. But because this procedure required infinitely many bisections, it was neither practical nor a solution to the classical problem. He also gave a method of computing two mean proportionals using a circle and a hyperbola.[7]

There was much to admire in this ambitious treatise, but the title seemed to imply that Gregory had solved the classical problem of squaring the circle. The frontispiece, shown in figure 17.1, did so as well; it features a cherub holding a square frame. The sun's rays pass through it and produce a circle on the ground that another cherub is tracing with a compass. This presupposition was enough to cast doubt on the entire work. But he made no claim to have squared the circle in a Euclidean fashion. Instead, he used ideas that were early precursors to integration to find the area of the circle.

Unfortunately, in 1651 Huygens found a subtle error in Gregory's area computation. This did not help the reputation of this otherwise excellent work. Despite finding the error, Huygens recommended the book to Leibniz in 1672 when he was in Paris trying to learn mathematics. Later Leibniz wrote, "More substantial help came from the famous triumvirs: from Fermat by his invention of a method pro maximis et minimis, from Descartes by his showing how to describe curves of usual Geometry by means of equations, and from Father Gregory of Saint Vincent by his numerous bright inventions."[8]

James Gregory

In his 20s James Gregory left Scotland, which was not known for its scientific education, for London and then Italy to study mathematics, mechanics, and astronomy. While he was in Padua, he thought

FIGURE 17.1. Frontispiece of Gregory of St. Vincent's *Geometric Work of Squaring the Circle and Conic Sections*. (1647, Antwerp)

he had proved it impossible to square the circle, and in 1667 he wrote *Vera circuli et hyperbolae quadratura* (*The True Squaring of the Circle and Hyperbola*). He generalized Archimedes's idea of using polygons to approximate the circumference of a circle to using them to approximate the areas of sectors of the circle, ellipse, and hyperbola. For instance, given the unit circle, and taking I_n and C_n to denote the areas of inscribed and circumscribed regular n-gons, respectively, Gregory obtained the formulas[9]

$$I_{2n} = \sqrt{C_n I_n} \quad \text{and} \quad C_{2n} = \frac{2C_n I_{2n}}{C_n + I_{2n}}.$$

Recall that I_{2n} is the geometric mean of C_n and I_n, and C_{2n} is called the *harmonic mean* of C_n and I_{2n}. The formulas are recursive and intertwined. The procedure entails finding the area of an inscribed polygon, followed by the area of a circumscribed polygon with the same number of sides, followed by the area of an inscribed polygon with twice as many sides, and so on. The sequences converge to π from above and below. (It was in this work that Gregory coined the term "convergent.")

Let's see how this algorithm works. Suppose we start with the unit circle and an inscribed and circumscribed square. The area of the inscribed square is $I_4 = 2$, and the area of a circumscribed square is $C_4 = 4$. We then use these two values to obtain the area of an inscribed octagon: $I_8 = \sqrt{C_4 I_4} = \sqrt{8} = 2.828\ldots$ We use C_4 and I_8 to obtain the area of the circumscribed octagon:

$$ C_8 = \frac{2C_4 I_8}{C_4 + I_8} = \frac{8\sqrt{8}}{4 + \sqrt{8}} = 4\sqrt{8} - 8 = 3.313\ldots $$

We use I_8 and C_8 to obtain the bounds $I_{16} = 3.061$ and $C_{16} = 3.182\ldots$ for π. The procedure continues in this way, getting ever-tighter bounds.

These formulas were correct, but it was the next step that was problematic. Gregory argued that this limiting value, π, cannot be expressed using the four arithmetic operations and nth roots. And, because of this, he asserted that we cannot square the circle.

James Gregory knew of the work of Gregory of Saint Vincent and of Huygens's criticisms of it, so he sent a complimentary copy of his book to Huygens. Unfortunately, Huygens discovered a flaw in Gregory's argument, and instead of replying, he published his criticisms.[10] He also asserted that he proved some of Gregory's results before Gregory. This kicked off an often-heated public debate between the mathematicians.

Although Gregory's argument about the impossibility of squaring the circle was flawed, his argument was deep and clever. As Max Dehn and E. D. Hellinger wrote, "A modern mathematician will highly admire Gregory's daring attempt of a 'proof of impossibility' even if Gregory could not attain his aim."[11] Moreover, as we shall see, his procedure of using two different means would later be exploited by Gauss, and thus it was ultimately a fruitful method of analysis.

The Gregory–Huygens feud illustrated that there was tension between the old guard—like Huygens, who was a proponent of the classical geometric approach to mathematics—and the new analysts. As Scriba noted, Gregory "was one of the wild young men who wanted to tear down the barriers of traditional mathematics at almost any price, who wanted to view hitherto uncultivated areas. Inspired by hopes for as yet unheard-of results, he freely introduced new methods while at times he neglected necessary care for details and exactness."[12]

Longomontanus

These episodes also showed that the problem of squaring the circle was a subtle and tricky one; even good mathematicians can be fooled into thinking they have a proof of its possibility or impossibility. Of course, the seventeenth century was not free of run-of-the-mill circle squarers such as the Dutch astronomer Christian Longborg (1562–1647), who went by his Latinized name Longomontanus.

Longomontanus, the sole disciple of Tycho Brahe (1546–1601), "found himself looking backward to Tycho, instead of forward into the seventeenth century. Although [Longomontanus] worked and wrote in the era of Kepler and Galileo, he denounced ellipses, denied heliocentrism, denigrated the telescope, and ignored logarithms."[13]

Longomontanus wrote a number of articles in which he claimed to have squared the circle. In 1644 he gave a faulty geometric argument that implied that $\pi = \sqrt{18{,}252}/43 \approx 3.1418596\ldots$ at a time when π was known to 35 digits.[14] Longomontanus was aware that his value did not fall between these very tight bounds and argued that the bounds were incorrect. As justification, he observed that his value was much closer to the average of Archimedes's bounds of 22/7 and 223/71 than was the current best value—which was both true and irrelevant. The British mathematician John Pell (1611–1685), who was in the Netherlands at the time, wrote a response critical of Longomontanus's so-called proof. He was supported by the likes of Descartes, Cavalieri, Roberval, and even the future circle squarer Hobbes, none of whom were able to convince Longomontanus of his errors.[15]

Huygens wrote, "Truly, I have neither discerned nor prescribed by what means a circle may be squared; but I urge this, that he who contends to have found that means should demonstrate it to be in truth useful and effective."[16]

The Mādhava–Leibniz Series

There are many formulas for π, but perhaps there are none as beautiful and as useless as the alternating series

$$\frac{\pi}{4} = 1 - \frac{1}{3} + \frac{1}{5} - \frac{1}{7} + \frac{1}{9} - \frac{1}{11} \cdots.$$

Its elegance is obvious at a glance, and one might think that this was the expression that digit hunters were waiting for—no polygons, no square roots, no infinite products, no trigonometric functions. But alas, the series converges extremely slowly. The sum of the first 10 terms yields an approximation of 3.04 for π. We have to sum 10^n terms to get n digits of accuracy! The series is based on an infinite series expansion for the inverse tangent (or arctangent) function, namely

$$\tan^{-1} x = \arctan x = x - \frac{x^3}{3} + \frac{x^5}{5} - \frac{x^7}{7} + \cdots.$$

Because $\tan(45°) = \tan(\pi/4) = 1$, we can plug 1 into the inverse tangent formula to get the series (because $\arctan(1) = \pi/4$).

We believe the series was discovered by the Indian astronomer and mathematician Mādhava of Sangamagrama (ca. 1340–ca. 1425), predating the discovery of calculus by over two centuries. None of Mādhava's works survive, but we know about his contributions from his students and followers.[17] The inverse tangent series was rediscovered by James Gregory, who published it in 1671, so it is often called the Gregory series. Gregory did not take the next step of plugging in 1 to obtain $\pi/4$. The series was rediscovered by Leibniz in 1673. He described it as finding "the quadrature of the circle by arithmetical way."[18]

In a rare display of praise for Leibniz, his rival Isaac Newton wrote, "Leibniz's method for obtaining convergent series is certainly very elegant, and it would have sufficiently revealed the genius of its author, even if he had written nothing else."[19] As we shall see, Newton also discovered a series representation of π. It was much less elegant, but it converged to the value of π significantly faster.

Isaac Newton

It is no hyperbole to say that Isaac Newton was one of history's greatest thinkers. His work in mathematics included codiscovering many of the ideas that form calculus; expanding our understanding of infinite series, including generalizing the binomial theorem for exponents other than positive integers; discovering an iterative method for approximating the roots of a function; classifying cubic curves; giving a new neusis construction of two mean proportionals; and helping formalize our modern concept of number. In physics he did foundational work in classical mechanics; introduced the notion of universal gravitation, which formed the basis for justifying Kepler's laws of planetary motion; put forth his three laws of motion; made the accurate theoretical prediction that the earth is an oblate spheroid; developed a deep theory of optics, light, and color; built the first reflecting telescope; investigated so-called Newtonian fluids; devised a law of heating and cooling; and studied the speed of sound. He was also interested in alchemy and in biblical chronology.

Newton was the second Lucasian Professor of Mathematics at the University of Cambridge, succeeding Isaac Barrow. Later, he became Master of the Royal Mint. He was elected a Fellow of the Royal Society in 1672 for inventing the reflecting telescope and served as its president from 1705 until his death in 1727. He was knighted by Queen Anne in 1705, although not for his scientific contributions, but for an unsuccessful run for Parliament. He died wealthy—he was well paid at the Mint—and is buried in Westminster Abbey under a grand memorial.

Newton famously wrote that he stood "on the shoulders of giants." Descartes was one of the giants. Abraham de Moivre (1667–1754) wrote,[20]

[The young Newton] took Descartes's Geometry in hand, tho he had been told it would be very difficult, read some ten pages in it, then stopt, began again, went a little farther than the first time, stopt again, went back again to the beginning, read on till by degrees he made himself master of the whole, to that degree that he understood Descartes's Geometry better than he had done Euclid.

In 1664 and 1665 Newton immersed himself in the mathematics of Barrow, Oughtred, Descartes, Schooten, Viète, and Wallis. It is notable who is absent from this list: Napier, Briggs, Harriot, Desargues, Pascal, Fermat, Stevin, Kepler, Cavalieri, and Torricelli. Although he read Euclid, he did not read Archimedes or Apollonius. "By mid-1665, Newton's urge to learn from others seems to have abated."[21]

In his career, Newton mastered, used, and created a wide range of mathematical techniques. Guicciardini wrote,[22]

> Indeed, it is often the case that in tackling a problem Newton made recourse to a baroque repertoire of methods: one encounters in the same folios algebraic equations, geometrical infinitesimals, infinite series, diagrams constructed according to Euclidean techniques, insights in projective geometry, quadratures techniques equivalent to sophisticated integrations, curves traced via mechanical instruments, numerical approximations. Newton's mathematical toolbox was rich and fragmentary; its owner mastered every instrument it contained with versatility.

The Plague spread to Cambridge in 1665 and Trinity College was closed. Newton went home to his family's farm in Woolsthorpe. These two years, now known as his *anni mirabiles* (wonderful years), rival only Einstein's *annus mirabilis* in 1905 for a period of remarkably brilliant scholarly productivity. Newton discovered the method of fluxions (differential calculus), the inverse method of fluxions (integral calculus), the generalized binomial theorem, his theory on light and colors, and the beginnings of his theory of gravitation.

He also employed some of his newly discovered mathematics to obtain a series for π.[23] His idea was simple: compute the area of the shaded region in figure 17.2 in two different ways—using geometry, which involves π, and using calculus—and set them equal.

The region is bounded by the x-axis, the line $x = 1/4$, and the circle with center $(1/2, 0)$ and radius $1/2$. So the area of the region is the area of a 60° sector (1/6 of the full circle) minus the area of a triangle with base 1/4 and height $\sqrt{3}/4$. The area of the sector is $\frac{1}{6}\pi\left(\frac{1}{2}\right)^2 = \frac{\pi}{24}$, and the area of the triangle is $\frac{1}{2} \cdot \frac{1}{4} \cdot \frac{\sqrt{3}}{4} = \frac{\sqrt{3}}{32}$. So the area of the shaded region is $\pi/24 - \sqrt{3}/32$. Now let us see what we get using calculus. The equation for the circle is $(x - 1/2)^2 + y^2 = (1/2)^2$. If we solve for y, we find that for the top half of the circle, $y = \sqrt{x - x^2}$. So the area of the

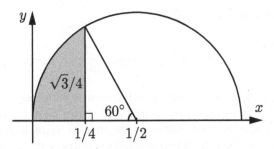

FIGURE 17.2. Newton computed the area of the shaded region using calculus and geometry.

shaded region is

$$\int_0^{1/4} \sqrt{x - x^2}\, dx.$$

Setting these expressions equal and solving for π we obtain

$$\pi = \frac{3\sqrt{3}}{4} + 24 \int_0^{1/4} \sqrt{x - x^2}\, dx.$$

Using this expression to obtain a good approximation of π relies on two things: obtaining a good approximation for $\sqrt{3}$ and evaluating the integral. The secret to both of these is Newton's binomial theorem.

The binomial theorem details how to expand expressions of the form[24] $(a+b)^n$, such as $(a+b)^4 = a^4 + 4a^3b + 6a^2b^2 + 4ab^3 + b^4$. Newton discovered a way to expand expressions of the form $(a+b)^n$ when n is negative or a fraction. But the price we pay is that the sum is infinite.

For instance, when x is between -1 and 1,

$$\sqrt{1-x} = (1-x)^{1/2} = 1 - \frac{1}{2}x - \frac{1}{8}x^2 - \frac{1}{16}x^3 - \frac{5}{128}x^4 - \frac{7}{256}x^5 - \cdots.$$

We can use this series to obtain a series for $\sqrt{3}$:

$$\sqrt{3} = 2\sqrt{1 - \frac{1}{4}} = 2\left(1 - \frac{1}{2\cdot4} - \frac{1}{8\cdot4^2} - \frac{1}{16\cdot4^3} - \frac{5}{128\cdot4^4} - \cdots\right).$$

The first five terms yield the approximation $\sqrt{3} \approx 1.7321167$, which differs from the true value by less than 0.00007. Newton used the same series to solve the integral:[25]

$$\int_0^{1/4} \sqrt{x - x^2}\, dx = \frac{1}{3 \cdot 2^2} - \frac{1}{5 \cdot 2^5} - \frac{1}{28 \cdot 2^7} - \frac{1}{72 \cdot 2^9} - \cdots .$$

At this point, it is straightforward to sum as many terms of the two quickly converging series as desired and obtain an approximation of π. In his article, Newton summed the first 20 terms to find π to 16 decimal places. The calculation was not the purpose of the work, it was an afterthought, an application. He added sheepishly, "I am ashamed to tell you how many places of figures I carried these computations, having no other business at the time."

As Petr Beckmann wrote, "The crumbs dropped by giants are big boulders."[26]

TANGENT
Digit Hunters

If we take the world of geometric relations, the thousandth decimal digit of π sleeps there, tho no one ever may try to compute it.
—William James, *The Meaning of Truth* (1909)[1]

IF WE USED a 50-digit approximation of π to compute the circumference of a circle through the North Star with earth as the center, the error would be significantly less than the radius of a proton. So today's 31 trillion digits seems a little ridiculous.

Why this obsession with finding more and more digits now that we have far exceeded the number of digits useful for practical applications? The mountaineer's response is one answer to this question: because it is there. Certainly, many digit hunters embarked on their mission just as a challenge—to see if they could extend the boundaries of knowledge.

Mathematicians asked important questions about the nature of π. Is it rational? The rational numbers are precisely those numbers that have terminating decimal expansions or decimal expansions that eventually repeat. Perhaps an investigation of the digits would give us a clue.

Once we knew (or suspected) it was irrational, we could ask whether the digits are evenly distributed. Are 1/10th of the digits 9s, 1/100th of the pairs 61s, 1/10,000th of the quads 9481s, and so on? In 1909 Émile Borel (1871–1956) formalized this notion of equidistributed digits and coined the term *normal* for such a number. Is π normal? Perhaps not—the first 0 does not appear until the 33rd digit. De Morgan observed that 7 appeared only 44 times in the first 608 digits of π, rather than the expected 61. He wrote, tongue in cheek, [2]

There is but one number which is treated with an unfairness which is incredible as an accident: and that is the mystic number *seven*! If the cyclometers and the apocalyptics would lay their heads together until they come to a unanimous verdict on this phenomenon, and would publish nothing until they are of one mind, they would earn the gratitude of their race.

As it turned out, De Morgan was using William Shanks's (1812–1882) digits, which were correct only up to the 527th digit. The digit 7 actually appears 50 times in the first 608 digits, and a large number of 7s come soon after; by the 650th place, 7 has appeared a more reasonable 62 times. We still don't know whether π is normal, although it looks as if it is. Computing digits of π may help answer such questions.[3]

Lastly, there are many examples of seemingly useless investigations in pure mathematics having surprising practical applications—the study of prime numbers made secure internet shopping possible, ideas from linear algebra enable web search engines to rank sites on the internet, and non-Euclidean geometries became the right framework for Einstein's theory of general relativity. Likewise, the race to obtain digits of π has pushed forward mathematics, computer science, and engineering. As a practical example, the problem of computing digits of π can be used to test computer hardware. In 1986 the results of two different π-generating programs yielded different outputs on a Cray-2 supercomputer. The cause was an obscure hardware problem. Cray subsequently added these algorithms to their test suites.[4]

In this chapter, we give a brief overview of the race to discover the digits of π. We will not discuss every broken record and record breaker, but we will tell some of the stories.[5]

Figure T.30 shows, at a glance, the race to find the digits. One cannot help but notice the rapid increase in the twentieth century after the invention of computers. In fact, it is an even quicker rise than it appears. Notice that the scale on the vertical axis does not increase linearly—it goes up by powers of 10! In other words, what appears to be linear growth in the pre- and post-computer eras is actually an exponential growth of digits. From 1400 to 1949 the number of digits increased by approximately 0.8% per year, and from 1949 onward they increased by approximately 45% annually.

The first stage in the quest for accuracy is best called the prehistory of π. There was no rigorous understanding about the geometry of the circle. The value of π (such as $3\frac{1}{8}$, $256/81$, $54 - 36\sqrt{2}$, $\sqrt{10}$) was based on approximations or faulty mathematical logic.

The second stage began with Archimedes. He gave the first exact method of computing π, which was based on approximating a circle

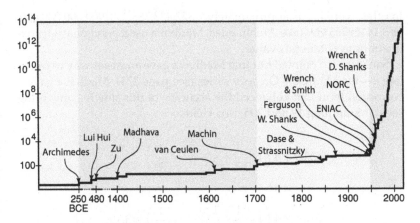

FIGURE T.30. The number of known digits of π by year. The shading corresponds to the prehistory of π, the polygon era, the calculus era, and the computer era.

by inscribed and circumscribed polygons. By repeatedly doubling the number of sides of the polygons we can obtain tighter bounds. Archimedes used 96-gons to place π between $25{,}344/8069 = 3.1409\ldots$ and $29{,}376/9347 = 3.1428\ldots$. For almost 2000 years, every improvement in our estimates of π was obtained using Archimedes's method.

In the second century CE, Ptolemy, in his *Almagest*, computed an approximation for a chord of $\frac{1}{2}^\circ$. If we view this value as the length of the side of a 720-gon, then we can compute an approximation for π: $3 + 8/60 + 30/360 = 377/120 \approx 3.14166\ldots$.

In China, Liu Hui used 192-gons to obtain the bounds $3.141024 < \pi < 3.142704$. He may also have used a 3072-gon to obtain $\pi \approx 3.1416$. And Zu Chongzhi improved this to 3.1415926 (see page 87).

In India in approximately 1400, Mādhava gave an approximation much better than the traditional Indian value $335/113$. Śaṅkara Vāriyar (c. 1500–c. 1560) attributed the following to Mādhava:[6]

Gods [33], eyes [2], elephants [8], serpents [8], fires [3], three, qualities [3], Vedas [4], naksatras [27], elephants [8], arms [2] (2,827,433,388,233)—the wise said that this is the measure of the circumference when the diameter of a circle is nine *nikharva* [10^{11}].

In other words, π is approximately 2,827,433,388,233/900,000,000,000,
or 3.14159265359. Like Archimedes, Mādhava used polygonal approximations to obtain this value.

Śaṅkara also pointed out that Mādhava gave an easier way to calculate π—the Mādhava–Gregory series (see page 274). Mādhava gave a
correction term that enhanced the accuracy of this slowly converging
series.[7] Summing 19 and 20 terms yields

$$3.09 \approx 4\left(1 - \frac{1}{3} + \frac{1}{5} - \cdots - \frac{1}{39}\right)$$

$$< \pi$$

$$< 4\left(1 - \frac{1}{3} + \frac{1}{5} - \cdots + \frac{1}{37}\right) \approx 3.19.$$

If we add n terms, Mādhava's correction term is $(n^2 + 1)/(4n^3 + 5n)$; it
is added if the last term was subtracted, and subtracted if the last term
was added. The same 19 or 20 terms now yield much tighter bounds:

$$4\left(1 - \frac{1}{3} + \frac{1}{5} - \cdots - \frac{1}{39} + \frac{20^2 + 1}{4 \cdot 20^3 + 5 \cdot 20}\right) = 3.1415926540\ldots,$$

$$4\left(1 - \frac{1}{3} + \frac{1}{5} - \cdots + \frac{1}{37} - \frac{19^2 + 1}{4 \cdot 19^3 + 5 \cdot 19}\right) = 3.1415926529\ldots.$$

Ludolph van Ceulen was the most persistent computer to use Archimedes's technique. Van Ceulen was born in Germany but emigrated
to the Netherlands where he taught mathematics and fencing. He
spent much of his life doubling and doubling the number of sides of
the polygons and performing careful and tedious calculations. By his
death in 1610 he had computed π to 35 places using 2^{62}-gons!

Ludolph van Ceulen's name became so attached to the circle constant, thanks to this remarkable feat, that for a time π was called the
"Ludolphine number" or "van Ceulen's number," as William Jones
did when he introduced the symbol π in 1706 (see the quote on
page 82). Just as Archimedes was proud of his work on this constant
and had the sphere and cylinder inscribed on his tombstone, van
Ceulen's tombstone in Leiden was decorated with the digits of π. The
tombstone was eventually lost, but a replica was made in 2000.

Van Ceulen's most famous student was the astronomer and mathematician Willebrord Snellius (Snell) (1580–1626). Snell is most well known for the law of refraction of light that now bears his name. He was also the first mathematician to improve Archimedes's method for finding digits of π. In his 1621 *Cyclometricus* (*Measuring the Circle*), Snell unveiled a method of squeezing even more out of inscribed and circumscribed polygons.[8] For instance, applying Archimedes's method to hexagons puts π between 3 and 3.464, but applying Snell's modification to hexagons yields the much tighter bounds of 3.1402 and 3.1416. Archimedes obtained three decimal places of accuracy using 96-gons, but Snell obtained seven-place accuracy. Likewise, Snell used a 2^{30}-gon to get 34-place accuracy, whereas van Ceulen obtained half as many decimal places from the same polygon. Snell did not give a formal proof that his method achieved the accuracy he claimed (although it did reproduce the digits that van Ceulen had generated). The rigorous geometric justifications came from Huygens in 1654.[9]

Archimedes's method of computing π was *the method* for computing π until the seventeenth century, when calculus came on the scene and mathematicians discovered many quickly converging infinite series. This was the third stage in the hunt for digits.

The Gregory series for the arctangent function is very useful for obtaining infinite series expressions for π. The Mādhava–Leibniz series

$$\frac{\pi}{4} = \arctan(1) = 1 - \frac{1}{3} + \frac{1}{5} - \frac{1}{7} + \cdots$$

is an elegant example that converges slowly. But many arctangents converge very rapidly. In 1706 John Machin derived the expression

$$\pi/4 = 4\arctan(1/5) - \arctan(1/239),$$

which he used to compute π to 100 digits. William Jones presented the result in the same publication in which he introduced the symbol π.

There are many, many other arctangent formulas for π. For instance, in 1738 Euler discovered the general formula

$$\arctan\left(\frac{1}{p}\right) = \arctan\left(\frac{1}{p+q}\right) + \arctan\left(\frac{q}{p^2 - pq + 1}\right),$$

which he used to derive other formulas, such as[10]

$$\pi/4 = \arctan(1/2) + \arctan(1/3)$$

and

$$\pi/4 = 20\arctan(1/7) + 8\arctan(3/79).$$

Euler used the latter series to compute 20 digits of π in one hour!

In 1844 the 20-year-old German mental calculator Zacharias Dase (1824–1861) spent two months using Schulz von Strassnitzky's (1803–1852) relation

$$\pi/4 = \arctan(1/2) + \arctan(1/5) + \arctan(1/8)$$

to obtain 205 digits of π (the last 5 of which were incorrect).

In 1853 the British amateur mathematician William Shanks used Machin's formula to compute π to a staggering 607 digits, and 20 years later he computed 100 more. Many years later—in 1945—another digit hunter D. F. Ferguson, also using an arctangent formula, discovered that Shanks's value was incorrect from the 528th digit onward. Shanks's digits were very widespread at the time. For instance, for the great world exhibition in Paris in 1937, the Palais de la Découverte displayed his digits around the circumference of a circular room. The erroneous representation continued to be reproduced in books for many years. Fortunately, the Palais de la Découverte was able to correct the digits on display. Ferguson continued calculating π, eventually turning to a mechanical calculator. He obtained 808 digits by 1947.[11]

As one last pre-electronic hurrah, John Wrench, Jr. (1911–2009), who had verified Ferguson's work, joined up with Levi Smith. They used a mechanical desk calculator to sum the terms of Machin's formula. By 1949, they had computed 1120 digits.

The year 1949 was the year everything changed for digit hunters. With the encouragement of John von Neumann (1903–1957), a team of researchers[12] used the ENIAC—one of the earliest general-purpose electronic computers—to compute the digits of e and π during the computer's downtime. They did the programming in the evenings and used the computer over the long 4th of July (for e) and Labor Day (for π) weekends. It took 70 hours of computer time for them to calculate 2037 digits of π using an algorithm based on Machin's formula.

In 1954 the NORC (Naval Ordinance Research Calculator) computed 3093 digits of π—again using Machin's formula—in 13 minutes. In 1958 the IBM 704 computed 10,000 digits in one hour and

40 minutes. In 1961 Wrench and Daniel Shanks (1917–1996) used an IBM 7090 to compute 100,265 digits. There was no going back. Every future record involved computers. As engineers built faster, more powerful computers, and as computer scientists developed more efficient algorithms, the speed of computing the digits continued to increase.

Although digit hunters now had computers to speed up and keep track of their calculations, the basic idea for computing π had not changed in three centuries. They were adding up variations of Machin's arctangent series. In 1976 Richard Brent (1946–) and Eugene Salamin independently discovered a new expression for π—although, surprisingly, it was essentially a rediscovery of a formula due to Gauss from 1809.[13] This new series converged extraordinarily quickly. Each new term *doubled* the number of digits! Their formula,

$$\pi = \frac{4\left(\text{AGM}(1, 1/\sqrt{2})\right)^2}{1 - \sum_{k=1}^{\infty} \left(2^{k+1} \left(a_k^2 - b_k^2\right)\right)},$$

requires some explanation.

It relies on computing a type of average of two numbers called the *arithmetic–geometric mean* (which we denote AGM). The arithmetic mean of a and b is the usual, well-known average $(a+b)/2$, and the geometric mean, which we saw on page 70, is the multiplicative average $\sqrt{ab}$. Computing the arithmetic–geometric mean of a and b is an iterative process that combines both of these means. We begin with $a_0 = a$ and $b_0 = b$. Then for $k \geq 1$, a_k and b_k are the arithmetic and geometric means, respectively, of a_{k-1} and b_{k-1}: $a_k = (a_{k-1} + b_{k-1})/2$ and $b_k = \sqrt{a_{k-1} b_{k-1}}$. The sequences $a_0, a_1, a_2, \ldots$ and $b_0, b_1, b_2, \ldots$ both limit on the same value, and this is $\text{AGM}(a, b)$.

The Brent–Salamin formula contains $\text{AGM}(1, 1/\sqrt{2})$. Let's compute the first few terms. We begin with $a_0 = 1$ and $b_0 = 1/\sqrt{2}$. Then we compute the arithmetic mean

$$a_1 = \frac{1}{2}\left(1 + \frac{1}{\sqrt{2}}\right) = \frac{1}{4}\left(\sqrt{2} + 2\right)$$

and the geometric mean

$$b_1 = \sqrt{1 \cdot \frac{1}{\sqrt{2}}} = \frac{1}{\sqrt[4]{2}}.$$

The next two terms are

$$a_2 = \frac{1}{2}\left(\frac{1}{4}\left(\sqrt{2}+2\right)+\frac{1}{\sqrt[4]{2}}\right) = 0.84722\ldots,$$

$$b_2 = \sqrt{\frac{\sqrt{2}+2}{4\sqrt[4]{2}}} = 0.84720\ldots.$$

These are followed by $a_3 = 0.8472130848\ldots$ and $b_3 = 0.8472130847\ldots$, and a_4 and b_4, which agree to 20 digits.

When he was young, Gauss discovered the convergence of the arithmetic–geometric mean—by tedious hand calculations! He discovered many other properties of the arithmetic–geometric mean, but most of his work remained unpublished until after his death.

In 1985 brothers Jonathan (1951–2016) and Peter Borwein (1953–) produced formulas that tripled and quadrupled the numbers of digits with each iteration. While these and other "faster" convergences seem like improvements on the Brent–Salamin algorithm, it takes longer to carry out each iteration, so they do not necessarily lead to more efficient computations of digits.

Because these algorithms are so efficient, they are used by many of today's π-generating computer programs. At the time the Brent–Salamin algorithm was discovered, we knew one million digits of π. Three years later—thanks to this algorithm—we knew over 30 million.

These aren't the only modern methods of computing digits. In 1994 brothers David (1947–) and Gregory Chudnovsky (1952–) computed more than four billion digits using a formula that resembled the ones discovered by the Indian amateur mathematician Srinivasa Ramanujan (1887–1920):[14]

$$\frac{1}{\pi} = 12 \sum_{k=0}^{\infty} \frac{(-1)^k (6k)!(13{,}591{,}409 + 545{,}140{,}134k)}{(3k)!(k!)^3 640{,}320^{3k+3/2}}.$$

In this case, each iteration yields 14 digits of π.

There have been other interesting digit-generating discoveries in recent years that have not necessarily helped those digit hunters trying to break world records. In 1995 Stanley Rabinowitz (1947–) and Stan Wagon (1951–) discovered an algorithm that produced digits of π one at a time, unlike the typical algorithms in which all of the calculations are done before producing the full approximation of π at once. The authors wrote that "this algorithm is a 'spigot' algorithm: it pumps out digits one at a time and does not use the digits after they are computed."[15]

In 1996 a computer program discovered the formula

$$\pi = \sum_{k=0}^{\infty} \frac{1}{16^k} \left(\frac{4}{8k+1} - \frac{2}{8k+4} - \frac{1}{8k+5} - \frac{1}{8k+6} \right).$$

David Bailey (1948–), Peter Borwein, and Simon Plouffe (1956–) realized that it could be used to extract any digit of π (written in base-16) without computing any of the digits before it! This is now known as the BBP formula.

Since the 1980s, the record for the most digits has been broken more than 30 times, and the number of known digits has increased from one million to 31.4 trillion. (This most recent π-like record was broken by Emma Haruka Iwao, who announced it on π day, March 14, 2019.) The record may well be broken again by the time you are reading this book. To put this number in perspective, if we printed the 31.4 trillion digits on these book pages, it would require approximately 10 billion pages. That is more than one page for each person on Earth. Which page do you want?

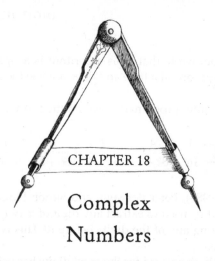

CHAPTER 18

Complex Numbers

We can make the same statement as in the case of negative numbers, that imaginary numbers made their own way into arithmetic calculation without the approval, and even against the desires of individual mathematicians, and obtained wider circulation only gradually and to the extent to which they showed themselves useful. Meanwhile the mathematicians were not altogether happy about it. Imaginary numbers long retained a somewhat mystic coloring, just as they have today for every pupil who hears for the first time about that remarkable $i = \sqrt{-1}$.

—Felix Klein, 1908[1]

IN 1900 PAUL PAINLEVÉ (1863–1933) wrote about the surprising usefulness of the complex numbers. He concluded that "between two truths of the real domain, the easiest and shortest path quite often passes through the complex domain."[2]

Indeed, it may seem strange that a book about Euclidean geometry would contain a chapter on the complex numbers. But as evidence of Painlevé's assertion, the solutions of two of our problems require complex numbers—the proofs that it is impossible to construct every regular polygon and that it is impossible to square the circle.

Imaginary Numbers in the Solution of the Cubic

Ask mathematics students (or even many mathematicians) why mathematicians created the notion of an imaginary number and what problem they were trying to solve, and almost surely their guess will be that it came from solving quadratic equations. After all, that is where many students first encounter complex numbers in their schooling: the roots of $ax^2 + bx + c$ are $\left(-b \pm \sqrt{b^2 - 4ac}\right)/(2a)$, and if $b^2 - 4ac$ is negative, the roots are complex.

However, when quadratic equations arise in practical situations, we are typically looking for real solutions, and $b^2 - 4ac$ is negative precisely when the equation has no real roots. So we can stop and say "the equation has no solution," and move on. The equation $x^2 + 1 = 0$ has no solutions, right? Thus, we have no incentive to invent the notion of an imaginary number. It would not help us solve any new problems.

In *Ars magna*, Cardano asked whether it is possible to find two numbers whose sum is 10 and product is 40. This problem is equivalent to solving the quadratic equation $x^2 - 10x + 40 = 0$. Cardano showed that it has roots $5 + \sqrt{-15}$ and $5 - \sqrt{-15}$. As he performed the multiplication to verify that $\left(5 + \sqrt{-15}\right)\left(5 - \sqrt{-15}\right) = 40$, he used the expression *dismissis incruciationibus*, which has two very different translations. In one translation, the phrase refers to the mathematics, "the imaginary parts being lost," while in another it describes his state of mind, "putting aside the mental tortures involved."[3] Perhaps this was an intentional play on words. He concluded the example by calling such expressions "as subtle as they are useless."[4] We cannot fault Cardano for not standing behind this new type of number. At the time, negative numbers were still not universally accepted; Cardano called them "numeri ficti" (fictitious numbers) and in 1637, nearly a century later, Descartes called them "faux" (false) numbers.

The first important use of imaginary numbers was hidden elsewhere in *Ars magna*, but that discovery came later and from someone else. As we mentioned in chapter 13 (see page 213), complex numbers arose out of necessity from solving cubic equations. When we apply the Tartaglia–Cardano formula to $x^3 = 15x + 4$ we obtain the root $x = \sqrt[3]{2 + \sqrt{-121}} + \sqrt[3]{2 - \sqrt{-121}}$, or, using the notation $i = \sqrt{-1}$ that Euler introduced 200 years later,[5] $x = \sqrt[3]{2 + 11i} + \sqrt[3]{2 - 11i}$. However, this equation, like all cubics fitting the casus irreducibilis form, has three real roots, and this is one of them.

Rafael Bombelli (1526–1572) was an Italian engineer known in his day for projects like draining and reclaiming the Val di Chiana marshes in central Italy and attempting to repair the Ponte Santa Maria bridge that crossed the Tiber. During the downtime from his day job, he wrote a book called *Algebra*.[6] Three parts of the book were published in 1572, and two more were published posthumously (in 1923).

Bombelli admired *Ars magna*, but thought it would be too difficult for students to learn from. He intended his book to be complete, up to date, accessible to read, and filled with real-world examples. Indeed, it was widely read and praised. Simon Stevin (1548–1620) called Bombelli a "great arithmetician of our time," and Leibniz referred to him as an "outstanding master of the analytical art."[7]

In his *Algebra*, Bombelli took a closer look at the casus irreducibilis, and in the process he constructed a theory of complex numbers. He did not understand what these numbers were; his approach was purely mechanical—formal rules for manipulating square roots of negatives.

He discovered a method for turning some expressions of the form $\sqrt[3]{a+bi}$ into the form $c+di$. He showed that $\sqrt[3]{2+11i}=2+i$ and $\sqrt[3]{2-11i}=2-i$, equalities that are unexpected, but not difficult to verify—just cube both sides.[8] These numbers are *complex conjugates* (a term that Cauchy introduced in 1821[9]); they have the same real part and their imaginary parts are negatives of each other. When we add them—or any pair of complex conjugates—the imaginary parts cancel and the real part doubles. Thus, the root of the cubic $x^3 = 15x + 4$, $\sqrt[3]{2+11i} + \sqrt[3]{2-11i} = (2+i) + (2-i) = 4$, is real after all.[10]

Despite his success, Bombelli was still trying to overcome his unease about working with these new mathematical entities. He wrote,[11]

> Although to many this will appear an extravagant thing, since even I held this opinion some time ago, because it appeared to me to be more sophistic than true, nevertheless I searched hard and found the demonstration, which will be set out below.

Although Bombelli was able to cleverly transform the nested radicals into the sum of complex conjugates whose sum yields a real number, this will not work in general. In 1843 Pierre Wantzel, who is the subject of chapter 20 and one of the heroes of this book, proved that if a cubic equation has three distinct real roots and none of them are rational, then it is impossible to express them in terms of radicals without using complex numbers.[12] In other words, Bombelli was fortunate that the root of the equation was the rational value 4.

Thus, mathematicians were forced to consider complex numbers not from quadratic equations, but from cubic equations. They were necessary to obtain real solutions. We are reminded again of Painlevé's assertion at the start of the chapter—we have to pass through the complex domain to obtain our real solutions.

So as much as mathematicians wanted to ignore complex numbers, they couldn't. While this was no doubt an important step forward in the introduction of complex numbers into mathematics, it was only a first step. There continued to be much resistance to them, and it took another three centuries for mathematicians to come to grips with the new realities of complex numbers and to fully embrace them as legitimate mathematical entities.

The Gradual Acceptance of Complex Numbers

Descartes coined the terms "real" and "imaginary" to separate these different classes of numbers; in *Geometry* he wrote, "Neither the true nor the false roots are always real: sometimes they are imaginary."[13] Descartes's use of the term imaginary, and the public's adoption of this term, has caused no end of trouble. Even today math-phobic students cringe and scoff at the study of "pretend" numbers. Arguably, "imaginary" is better than "impossible," which was used widely in the seventeenth century and goes back to Cardano who referred to the "impossible case" for quadratic equations in which the roots are complex.

Leibniz toyed with complex numbers. In a letter to Huygens in 1674 or 1675[14] he pointed out that $\sqrt{1+\sqrt{-3}}+\sqrt{1-\sqrt{-3}}=\sqrt{6}$, and in 1702 he factored x^4+a^4 as

$$\left(x+a\sqrt{-\sqrt{-1}}\right)\left(x-a\sqrt{-\sqrt{-1}}\right)\left(x+a\sqrt{\sqrt{-1}}\right)\left(x-a\sqrt{\sqrt{-1}}\right).$$

Leibniz saw their usefulness, but was uneasy nonetheless. He wrote,[15]

> From the irrationals are born the impossible or imaginary quantities whose nature is very strange but whose usefulness is not to be despised.

In 1702 he wrote the following colorful description of imaginary numbers straddling two worlds:[16]

In fact Nature, the mother of limitless variations, or rather, Divine Reason, certainly adheres so closely to her own grand diversity than to permit all things to be confined within a single cast. And so she finds an elegant and amazing escape in that wonder of analysis, that Platonic entity, almost living a double life between being and not-being, that we call an imaginary root.

Even Euler, who used complex numbers extensively, was not sure what to make of them. In his 1770 *Elements of Algebra*, he wrote,[17]

Since all numbers which it is possible to conceive are either greater or less than 0, or are 0 itself, it is evident that we cannot rank the square root of a negative number amongst possible numbers, and we must therefore say that it is an impossible quantity. In this manner we are led to the idea of numbers which from their nature are impossible, which numbers are usually called *imaginary quantities*, because they exist merely in the imagination.

However, Euler and his contemporaries were willing to push symbols around without having what we would today view as a rigorous justification. The historian of mathematics Judith Grabiner states that eighteenth-century mathematicians "placed great reliance on the power of symbols. Sometimes it seems to have been assumed that if one could just write down something which was symbolically coherent, the truth of the statement was guaranteed."[18]

Full acceptance of imaginary numbers did not come until Gauss lobbied on their behalf. Perhaps most important were his strong words in their defense in 1831.[19] He reminded us that earlier mathematicians balked when they had to add rational numbers to the set of integers, irrational numbers to the rational numbers, and negative numbers to the positive numbers. There are contexts in which rational numbers and negative number make no sense (like when counting objects), yet we still accept them as numbers. Likewise, we must accept complex numbers even though they are often irrelevant in particular situations.

Gauss also proposed the term "complex number" as a replacement for the terms "impossible" and "imaginary" numbers. He wrote,[20]

The imaginary quantities—which are contrasted with the real ones, and which were formerly, and are still occasionally (although improperly) called *impossible*—are still tolerated rather than fully accepted, and therefore appear more like a game with symbols, in itself empty of content, to which one unconditionally

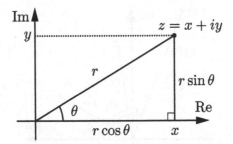

FIGURE 18.1. The polar representation of complex numbers.

denies a thinkable substrate—without, however, wishing to scorn the rewards which this game with symbols achieved for our understanding of the relationships of the real quantities.

Gauss also addressed another stumbling block in the acceptance of complex numbers—how to represent them geometrically—by presenting their now-standard planar representation. Each complex number $z = x + iy$ has coordinates (x, y) in the complex plane; the x- and y-axes represent the real and purely imaginary numbers, respectively. He was not the first to view complex numbers in this way, but it was from Gauss that many learned of it.[21]

The Polar Representation of Complex Numbers

We can represent complex numbers in another useful way: using polar coordinates. If a complex number $z = x + iy$ is r units from the origin and θ is the angle between z and the real axis (see figure 18.1), then we can represent the point as (r, θ). Furthermore, a little trigonometry shows that $x = r \cos \theta$ and $y = r \sin \theta$, and the Pythagorean theorem gives $r = \sqrt{x^2 + y^2}$. Thus $z = x + iy = r \cos \theta + i r \sin \theta$.

The polar representation gives a nice intuitive interpretation of complex multiplication. When we multiply the complex numbers $z_1 = r_1(\cos \theta_1 + i \sin \theta_1)$ and $z_2 = r_2(\cos \theta_2 + i \sin \theta_2)$ we obtain a mess of sines and cosines, which can be simplified using the angle sum formulas for the trigonometric functions:

$$z_1 z_2 = r_1(\cos \theta_1 + i \sin \theta_1) \cdot r_2(\cos \theta_2 + i \sin \theta_2)$$

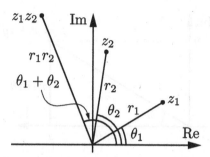

FIGURE 18.2. To multiply complex numbers, multiply their distances from the origin and add their angles.

$$= r_1 r_2 ((\cos\theta_1 \cos\theta_2 - \sin\theta_1 \sin\theta_2) + i(\cos\theta_1 \sin\theta_2 + \sin\theta_1 \cos\theta_2))$$
$$= r_1 r_2 (\cos(\theta_1 + \theta_2) + i\sin(\theta_1 + \theta_2)).$$

In other words, when we multiply z_1 and z_2 we multiply their distances from the origin and we add the angles (see figure 18.2).

In 1748 Euler considered the special case in which we square a complex number on the unit circle $(r = 1)$.[22] The formula gives

$$(\cos\theta + i\sin\theta)^2 = \cos(2\theta) + i\sin(2\theta).$$

More generally, if we raise it to the nth power we have

$$(\cos\theta + i\sin\theta)^n = \cos(n\theta) + i\sin(n\theta).$$

And this holds for any real value of n, not just for positive integers.

Today we call this formula de Moivre's formula after Abraham de Moivre. Although this formula bears his name, he never stated it in this form. It is likely that he knew this formula as early as 1707, and he used a number of variations of it. We do not know whether Euler knew of de Moivre's work.

To give an example of the usefulness of de Moivre's formula, let's see how we can use it to quickly generate trigonometric identities. In chapter 20 we will need a trigonometric identity for $\cos(3\theta)$. De Moivre's formula gives us $(\cos\theta + i\sin\theta)^3 = \cos(3\theta) + i\sin(3\theta)$. Cubing the expression on the left we obtain

$$(\cos\theta + i\sin\theta)^3 = \left(\cos^3\theta - 3\sin^2\theta\cos\theta\right) + i\left(3\sin\theta\cos^2\theta - \sin^3\theta\right).$$

The real parts of the two expressions must be equal and so must the imaginary parts. Setting the real parts equal gives us the identity $\cos(3\theta) = \cos^3 \theta - 3\sin^2 \theta \cos \theta$. Finally, to get this to the desired form we substitute the Pythagorean identity $\cos^2 \theta + \sin^2 \theta = 1$ and obtain $\cos(3\theta) = 4\cos^3 \theta - 3\cos \theta$.

Euler's Identity

We have encountered Leonhard Euler several times already. He solved the bridges of Königsberg problem, solved the Basel problem, worked on the problem of squarable lunes, discovered arctangent formulas for π, and did foundational work in complex analysis. Although he is not one of the major players in the story of the problems of antiquity, his fingerprints are everywhere. To those familiar with the history of mathematics, this is not a surprise. Euler was one of the mathematical giants—not just of the eighteenth century, but of all time.

Euler was born in Basel, Switzerland, in 1707. He studied under Johann Bernoulli, who was a family friend and was one of the major mathematical figures of the day. In 1726 Euler was recruited to move to the young city of St. Petersburg, Russia, to take a position in Peter the Great's new Academy of Sciences. After a decade and a half in Russia, Frederick the Great enticed Euler to move to Berlin to take a position at the Berlin Academy. He remained there for 25 years until Catherine the Great brought him back to St. Petersburg. He lived in St. Petersburg until his death in 1783.

Despite numerous personal hardships, including being mostly blind for years, Euler was one of the most—if not the most—prolific mathematicians of all time. His work was extremely diverse. He wrote influential books, long articles, short articles, technical papers, and books for general audiences. He played a major role in building up calculus after the early work of Newton and Leibniz. But he contributed to many other areas as well: topology, number theory, algebra, mechanics, fluid dynamics, optics, astronomy, logic, music theory, and more.

Ironically, Euler may be most well known for something he is not directly responsible for—or at least something that he did not put in writing—the identity $e^{i\pi} + 1 = 0$. Euler's identity has been repeatedly selected as the most beautiful mathematical expression.[23] It contains the mathematically significant constants 0, 1, π, e, and i and has graced countless posters, T-shirts, and tattooed body parts of geeky rebels who want to share their love of mathematics with the world.[24] And fittingly, as we shall see, Euler's identity is the key ingredient in the eventual solution of the problem of squaring the circle.

To understand Euler's identity, we need to learn about the number $2.71828\ldots$, now known as e, and its relationship to the complex numbers and to π.

The number e is much younger than its famous cousin π. The history of the discovery of e is much more complicated than that for π, which for many centuries was related only to the circumference, diameter, and area of a circle. The many other lives of π were discovered later. On the other hand, the nature and properties of e came out of a comparatively brief 150-year period and are intertwined with the discovery of logarithms, the birth of calculus (limits, derivatives, integrals, and infinite series), the development of the theory of complex numbers, and the introduction of functions. We will give only a brief outline of this fascinating history.

In 1614, after 20 years of research, John Napier (1550–1617) introduced logarithms in *Mirifici logarithmorum canonis descriptio* (*A Description of the Wonderful Law of Logarithms*).[25] His motivation for this work was to make it easier for astronomers to perform tedious computations. Rather than describing Napier's work (which would look unfamiliar to us), we recall our current understanding of the logarithm: $\log_a(b)$ is the exponent to which we raise a to obtain b. In other words, if $c = \log_a(b)$, then $a^c = b$. Today, schoolchildren learn base-10 logarithms, computer scientists use base-2 logarithms, but the most useful and mathematically elegant logarithm is the base-e logarithm ($\ln x = \log_e x$), which Nicolaus Mercator (1620–1687) called the *natural logarithm*.

Napier recognized that logarithms make arithmetic computations easier. They turn exponentiation into multiplication (because $\log_a(b^r) = r\log_a(b)$), roots into division, multiplication into addition ($\log_a(bc) = \log_a(b) + \log_a(c)$), and division into subtraction. So Napier and others after him created tables of logarithms to enable speedier calculations.

In 1647 Gregory of Saint Vincent published a work on geometry that contained results about the area under the hyperbola; essentially, he proved that the area was related to the logarithm. His student Alphonse Antonio de Sarasa (1618–1667) made the explicit connection to logarithms two years later. For instance, the area under the hyperbola $xy = 1$ between $x = 1$ and $x = a$, as shown in figure 18.3, is precisely

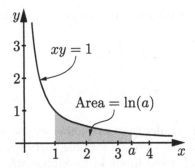

FIGURE 18.3. The area under the hyperbola is related to logarithms.

ln(a). Thus, in a sense this forms a connection between e and π—they are both related to the quadrature of conic sections. We could call π the circle constant[26] and e the hyperbola constant.

Jacob Bernoulli, in 1683, was probably the first person to define the number e (although he didn't call it e). He discovered it through the study of compound interest, not logarithms.[27]

Euler introduced the letter e for this value as early as 1727, when he was 20 years old.[28] But Euler's hugely influential 1748 precalculus textbook *Introductio in analysin infinitorum* (*Introduction to Analysis of the Infinite*),[29] which math historian Carl Boyer called the "foremost textbook of modern times,"[30] likely cemented its use. In the following excerpt, Euler introduced e, gave a decimal approximation, presented its infinite series representation, and gave the infinite series for the inverse function to the natural logarithm, e^z:[31]

> For the sake of brevity, for this number 2.718281828459 etc. we will use the symbol e, which will denote the base for the natural or hyperbolic logarithms,... and e represents the sum of the infinite series
>
> $$1 + \frac{1}{1} + \frac{1}{1 \cdot 2} + \frac{1}{1 \cdot 2 \cdot 3} + \frac{1}{1 \cdot 2 \cdot 3 \cdot 4} + \text{ &c. to infinity.}$$
>
> ...Therefore on putting e for the number found above there will be always
>
> $$e^z = 1 + \frac{z}{1} + \frac{z^2}{1 \cdot 2} + \frac{z^3}{1 \cdot 2 \cdot 3} + \frac{z^4}{1 \cdot 2 \cdot 3 \cdot 4} + \text{ &c.}$$

Today it is more common to express these series using factorials:[32]

$$e = 1 + \frac{1}{1!} + \frac{1}{2!} + \frac{1}{3!} + \cdots,$$

$$e^z = 1 + \frac{z}{1!} + \frac{z^2}{2!} + \frac{z^3}{3!} + \cdots.$$

In the *Introductio*, Euler gave us another way to represent complex numbers. He took de Moivre's formula, performed some algebraic manipulation, took some limits in his eighteenth-century way, and arrived at the gorgeously strange and unexpected[33] $e^{i\theta} = \cos\theta + i\sin\theta$. To the uninitiated, the left-hand side looks meaningless: What could it possibly mean to raise a number to an imaginary power? One way to think about e^z is not as an exponential function, but in terms of its infinite series. This is still mysterious, but at least we can imagine plugging an imaginary number into the infinite sum. In fact, a modern method of proving Euler's formula is to plug $z = i\theta$ into the series for e^z, rearrange terms, and see the series for sine and for cosine magically emerge.[34]

Thus, we now have a whole slew of ways to represent a complex number, each with its own pros and cons:

$$z = x + iy = r\cos\theta + ir\sin\theta = re^{i\theta}.$$

We now have what we need to produce Euler's identity, which Euler never wrote in print. Substitute $\theta = \pi$ into Euler's formula:

$$e^{i\pi} = \cos\pi + i\sin\pi = -1.$$

This expression, which at first glance seemed so wild and unfathomable, now nicely falls out of the framework that Euler built.

We can get other surprising expressions from Euler's formula. Consider, for example, i^i. If it was put to a vote, this number would probably be elected the most complex number in existence, and yet it is real! Euler pointed this out to Christian Goldbach in a 1746 letter. Notice that i makes a 90° angle ($\pi/2$ radians) with the real axis and it is one unit from the origin, so in exponential form, $i = e^{i\pi/2}$. Thus,

$$i^i = (e^{i\pi/2})^i = e^{i \cdot i\pi/2} = e^{-\pi/2} = 0.2078795763\ldots,$$

which is clearly a real number.[35]

Finally, with Euler's formula in hand, de Moivre's formula, which seemed so remarkable when we first saw it, is trivial to prove:

$$(\cos\theta + i\sin\theta)^n = (e^{i\theta})^n = e^{in\theta} = \cos(n\theta) + i\sin(n\theta).$$

The numbers e, π, and i are so special that the Harvard mathematician Benjamin Peirce (1809–1880) thought they deserved their own unique symbols. He stated that the existing notation for π and e were "for many reasons, inconvenient," and that "the close relation between these two quantities ought to be indicated in their notation."[36] In 1859 he introduced new symbols for π, first ◖ and later ◗, which he claimed looked like a C for circumference; for e, first ◖ and later ◗, which he said looked like a b, for the base of the natural logarithm; and ⌐ for i. He brought these together with the single formula[37] ⌐$^{-⌐} = \sqrt{◗◗}$, which we would recognize as $i^{-i} = \sqrt{e^\pi}$. Although his mathematician sons used his notation, it never caught on outside his family.

TANGENT
The τ Revolution

Futile is the labor of those who fatigue themselves
with calculations to square the circle.
—Michael Stifel[1]

IN 2001 ROBERT PALAIS wrote an article, entitled "π is wrong,"[2] that set in motion an underground movement to topple the famous number. Palais was not a circle squarer. No, when he said that π was wrong, he meant it was not the ideal choice for the circle constant. He argued that $2\pi = 6.283\ldots$ was better and even suggested a new symbol—an unsightly glyph resembling two πs that crashed into each other: $\pi\!\pi$.

In 2010 Michael Hartl published "The tau manifesto,"[3] arguing many of the same points that Palais raised. He called for mathematicians to abandon π and embrace 2π, which he called τ. The new name stuck.

Palais, Hartl, and others argued that τ is better because in many formulas in which π appears, there is also a 2 present (the definition of the normal distribution, the Gauss–Bonnet theorem, Cauchy's integral formula, and so on). But by far the most persuasive argument is that τ is the correct constant for circles. One revolution should not be twice a constant, 2π, but the constant itself, τ. Hartl chose τ because it represented *one turn* of the circle. So one-quarter of a full rotation would be the more intuitive $\tau/4$, and not $\pi/2$. This would alleviate much student confusion when they learned about radian measure.

Historically, π had the elegant definition C/d. But the radius is a more natural measure for a circle than the diameter; after all, a compass is set to the radius. Why not define the circle constant as C/r, which is τ? Sine and cosine have period τ. And we can replace Euler's lovely identity $e^{i\pi} = -1$ with its equally beautiful τ counterpart: $e^{i\tau} = 1$.

True, the area formula $A = \pi r^2$ looks nicer than $A = \frac{1}{2}\tau r^2$, with its unsightly $\frac{1}{2}$.[4] On the other hand, Archimedes proved that the area of a circle is the area of a triangle with height r and base C, where the $\frac{1}{2}$ appears naturally: $A = \frac{1}{2} \cdot \text{base} \cdot \text{height} = \frac{1}{2} \cdot \tau r \cdot r$.

The τ movement is gaining momentum, but it is too early to tell if π's days are numbered.

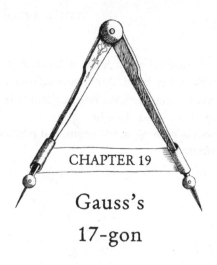

CHAPTER 19

Gauss's
17-gon

Mathematics is the queen of sciences and number
theory is the queen of mathematics.
—Carl Friedrich Gauss[1]

THE BELIEF THAT all great mathematicians were child geniuses is
a myth. There are many late bloomers when it comes to the queen
of the sciences. But not Carl Friedrich Gauss. His mathematical skills
were apparent at a young age. In his golden years, Gauss liked to tell
of the time his teacher asked his class to sum an arithmetic series to
keep them busy and out of trouble. The young prodigy deduced the
closed-form summation formula for such series and promptly brought
his slate to the teacher announcing that he was finished: *"ligget se"*
(there it lies).[2] But his early attempts at mathematics were not lim-
ited to impressing his teacher. He made deep mathematical discoveries
while still a teenager—one of which is a major plot point in our story.

Gauss was born in 1777 in Brunswick, now part of Lower Saxony,
Germany. Because of his early aptitude for mathematics, he caught
the eye of the Duke of Brunswick. The duke began giving Gauss a
stipend. This allowed him to attend the Brunswick Collegium Car-
olinum and then Göttingen University. He left Göttingen without a
diploma, but with a number of important mathematical discoveries in
hand. He received his PhD from the University of Helmstedt.

The duke continued to pay Gauss a stipend, so he was able to devote all of his time to his research career. This ended when the duke was killed fighting for the Prussian army. Needing work, Gauss took a position as the director of the new Göttingen Observatory. He continued his research into old age, and died in Göttingen in 1855.

Unlike Euler who published relentlessly, Gauss was a perfectionist. He held on to his discoveries, refined the ideas, and waited until he was able to publish a masterpiece. They came out fully formed, extremely deep, and challenging for others to grasp; it often took years for the mathematical community to fully understand his ideas. His personal motto *pauca sed matura* (few, but ripe) was apt. Mathematicians repeatedly learned, either from Gauss directly or from the notebooks he left behind, that some new mathematical advancement—even decades or centuries later—was a rediscovery of an idea Gauss already had.

Gauss's body of work is too broad to fully describe. He made important contributions in algebra, analysis, number theory, geometry, topology, complex analysis, linear algebra, statistics, and many areas of physics and astronomy.

The 17-gon

In 1796 Gauss started his first mathematical journal. This first year it was nothing more than a list of his mathematical accomplishments and observations from that year (there are 49 of them). Although many of the entries are cryptic (not only to us, but apparently to Gauss when he was older too), the first entry, dated March 30, 1796, is clear:[3]

[1] The principles upon which the division of the circle depend, and geometrical divisibility of the same into seventeen parts, and so on.

He didn't include anything more at this time, just this single sentence. Still one month before his 19th birthday, this teenager discovered that it was possible to construct a regular 17-gon using only a straightedge and compass, and moreover, he provided general rules for which regular polygons are constructible. Later that year he wrote,[4]

It seems that one has persuaded oneself ever since [Euclid's time] that the domain of elementary geometry could not be extended;

at least I do not know of any successful attempts to enlarge its boundaries on this side. It seems to me then to be all the more remarkable that *besides the usual polygons there is a collection of others which are constructible geometrically, e.g. the 17-gon.*

He went on to write, "This discovery is properly only a corollary of a not quite completed discovery of greater extent which will be laid before the public as soon as it is completed." He did indeed have more to say. This became one piece of his book *Disquisitiones arithmeticae*, a masterful treatment of number theory, which Gauss published in 1801.

Olaf Neumann wrote that *Disquisitiones* "was quickly recognized by contemporary experts . . . as a masterpiece of unprecedented organization, rigour, and extent, which transformed number theory from a scattering of islands into an established continent in mathematics. . . . This work belongs to the 'eternal canon' of mathematics, and thus of human culture."[5]

Gauss was always extremely proud of his discovery of the constructibility of these polygons. By some accounts it is what inspired him to become a mathematician. He even requested that the 17-gon be inscribed on his tombstone. This wish was not honored, but there is now a statue of Gauss in his hometown of Brunswick emblazoned with a 17-pointed star.

Interestingly, Gauss did not, at least in print, give a step-by-step construction of the 17-gon using Euclidean tools, but he proved that such a construction is possible. We know that $\cos(2\pi/17)$ is the x-coordinate of a vertex of a regular 17-gon inscribed in the unit circle. Gauss wrote $\cos(2\pi/17)$ explicitly using only integers, the four arithmetic operations, and square roots.[6] By Descartes's theorem, it is constructible, and hence so is the 17-gon.

The literature now contains many constructions of the 17-gon. In 1915 Robert Goldenring wrote a book containing more than 20,[7] and one reviewer pointed out that this collection is "far from being anything like complete."[8] Figure 19.1 shows Herbert Richmond's streamlined 1893 construction.[9] Begin with a circle with center O and points on the circle A and B such that $\angle AOB$ is a right angle (see figure 19.1). Let the point C be 1/4 the distance from O to B. Draw segment AC. Find the point D on AO such that angle $\angle DCO = \frac{1}{4}\angle ACO$. Find the point E on AO so that $\angle DCE = 45°$. Now construct a circle that has diameter AE. It crosses BO at a point F. Construct a circle with center D that passes through F. It intersects AO at G and H. Then lines

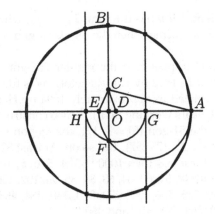

FIGURE 19.1. A construction of the 17-gon.

passing through G and H and perpendicular to AO intersect the circle at four points. These are four vertices of the 17-gon (A is a fifth vertex). Use these five vertices to construct the remaining 12 vertices of the 17-gon.

Gauss's Theorem

Gauss's proof that the 17-gon is constructible was the icing on the cake. It was a result that was easy to parade around (or inscribe for eternity on a memorial). But it was the showpiece for a much deeper body of work.

Recall that the Greeks knew how to construct an equilateral triangle, a square, and a regular pentagon, and from these they were able to construct a regular n-gon provided n had the form 2^j, $2^j \cdot 3$, $2^j \cdot 5$, or $2^j \cdot 3 \cdot 5$. Gauss discovered other primes p, besides 3 and 5—like 17—for which the regular p-sided polygon is constructible.

What is so special about the primes 3, 5, and 17? The key, it turns out, is that in each case $p - 1$ is a power of 2: $3 - 1 = 2^1$, $5 - 1 = 2^2$, and $17 - 1 = 2^4$. And in fact, the exponents themselves are powers of 2: $3 - 1 = 2^{2^0}$, $5 - 1 = 2^{2^1}$, and $17 - 1 = 2^{2^2}$. We call a number of the form $2^{2^m} + 1$ a *Fermat number*, and if a Fermat number is prime, we call it a *Fermat prime*.[10] The next two Fermat numbers after 3, 5, and 17 are also prime: $257 = 2^{2^3} + 1$ and $65{,}537 = 2^{2^4} + 1$. In *Disquisitiones*, Gauss proved the following remarkable theorem.

Gauss's theorem.[11] If n has the form $2^j p_1 \cdots p_k$, where $j \geq 0$ and the p_i are distinct Fermat primes, then the regular n-gon is constructible.

Gauss's theorem implies that the regular 257-gon is constructible (it was constructed in 1832 by F. J. Richelot), as is the regular 65,537-gon (it was constructed in approximately 1894 by Hermes of Lingen after spending 10 years working on it).[12] Also constructible are the 34-gon ($34 = 2 \cdot 17$), the 51-gon ($51 = 3 \cdot 17$), the 68-gon ($68 = 2^2 \cdot 17$), the 69,904-gon ($69,904 = 2^4 \cdot 17 \cdot 257$), and so on. At least 52 regular n-gons are constructible for n less than 1000:[13] 3, 4, 5, 6, 8, 10, 12, 15, 16, 17, 20, 24, 30, 32, 34, 40, 48, 51, 60, 64, 68, 80, 85, 96, 102, 120, 128, 136, 160, 170, 192, 204, 240, 255, 256, 257, 272, 320, 340, 384, 408, 480, 510, 512, 514, 544, 640, 680, 768, 771, 816, and 960.

As remarkable as it is, Gauss's theorem is only a partial answer to the question of which regular polygons are constructible. First of all, a complete answer would require, at the very least, information about which Fermat numbers are prime. This question has an interesting history, to which we will return shortly.

A more significant gap is the question of whether n-gons are constructible for n-values not of this form. Gauss's theorem does not address the heptagon or the nonagon. They are not of the form addressed by Gauss's theorem—7 is not a Fermat prime, and $9 = 3^2$ is the square of a Fermat prime. His theorem leaves open the possibility that they are constructible. In fact, they are not constructible and according to Gauss, he *knew* this was the case. He wrote that he could prove the converse "with all rigor," but went on to write,[14]

> The limits of the present work exclude this demonstration here, but we issue this warning lest anyone attempt to achieve geometric constructions for sections other than the ones suggested by our theory (e.g., sections into 7, 11, 13, 19, etc. parts) and so spend his time uselessly.

In other words, "It is not worth using this space to give a justification. But trust me, I know that the rest are not constructible. Don't waste your time trying." Pierre Wantzel gave a rigorous proof of this in 1837 (we say more about this in chapter 20).[15]

Putting together Gauss's and Wantzel's theorems, we obtain a new result.

Gauss–Wantzel theorem. A regular n-gon is constructible if and only if $n = 2^j p_1 \cdots p_k$, where $j \geq 0$ and the p_i are distinct Fermat primes.

Fermat Primes

Thanks to the Gauss–Wantzel theorem, we know exactly which regular polygons are constructible by a compass and straightedge—at least in principle. In practice, we need to know which Fermat numbers are prime. This is an open problem.

The investigation dates back to the first half of the seventeenth century. Fermat observed that when $j = 0, 1, 2, 3, 4$, the numbers $2^{2^j} + 1$ are prime (3, 5, 17, 257, and 65,537), and he conjectured that all such numbers are prime. In an August 1640 letter to Bernard Frénicle de Bessy (ca. 1604–1674), Fermat listed the first seven of these numbers (the seventh has 20 digits) and wrote,[16]

> I do not have the exact proof, but I have excluded such a large number of divisors by infallible proofs, and I have such a strong insight, which is the foundation for my thought, that it would be difficult for me to retract it.

Fermat revisited these numbers repeatedly in the 1640s and '50s in correspondences with the mathematicians, always asserting that this was a sequence of primes, but admitting that the proof was out of his reach. If it were true, it would be a great theorem. It had been known since the Greeks that there are infinitely many primes, but there was still no method of generating primes. In Fermat's formula was the hope of a prime generating function.

The conjecture was still unproved more than half a century later. Then came the 22-year-old Euler. He was settling in to his new job at the Academy of Sciences in the young city of St. Petersburg. Christian Goldbach, who had recently moved from St. Petersburg to Moscow to tutor the young Peter II, began a correspondence with Euler that would consist of nearly 200 letters and last for over 30 years.

On December 1, 1729, in a postscript to his first letter to Euler, Goldbach introduced Fermat's problem:[17]

> P.S. Note the observation of Fermat that all numbers of this form $2^{2^{x-1}} + 1$, that is 3, 5, 17, and so on, are primes, which he himself

admits that he was not able to prove, and, as far as I know, nobody else has proved either.

At first Euler was uninterested in the problem, but Goldbach continued to prod him. Finally, his interest piqued, Euler turned his attention to Fermat's sequence. On September 26, 1732, Euler presented his conclusion to the St. Petersburg Academy: Fermat's conjecture is false. In the accompanying article he wrote,[18]

> But I do not know by what fate it turned out that the number immediately following, $2^{2^5} + 1$, ceases to be a prime number; for I have observed after thinking about this for many days that this number can be divided by 641, which can be seen at once by anyone who cares to check.

And just like that Euler destroyed Fermat's hopes of a sequence of primes by showing that $2^{2^5} + 1 = 4{,}294{,}967{,}297$ has factors 641 and 6,700,417. How did Euler find these factors? The method of brute force is not completely out of the question. Although there are approximately 6500 possible prime divisors of $2^{2^5} + 1$, a check starting with 2, then 3, then 5, and so on would land on a factor after 116 attempts. (In fact, when the eight-year-old human calculator Zerah Colburn (1804–1840) was paraded around the United States to show off his powers of mental calculation, "the same number $[2^{2^5} + 1]$ was proposed to this child, who found out the factors by mere operations of his mind."[19]) This, of course, was not Euler's technique.

Euler did not explain how he arrived at this factorization in the first article, but 15 years later, in another article, he explained how he found 641.[20] Euler applied "Fermat's little theorem," which was stated without proof by Fermat, and was eventually proved by Euler in 1736,[21] to show that if there is a prime divisor of $2^{2^j} + 1$ ($j \geq 2$), then it must have the form $2^{j+1}k + 1$ for some whole number k. Thus, Euler needed to check prime numbers of the form $64k + 1$ only. Four of the first nine such numbers are prime—193, 257, 449, and 577—but none divide $2^{2^5} + 1$. Euler struck gold with the tenth number: $64 \cdot 10 + 1 = 641$.

It is surprising that Fermat was unable to factor $2^{2^5} + 1$ because he used a technique very similar Euler's to factor $2^{37} - 1$. But, as André Weil (1906–1998) wrote, "One may imagine that, when [Fermat]

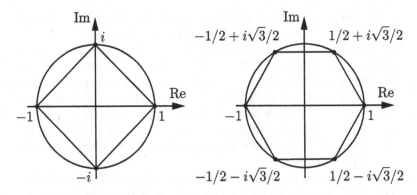

FIGURE 19.2. Roots of unity form regular polygons.

first conceived the conjecture, he was so carried away by his enthusiasm that he made a numerical error, and then never checked his calculations again."[22]

We have found no new Fermat primes in the 370 years since Fermat discovered the first five. Currently, we know that the next 28 Fermat numbers are composite. The gargantuan 2.5 billion digit number $2^{2^{33}} + 1$ is the first Fermat number whose primality is unknown.

Roots of Unity and Regular Polygons

Before we discuss the proof of Gauss's theorem, we must return to the realm of complex numbers. We start with a seemingly simple question: What is $\sqrt[4]{1}$? It's 1, right? Yes—if we raise 1 to the 4th power we obtain 1. But 1 is not the only 4th root of 1. There are three others, -1, i, and $-i$, because $(-1)^4 = i^4 = (-i)^4 = 1$. In general, an *nth root of unity* is any complex solution to the equation $z^n - 1 = 0$. There are four 4th roots of unity: 1, -1, i, and $-i$.

In figure 19.2 we see that the 4th roots of unity are the vertices of a square inscribed in the unit circle in the complex plane. It is tedious but not difficult to check that ± 1 and $\pm 1/2 \pm i\sqrt{3}/2$ are the six 6th roots of unity. They are the vertices of a regular hexagon inscribed in the unit circle. So the problem of inscribing regular n-gons in a circle is intimately related to the values of the nth roots of unity—they are the vertices of the polygon.

We can use Euler's formula to verify this assertion. Suppose $z = r \cos \theta + r \sin \theta = re^{i\theta}$ is an nth root of unity. Then $1 = z^n = (re^{i\theta})^n = r^n e^{in\theta}$. For this to be true, we must have $r^n = 1$, and because r is a non-negative real number, $r = 1$. So $1 = e^{in\theta} = \cos(n\theta) + i \sin(n\theta)$, and hence $\cos(n\theta) = 1$ and $\sin(n\theta) = 0$. The only way this will happen is if $n\theta$ is an integer multiple of $360°$ (or 2π, using radians), or equivalently $\theta = 2k\pi/n = k \cdot 360°/n$ for some integer k. So the nth roots of unity are $\cos(2k\pi/n) + i \sin(2k\pi/n)$, for $k = 0, \ldots, n - 1$. Viewed in this way, we can see that the nth roots of unity are n points evenly spaced about the unit circle.

If we apply this to $n = 6$, we find that the 6th roots of unity are numbers of the form $\cos(k \cdot 60°) + i \sin(k \cdot 60°)$. For instance, when $k = 1$, we have $\cos(60°) + i \sin(60°) = 1/2 + i\sqrt{3}/2$, and when $k = 3$, we have $\cos(180°) + i \sin(180°) = -1$.

Thus, to solve the polygon question, Gauss realized that he must find the roots of $z^n - 1 = 0$. If the real and imaginary parts of $\cos(2k\pi/n) + i \sin(2k\pi/n)$ can be expressed in terms of the four arithmetic operations and square roots, then the polygon is constructible. In fact, if $\cos(2k\pi/n)$ is constructible, so is $\sin(2k\pi/n)$.

Gauss was not the first mathematician to look at $z^n - 1$. In 1740 Euler showed that for $n = 1, \ldots, 10$, the equation $z^n - 1 = 0$ is solvable by radicals; that is, all of the roots of the polynomial can be expressed in terms of the four arithmetic operations plus square roots, cube roots, 4th roots, and so on. In 1770 Alexandre-Théophile Vandermonde (1735–1796) did the same for $n = 11$.[23] These were important mathematical results, especially in retrospect, because not all polynomials of degree 5 or larger can be solved by radicals. But it did not answer the question of whether the polygons are constructible. To do so, we must show that square roots are the only roots required to express the coordinates of the vertices. And this was the teenage Gauss's discovery.

Idea of Gauss's Proof

The seventh, and final, section of Gauss's *Disquisitiones* was devoted to the problem of constructing regular polygons. His proof relied on number theory, the theory of equations, and properties of complex numbers. Viewed in one way, it was about the constructibility of polygons. But really, it was a study of polynomials of the form $z^n - 1$.

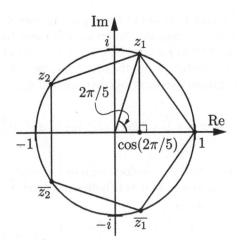

FIGURE 19.3. A regular pentagon in the complex plane.

Every mathematician knows and relishes that wonderful aha moment that comes at the most unexpected time—in the shower, driving to work, walking the dog, at the dinner table, or lying in bed. Days, weeks, months, or years of thinking intently about a seemingly intractable problem initiates a subconscious attack by the brain. Often in these moments of relaxation, the answer springs out, as if by divine revelation. Gauss described just such an experience:[24]

> By thinking with great effort about the relation of all the roots [of $z^p - 1$] to each other with respect to their arithmetic properties, I succeeded, while I was on vacation in Braunschweig, on that day (before I got out of bed) in seeing this relation with utmost clarity, so that I was able to make on the spot the special application to the 17-gon and to verify it numerically.

The full proof of Gauss's theorem is beyond the scope of this book, but we can give the general idea by showing how his argument would proceed for a regular pentagon, then sketching it for the 17-gon and more general regular polygons.

We view the vertices of the pentagon as the five 5th roots of unity; that is, the five roots of $z^5 - 1$. One root is $z = 1$, so the polynomial factors are $z^5 - 1 = (z - 1)(z^4 + z^3 + z^2 + z + 1)$, and it suffices to focus on the four roots of $z^4 + z^3 + z^2 + z + 1$. As we see in figure 19.3, these come in complex conjugate pairs, z_1 and $\overline{z_1}$, and z_2 and $\overline{z_2}$.

There are two nice facts about these pairs of complex conjugates. First, like any pair of conjugates, when we add them, their imaginary parts cancel and the real part doubles. So $z_k + \overline{z_k} = 2\operatorname{Re}(z_k) = 2\operatorname{Re}(\overline{z_k})$. And second, because they are on the unit circle, they are reciprocals:[25] $\overline{z_k} = 1/z_k$. Putting these two facts together, we find that

$$z_k + \frac{1}{z_k} = z_k + \overline{z_k} = 2\operatorname{Re}(z_k) = 2\operatorname{Re}(\overline{z_k}) = \overline{z_k} + \frac{1}{\overline{z_k}}.$$

We would like to find the z-values for which $z^4 + z^3 + z^2 + z + 1 = 0$. But we know $z \neq 0$, so we can divide both sides by z^2. This gives us

$$0 = \frac{1}{z^2}(z^4 + z^3 + z^2 + z + 1)$$

$$= \left(z + \frac{1}{z}\right)^2 + \left(z + \frac{1}{z}\right) - 1$$

$$= w^2 + w - 1,$$

where $w = z + \frac{1}{z}$. By the quadratic formula, $w = \frac{1}{2}(-1 \pm \sqrt{5})$.

We could stop here; we have enough information to construct our pentagon with a compass and straightedge. These two real numbers are twice the real part of the four remaining vertices of the pentagon. So $\operatorname{Re}(z_1) = \frac{1}{4}(-1 + \sqrt{5}) = \cos(2\pi/5)$. As we described in chapter 1, we can construct the point $\cos(2\pi/5)$ in the complex plane— or $(\cos(2\pi/5), 0)$ in the Cartesian plane—then draw the line that goes through the point and is perpendicular to the real axis. It intersects the unit circle at z_1 and z_4. Then we can use the compass to find the remaining vertices.

However, let's not take this approach. Let's carry the procedure through to the bitter end and find the coordinates of the vertices. Because $w = z + \frac{1}{z}$, we can find z by solving the quadratic equation $0 = z^2 - wz + 1$, where we replace w with the two roots $\frac{1}{2}(-1 \pm \sqrt{5})$. The points z_1 and $\overline{z_1}$ are the roots of the quadratic equation $0 = z^2 + \frac{1}{2}(1 - \sqrt{5})z + 1$, which by the quadratic formula are $\frac{1}{4}(-1 + \sqrt{5}) \pm i\frac{1}{4}\sqrt{10 + 2\sqrt{5}}$. So the coordinates are $\left(\frac{1}{4}(-1 + \sqrt{5}), \pm\frac{1}{4}\sqrt{10 + 2\sqrt{5}}\right)$. A similar procedure yields the other vertices: $\left(-\frac{1}{4}(1 + \sqrt{5}), \pm\frac{1}{4}\sqrt{10 - 2\sqrt{5}}\right)$.

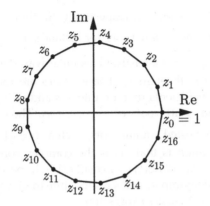

FIGURE 19.4. A regular 17-gon in the complex plane.

The takeaway point from these messy calculations is that to find the roots of the 4th-degree equation, we were able to reduce the problem to finding the roots of three quadratic equations—one with integer coefficients and the two whose coefficients included the roots of the first. And in the end, the coordinates of our points consisted of nested square roots of rational numbers.

Gauss's analysis of the 17-gon is similar, but it requires solving more quadratic equations, which yields roots with even more nested square roots. We give a brief sketch of the mathematics involved.

Suppose $z_0, z_1, z_2, \ldots, z_{16}$ are the vertices of the regular 17-gon, as in figure 19.4. Gauss proved a theorem about prime numbers that allowed him to list the vertices (except $z_0 = 1$) in a very specific order:[26] $z_1, z_3, z_9, z_{10}, z_{13}, z_5, z_{15}, z_{11}, z_{16}, z_{14}, z_8, z_7, z_4, z_{12}, z_2, z_6$. He showed that the sums of every other term yield the two complex numbers

$$w_1 = z_1 + z_9 + z_{13} + z_{15} + z_{16} + z_8 + z_4 + z_2,$$
$$w_2 = z_3 + z_{10} + z_5 + z_{11} + z_{14} + z_7 + z_{12} + z_6,$$

which are roots of the quadratic equation $z^2 + z - 4$. Thus w_1 and w_2 can be written in terms of square roots; specifically $w_1 = \frac{1}{2}\left(-1 + \sqrt{17}\right)$ and $w_2 = \frac{1}{2}\left(-1 - \sqrt{17}\right)$. Next, take every other term in w_1 to create complex numbers $u_1 = z_1 + z_{13} + z_{16} + z_4$ and $u_2 = z_9 + z_{15} + z_8 + z_2$, and similarly every other term in w_2 to obtain u_3 and u_4. These two pairs of complex numbers are roots of the quadratic equations

$z^2 - w_1 z - 1$ and $z^2 - w_2 z - 1$, respectively. So they can also be written in terms of nested square roots; for instance, $u_1 = \frac{1}{2}\left(w_1 + \sqrt{w_1^2 + 4}\right)$ and $u_3 = \frac{1}{2}\left(w_2 + \sqrt{w_2^2 + 4}\right)$. Lastly, in a similar way, define $t_1 = z_1 + z_{16}$, $t_2 = z_{13} + z_4$, and t_3 through t_8. These values are roots of quadratic equations with coefficients that are integers and u_1, u_2, u_3, and u_4. In particular, t_1 and t_2 are roots of $z^2 - u_1 z + u_3$. Thus, these roots too can be written as nested square roots, such as $t_1 = \frac{1}{2}\left(u_1 + \sqrt{u_1^2 - 4u_3}\right)$. The final observation is that z_{16} is the complex conjugate of z_1, so $t_1 = z_1 + z_{16} = z_1 + \overline{z_1} = 2\,\mathrm{Re}(z_1) = 2\cos(2\pi/17)$. In particular, when we substitute w_1 and w_2 into u_1 and u_3, substitute these values into t_1, and simplify, we obtain Gauss's expression

$$\cos\left(\frac{2\pi}{17}\right) = \frac{1}{2}t_1 = \frac{1}{16}\left(-1 + \sqrt{17} + \sqrt{34 - 2\sqrt{17}}\right.$$
$$\left. + 2\sqrt{17 + 3\sqrt{17} + \sqrt{34 - 2\sqrt{17}} - 2\sqrt{34 + 2\sqrt{17}}}\right),$$

which is a constructible number!

This shows the general approach that Gauss used. He wanted to find the roots of $z^p - 1$, where p is a prime number. Because $z = 1$ is a root of this polynomial, it sufficed to find the roots of $z^{p-1} + z^{p-2} + \cdots + z + 1$. Now, p is prime, but $p - 1$ need not be. Suppose it has the prime factorization $p - 1 = p_1 p_2 \cdots p_m$. Gauss reduced the problem to finding the roots of polynomials of degree $p_1, p_2, \ldots, p_m$, which are solved sequentially. The first one has integer coefficients. Then the coefficients of subsequent polynomials are built from the roots of the previously solved polynomials.

Gauss observed that if $p - 1$ is a power of 2, then $p_1 = p_2 = \cdots = p_m = 2$. So we can find the roots of f by solving a sequence of quadratic equations, and thus the roots are constructible. In short, if $p = 2^l + 1$ is prime, then the regular p-gon is constructible.

This p-value does not yet have the form of a Fermat prime. However, if $p = 2^l + 1$ is prime, then the exponent l must be a power of 2. Suppose it isn't; then l has an odd factor $s \neq 1$. Then $l = rs$ (with $r = 1$ a possibility). In this case,

$$p = 2^l + 1 = 2^{rs} + 1 = (2^r + 1)((2^r)^{s-1} - (2^r)^{s-2} + \cdots - 2^r + 1),$$

which implies that it is not prime.[27]

Lastly, suppose $n = 2^j p_1 \cdots p_k$, where $j \geq 0$ and the p_i are distinct Fermat primes. Then we can construct the regular p_i-gons, and using the technique that we used to create a regular 15-gon from an equilateral triangle and a regular pentagon, we can construct a regular $(p_1 \cdots p_k)$-gon. Finally, double the number of sides j times to obtain the regular n-gon. This yields Gauss's theorem.

TANGENT
Mirrors

Mirror, mirror, on the wall, who in this land is fairest of all?
— *Jacob and Wilhelm Grimm*, Little Snow-White

A MIRA IS a teaching tool invented by George Scroggie and N. J. Gillespie to help students learn about reflections and lines of symmetry.[1] It is a piece of tinted acrylic glass with legs that allow it to stand vertically. The Mira's key feature is that it is both transparent and reflective, so it acts simultaneously as a window and mirror.

To reflect a point P across a line l, align the Mira along l, and mark the point that lines up with the reflection of P on the opposite side of the glass (see figure T.31). To find the line of reflection between P and another point Q, situate the Mira so the reflection of P lines up with Q, then trace along the glass. We can also draw a line through a point that reflects a second point to a given line, if such a line exists. And figure T.32 shows the fourth basic action: given points P and R and lines l_1 and l_2, place the Mira so the reflections of P and R lie on l_1 and l_2, respectively.[2]

These four actions are surprisingly powerful. We can imagine abandoning the compass and straightedge and using only the Mira for our geometric constructions. We cannot draw a circle but, using only the first three of the Mira actions, we can construct every Euclidean-constructible point. Conversely, the first three actions can be carried out with the Euclidean tools. Thus, the points constructible with the first three actions are exactly the compass-and-straightedge points. The fourth action allows us to construct points that are not Euclidean constructible. Let's see why.

A parabola is the set of points that are equidistant from a line (the *directrix*) and a point not on the line (the *focus*). In figure T.33, Q is equidistant from the focus P and directrix l_1. In fact, the tangent line to the parabola at Q, which we denote l, bisects $\angle PQS$. When we reflect P across l, we obtain S. In other words, if we place the Mira along l, it

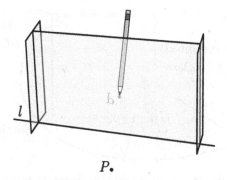

FIGURE T.31. The Mira reflects P across l.

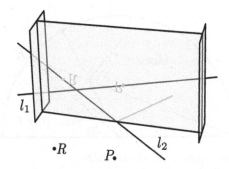

FIGURE T.32. The Mira reflects P and R to l_1 and l_2, respectively.

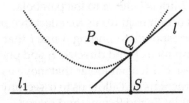

FIGURE T.33. If we reflect P onto l_1, the Mira is a tangent line to the parabola with focal point P and directrix l_1.

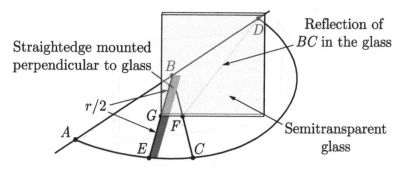

FIGURE T.34. We can use a straightedge mounted to a semitransparent piece of glass to trisect an angle.

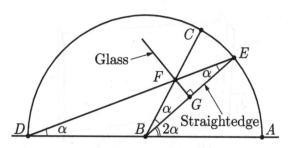

FIGURE T.35. The segment BE trisects $\angle ABC$.

will reflect P to S. Conversely, if we place the Mira so the reflection of P is on l_1, then the Mira is tangent to the parabola.

The fourth Mira action requires us to reflect two points to two lines. This procedure corresponds to finding a line l that is tangent to two parabolas—one with focus P and directrix l_1 and one with focus R and directrix l_2. It is these hidden parabolas that move us outside the realm of Euclidean constructions and allow us to trisect any angle, double the cube, construct regular 7- and 9-gons, and more.[3]

In 1963—before the invention of the Mira—A. E. Hochstein invented a mirrored device that could be used with a compass and straightedge to trisect an angle.[4] It is an ordinary straightedge of length r attached to a semitransparent piece of glass. The glass is attached so that it is perpendicular to the straightedge and so that the reflective face divides the straightedge in half (see figure T.34).

Here's how to use the tool. We wish to trisect $\angle ABC$. For simplicity, assume $AB = BC = r$. Use a straightedge to extend AB, and use a compass to draw a circle with center B of radius r, meeting AB at D. Next put the straightedge so that one corner sits on B and the corner at the other end sits on the circle. We see the reflection of BC in the glass. Move the straightedge until the reflection passes through D. Then draw along the edge to obtain BE; this trisects the angle.

We now prove this result. First, notice that because $BG = EG = r/2$, triangles BFG and EFG are congruent, and hence BEF is isosceles (see figure T.35). Moreover, because $BE = BD = r$, triangle BDE is isosceles and similar to BEF. Thus $\angle EBC = \angle BED = \angle BDE$, which we will call α. Because $\angle ABE$ is an exterior angle of triangle BDE, then $\angle ABE = \angle BDE + \angle BED = 2\alpha$. Thus $\angle CBE = \frac{1}{3}\angle ABC$.

CHAPTER 20

Pierre
Wantzel

God created the integers, all else is the work of man.
—Leopold Kronecker, 1886[1]

IN 1837 QUEEN VICTORIA acceded to the throne, Michigan became
the 26th US state, Chicago and Houston were incorporated as cities,
the electric telegraph was born, and Descartes's *Geometry* celebrated
its 200th birthday. In France, 23-year-old Pierre Wantzel published
a seven-page article that put to an end a millennia of speculation
about three of the four problems of antiquity. In these few short
pages, Wantzel proved that it was impossible to trisect every angle,
to construct every regular polygon, and to double the cube.

Surely the news was greeted with fanfare, banner headlines in the
major news publications, and fame for the young mathematician.
Right? Wrong! The result came with a deafening silence. Not only
was it not publicized at the time, prominent mathematicians even
a century later did not know who proved the impossibility results.
The historical record has been set straight, but even today Wantzel
is largely unknown and underappreciated. His Wikipedia page fits on
one screen. He has no entry in the 27-volume *Complete Dictionary of
Scientific Biography*, a scholarly work that profiles the lives and work of
influential scientists and mathematicians. It skips from chemist James
Wanklyn to physicist Emil Warburg without a mention of Wantzel.

As Brian Hayes wrote, Wantzel "is hardly a household name, even in mathematical households."[2]

Pierre Wantzel

Pierre Wantzel's father came from a German banking family. He moved to Paris for work at the end of the eighteenth century. He entered the French army three months before his son was born on June 5, 1814. After serving for seven years, he returned to his family and became a professor of applied mathematics.

It was apparent at a young age that Wantzel inherited his father's mathematical aptitude. When he was nine years old, his teacher—who was also a surveyor—asked the young boy for help with a difficult surveying problem. When he was 15, Antoine Reynaud (1771–1844) asked him to proofread the latest edition of his popular text *Traité d'arithmétique*. But Reynaud got more than he asked for. Wantzel provided a proof for a commonly used, but previously unjustified, method of computing square roots. He excelled in school and was known for "dazzling his classmates with the superiority of his intellect, as he charmed them with his frankness and nobility of character."[3]

Wantzel had a broad set of academic interests and strengths. He had a facility for languages, such as Latin, Greek, and German. As Gaston Pinet wrote, Wantzel was "endowed with extreme vivacity of impressions and with truly universal aptitudes.... He threw himself into mathematics, philosophy, history, music, and into controversy, exhibiting everywhere equal superiority of mind."[4]

At 18, Wantzel scored first place in the entrance exams to both the École Polytechnique and the science department of the École Normale Supérieure—a first for any applicant. He attended the École Polytechnique.

When he was 20, he began training as an engineer in the École des Ponts et Chaussées (School of Roads and Bridges). It was at this time that Wantzel proved the impossibility theorems. Perhaps because of this achievement, he "said cheerfully to his friends that he would be but a mediocre engineer," and that "he preferred the teaching of mathematics."[5] Although he completed his engineering program, he ended up teaching at the École des Ponts et Chaussées, the École Polytechnique, and elsewhere and to numerous private students.

His collaborator Jean Claude Saint-Venant (1797–1886) wrote,[6]

> His teaching ... carried a particular cachet of clearness, steadiness, lucidity, and charm. No one knew as he did how, with gentleness and patience, to obtain from his listeners a most attentive silence. He was always understood, in spite of the rapidity of his exposition, the originality of his methods, and the volubility of his speech, the tone of which was never raised.... His students adored and venerated him.

But Wantzel was overextended. In addition to his teaching, he had a wide-ranging research career. He worked on the compass-and-straightedge problems, the theory of radicals, the solvability of polynomial equations, the case of the irreducible cubic, the flow of air, and the curvature of elastic rods. He was also in charge of the entrance exams for the École Polytechnique, and he spent 10 to 12 hours per day administering the exams, even when he was ill.

The hard-driving mathematician died when he was merely 33 years old. Saint-Venant wrote,[7]

> He was blameworthy for having been too rebellious to the counsels of prudence and of friendship. Ordinarily he worked evenings, not lying down until late; then he read, and took only a few hours of troubled sleep, making alternately wrong use of coffee and opium, and taking his meals at irregular hours until he was married. He put unlimited trust in his constitution, very strong by nature, which he taunted at pleasure by all sorts of abuse. He brought sadness to those who mourn his premature death.

Wantzel was survived by his wife—the daughter of one of his former teachers whom he had married six years earlier—two daughters, and his father. He also left behind all the mathematics he did not discover.

Again, we quote Saint-Venant:[8]

> [His research accomplishments] are not equal to what he could have attained with his active imagination, extreme facility, and his extended and profound knowledge of pure mathematics.... I believe that this is mostly due to the irregular manner in which he worked, to the excessive number of occupations in which he was engaged, to the continual movement and feverishness of his

thoughts, and even to the abuse of his own faculties. Wantzel improvised more than he elaborated: He probably did not give himself the leisure nor the calm necessary to linger long on the same subject. All this leads us to think that if he had lived a few more years, had modified this routine, and, finally applying himself seriously to work on these accumulated materials, he would have, in ensuing works of great importance, taken the place in the intellectual world that his mathematical genius deserved.

As Albert Lapparent so eloquently wrote, Wantzel left "a luminous trace, unfortunately too similar to that which the meteors draw in the sky that fade as soon as glimpsed."[9] Fortunately, we are able to glimpse the luminous trace: Wantzel's impossibility theorems for the famous compass-and-straightedge problems. However, before we discuss these lovely results, we must introduce some necessary algebra.

Irreducible and Minimal Polynomials

In our algebra classes we learned to factor polynomials such as

$$x^2 - 9 = (x - 3)(x + 3),$$
$$5x^4 - 19x^3 - 3x^2 + 57x - 36 = (x - 3)(5x - 4)(x^2 - 3),$$

and

$$x^4 - 6x^3 + 13x^2 - 24x + 36 = (x - 3)^2(x^2 + 4).$$

We know—and in fact Descartes knew—that a is a root of a polynomial if and only if $x - a$ is a factor of the polynomial. So $x = 3$ is a root of all three polynomials. The first also has $x = -3$ as a root. If we factor the second polynomial further, as $5(x - 3)(x - 4/5)(x - \sqrt{3})(x + \sqrt{3})$, we see three more roots: $4/5$, $\sqrt{3}$, and $-\sqrt{3}$.

Descartes knew that the number of roots of a polynomial cannot be greater than the degree of the polynomial. These three polynomials of degrees 2, 4, and 4 have two, four, and one real roots, respectively. But we can do better than that. Gauss's 1799 PhD thesis gave a proof of the so-called *fundamental theorem of algebra*, which states that a polynomial of degree n has *exactly* n roots, provided we count complex roots and we count the roots with multiplicity.[10] For instance, we can factor the third polynomial as $(x-3)(x-3)(x-2i)(x+2i)$. So it has two real roots 3 and 3 (or 3 with multiplicity 2) and the complex roots $2i$ and $-2i$. Gauss believed this was such an important theorem that he proved it three more times—twice in 1816 and once in 1849.

When we are required to factor a polynomial, we may reasonably ask: How far is far enough? All three factorizations at the start of the section have integer coefficients. We could have gone further, but doing so would have required using irrational values (in the second polynomial) or complex values (in the third).

If we restrict our coefficients to some set other than the full set of complex numbers, some polynomials can't be factored at all; these polynomials are called *irreducible*. A polynomial $f(x)$ is *irreducible over the rational numbers* if there do not exist nonconstant polynomials $g(x)$ and $h(x)$ with rational coefficients such that $f(x) = g(x)h(x)$. We can define irreducibility over other types of numbers as well—the integers, real numbers, complex numbers, and so on.

The polynomials $2x+4$, x^2+1, and x^2-3 are all irreducible over the integers and the rational numbers. The first two are irreducible over the real numbers, and only the first is irreducible over the complex numbers. The three polynomials at the start of the section are reducible over the rational numbers, and we have reduced them to a product of irreducible factors. From here onward, if we don't say otherwise, irreducible means irreducible over the rational numbers.

It is generally difficult to determine whether a polynomial is irreducible. But there are a number of techniques to help decide.[11] We begin with an important result due to Gauss.

Gauss's lemma.[12] If a polynomial with integer coefficients is irreducible over the integers, it is irreducible over the rational numbers.

Suppose we want to show that $3x^2+2x+1$ is irreducible over the rational numbers. We would have to show that there do not exist rational numbers a, b, c, and d such that $3x^2+2x+1 = (ax+b)(cx+d) = acx^2 + (ad+bc)x + bd$, or equivalently, $ac=3$, $ad+bc=2$, and $bd=1$. That's a lot of work! Fortunately, Gauss's lemma says that because the original polynomial has integer coefficients, we need only check integer values of a, b, c, and d, and because all the coefficients are positive, we may assume that these four values are positive. Because $ac=3$ and $bd=1$, the only possible values for a and c are 1 and 3 and the only possible value for b and d is 1. Because $ad + bc = 3 \cdot 1 + 1 \cdot 1 = 4 \neq 2$, the polynomial doesn't factor; it is irreducible over the rational numbers.

Let us look a little closer at polynomials with integer coefficients and rational roots. Suppose $f(x) = a_n x^n + a_{n-1} x^{n-1} + \cdots + a_1 x + a_0$ has a rational root r/s, where r and s have no common factors. Then

$$f(r/s) = a_n (r/s)^n + a_{n-1}(r/s)^{n-1} + \cdots + a_1(r/s) + a_0 = 0.$$

Multiplying through by s^n we have

$$a_n r^n + a_{n-1} r^{n-1} s + \cdots + a_1 r s^{n-1} + a_0 s^n = 0.$$

And rearranging terms we obtain

$$s(a_{n-1} r^{n-1} + \cdots + a_1 r s^{n-2} + a_0 s^{n-1}) = -a_n r^n.$$

Because all the values in the expression are integers, s divides $a_n r^n$. But by assumption, s and r have no common factors, so it must be the case that s divides a_n. A similar argument shows that r divides a_0. These observations form the rational root theorem.

Rational root theorem. Suppose $a_n x^n + \cdots + a_1 x + a_0$ is a polynomial with integer coefficients and r/s is a rational root (written in lowest terms). Then r is a factor of a_0 and s is a factor of a_n.

This theorem gives us a list of candidate roots to check. For example, if $16x^4 - 24x^3 - 12x^2 + 16x + 3$ has a rational root r/s, then r divides 3 and s divides 16. This implies that there are only 18 candidates to check: ± 1, $\pm 1/2$, $\pm 1/4$, $\pm 1/8$, ± 3, $\pm 3/2$, $\pm 3/4$, $\pm 3/8$, and $\pm 3/16$. This is a little bit of a hassle, but certainly doable. In this case, $3/2$ is the only rational root. So we can factor the polynomial as $(x - \frac{3}{2})(16x^3 - 12x - 2)$ or $(2x - 3)(8x^3 - 6x - 1)$ if we prefer integer coefficients.[13] Consider the cubic factor $8x^3 - 6x - 1$. It is either irreducible or it can be factored into the product of integer polynomials of degrees 1 and 2. But because it has no rational root, it has no degree-1 factor. It must be irreducible.

Here's another example. Consider the polynomial $x^m - n$ in which m and n are positive integers. This polynomial has one real root, $\sqrt[m]{n}$. By the rational root theorem, $\sqrt[m]{n}$ is either irrational or it is a rational number r/s in which r divides n and s divides 1. But the only way this latter case can occur is if $s = 1$, which implies that the root is an integer. So these roots are either irrational, like $\sqrt{2}$ and $\sqrt[3]{2}$, or integers, like $\sqrt{9} = 3$ and $\sqrt[5]{32} = 2$. In other words, $\sqrt[m]{n}$ cannot be a nonintegral rational number. This argument also shows that if n is not a perfect mth power, then $x^m - n$ is irreducible.

Now, let's change our point of view from a polynomial-centric view to a number-centric view. Let's start with a number, such as $\sqrt[3]{2}$. It is the root of the polynomial $x^3 - 2$, which we have just seen is irreducible. We now ask: Is it the root of any *other* irreducible polynomial? One with a lesser degree perhaps? A greater degree? The same degree?

The quick answer is yes. It is a root of $2x^3 - 4$, which is irreducible. But this is a technicality; this polynomial is just a constant multiple of the original one. If we insist that the coefficient of the largest power of x is 1—such polynomials are called *monic*—then the answer is no. In 1829 Abel proved the following result.

Abel's irreducibility theorem. If polynomials $f(x)$ and $g(x)$ have a common root, and $f(x)$ is irreducible, then $f(x)$ is a factor of $g(x)$.

For instance, $\sqrt[3]{2}$ is a root of $g(x) = x^5 + 2x^4 - 15x^3 - 2x^2 - 4x + 30$ and the irreducible polynomial $f(x) = x^3 - 2$. And indeed, $g(x) = (x^3 - 2)(x^2 + 2x - 15)$. Now, suppose there is another irreducible monic polynomial $p(x)$ with root $\sqrt[3]{2}$. By Abel's theorem, there exists a polynomial $h(x)$ such that $p(x) = (x^3 - 2)h(x)$. But because $p(x)$ is irreducible, it cannot be factored. So the only possibility is that $h(x)$ is the constant polynomial 1, and hence $p(x) = f(x)$.

Thus, we may define the *minimal polynomial* of a to be the unique irreducible monic polynomial having root a. In this case, $x^3 - 2$ is the minimal polynomial of $\sqrt[3]{2}$.

Wantzel's Theorem

Descartes proved that the constructible numbers are those that can be written using the integers, the four arithmetic operations, and square roots. Unfortunately, as important as this result was, because there are multiple ways of representing numbers it did not yield a checkable criterion to show that a number is not constructible. The checkable criterion came from Wantzel, and it was key to proving the impossibility of the problems of antiquity.

In his 1837 article "Recherches sur les moyens de reconnaître si un problème de géométrie peut se résoudre avec la règle et le compas" ("Research on the means of knowing whether a problem in geometry can be solved with ruler and compass")[14] Wantzel proved the following result.[15]

Wantzel's theorem. The degree of the minimal polynomial of a constructible number is 2^n for some n.

In particular—and here is the key observation—if a number is a root of an irreducible polynomial whose degree is *not* a power of 2, then it is not a constructible number.

Of course, Wantzel's proof applies to every constructible number. But rather than presenting that demonstration, we give a simplified, concrete example that illustrates the idea behind the proof.[16] We will find an irreducible polynomial with rational coefficients of degree 2^n for some n having the following constructible number as a root:

$$a = \frac{4}{\sqrt{5}+\sqrt{3}} + \sqrt{3} - \sqrt{2+\sqrt{5}}.$$

The first step is to simplify a as much as possible; in this case we want to rationalize the denominator. Standard simplification procedures yield $a = 2\sqrt{5} - \sqrt{3} - \sqrt{2+\sqrt{5}}$. We now want to find a quadratic equation with a as a root. To do so, set it equal to x and bring all terms to one side, $x - 2\sqrt{5} + \sqrt{3} + \sqrt{2+\sqrt{5}} = 0$. Now the trick: to get rid of the nested square roots, multiply the polynomial by the identical polynomial, but with the sign changed for $\sqrt{2+\sqrt{5}}$:

$$0 = \left(x - 2\sqrt{5} + \sqrt{3} + \sqrt{2+\sqrt{5}}\right)\left(x - 2\sqrt{5} + \sqrt{3} - \sqrt{2+\sqrt{5}}\right)$$

$$= x^2 + \left(2\sqrt{3} - 4\sqrt{5}\right)x + 21 - 4\sqrt{3}\sqrt{5} - \sqrt{5}.$$

Now we no longer have nested square roots, which is an improvement, but there are still square roots in the coefficients: $\sqrt{3}$ and $\sqrt{5}$. To get rid of the $\sqrt{3}$, perform the trick again: multiply the polynomial by the identical polynomial, but with the sign changed for each instance of $\sqrt{3}$. We obtain a 4th-degree polynomial with no $\sqrt{3}$ terms:

$$\left(x^2 + (2\sqrt{3} - 4\sqrt{5})x + 21 - 4\sqrt{3}\sqrt{5} - \sqrt{5}\right)$$
$$\cdot \left(x^2 + (-2\sqrt{3} - 4\sqrt{5})x + 21 + 4\sqrt{3}\sqrt{5} - \sqrt{5}\right)$$
$$= x^4 - 8\sqrt{5}x^3 + (110 - 2\sqrt{5})x^2 + (40 - 120\sqrt{5})x - 42\sqrt{5} + 206.$$

Now do the same with this polynomial, but change the signs of each $\sqrt{5}$. In this way, we obtain an 8th-degree polynomial with rational coefficients (which, in this case, all happen to be integers) with root a:

$$x^8 - 100x^6 - 80x^5 + 2892x^4 + 3040x^3 - 25{,}920x^2 - 33{,}920x + 33{,}616.$$

In general, each iteration of this procedure doubles the degree of the polynomial, and we end up with a polynomial of degree 2^n for some n. Wantzel argued that this procedure will yield an irreducible polynomial; we will not give his argument here.

Although Wantzel's theorem gives a way of proving that a number is not constructible, it isn't perfect. Being the root of an irreducible polynomial of degree 2^n is a necessary but not sufficient condition. For instance, the irreducible polynomial $x^4 - x^3 - 5x^2 + 1$ has four real roots, none of which are constructible.[17]

The Impossibility Theorems

At this point, Wantzel turned his attention to the problems of antiquity. In what is arguably the greatest single page in the history of mathematics (shown in figure 20.1), he proved that three of the problems are impossible.[18] Newton had his *anni mirabiles* and Einstein had his *annus mirabilis*. Perhaps we should call page 396 of Wantzel's article the *pagina mirabilis*—the miraculous page.

Near the top of the page, Wantzel attacks the first problem: doubling the cube. To prove that this problem is impossible, we need to show that $\sqrt[3]{2}$ is not a constructible number. We proved that $x^3 - 2$ is the minimal polynomial for $\sqrt[3]{2}$. The polynomial has degree 3, which is not a power of 2, so $\sqrt[3]{2}$ is not constructible. That's it! Thus, it is impossible

PURES ET APPLIQUEES.   369

et continuant de cette manière on conclura que $F(x_a)$ s'annulera pour les 2^n valeurs de x_a auxquelles conduit le système de toutes les équations (A) ou pour les 2^n racines de $f(x) = 0$. Ainsi une équation $F(x) = 0$ à coefficients rationnels ne peut admettre une racine de $f(x) = 0$ sans les admettre toutes; donc l'équation $f(x) = 0$ est irréductible.

IV.

Il résulte immédiatement du théorème précédent que tout problème qui conduit à une équation irréductible dont le degré n'est pas une puissance de 2, ne peut être résolu avec la ligne droite et le cercle. Ainsi *la duplication du cube*, qui dépend de l'équation $x^3 - 2a^3 = 0$ toujours irréductible, ne peut être obtenue par la Géométrie élémentaire. Le problème *des deux moyennes proportionnelles*, qui conduit à l'équation $x^3 - a^2b = 0$ est dans le même cas toutes les fois que le rapport de b à a n'est pas un cube. *La trisection de l'angle* dépend de l'équation $x^3 - \frac{3}{4} x + \frac{1}{4} a = 0$; cette équation est irréductible si elle n'a pas de racine qui soit une fonction rationnelle de a et c'est ce qui arrive tant que a reste algébrique; ainsi le problème ne peut être résolu en général avec la règle et le compas. Il nous semble qu'il n'avait pas encore été démontré rigoureusement que ces problèmes, si célèbres chez les anciens, ne fussent pas susceptibles d'une solution par les constructions géométriques auxquelles ils s'attachaient particulièrement.

La division de la circonférence en parties égales peut toujours se ramener à la résolution de l'équation $x^m - 1 = 0$, dans laquelle m est un nombre premier ou une puissance d'un nombre premier. Lorsque m est premier, l'équation $\frac{x^m - 1}{x - 1} = 0$ du degré $m - 1$ est irréductible, comme M. Gauss l'a fait voir dans ses *Disquisitiones arithmeticæ*, section VII; ainsi la division ne peut être effectuée par des constructions géométriques que si $m - 1 = 2^n$. Quand m est de la forme a^n, on peut prouver, en modifiant légèrement la démonstration de M. Gauss que l'équation de degré $(a-1)a^{n-1}$, obtenue en égalant à zéro le quotient de $x^{a^n} - 1$ par $x^{a^{n-1}} - 1$, est irréductible; il faudrait donc que $(a-1)a^{n-1}$ fût de la forme 2^n en même temps que $a-1$, ce qui est impossible à moins que $a = 2$. Ainsi, *la division de la circonférence en N parties ne peut être effectuée avec la règle et le compas que si les facteurs premiers de N différents de 2 sont de la forme $2^n + 1$ et s'ils entrent seulement à la première puissance dans ce nombre.* Ce

FIGURE 20.1. The single greatest page in mathematics? (P. L. Wantzel, 1837, Recherches sur les moyens de reconnaître si un problème de géométrie peut se résoudre avec la règle et le compas, *J. Math. Pures Appl.* 2(1), 369; highlights added)

to double the cube, and more generally it is impossible to find two mean proportionals using only a compass and straightedge.

In fact, there is nothing special about doubling the cube. Because $x^3 - m$ is irreducible whenever m is not a perfect cube, it is impossible to triple a cube, quadruple a cube, and so on.

A few lines later, Wantzel turned his attention to the angle trisection problem. Recall that we can trisect an angle θ if and only if $\cos(\theta/3)$

is a constructible number. Also, recall that we *can* trisect some angles. To prove that the problem is impossible, all we must do is find some angle θ so that $\cos(\theta/3)$ is not constructible.

Our aim is to show that a 60° angle cannot be trisected. We must find the minimal polynomial for $\cos(20°)$. To do this, we use the trigonometric identity we derived in chapter 18: $\cos\theta = 4\cos^3(\theta/3) - 3\cos(\theta/3)$. Plugging in $\theta = 60°$, we see that $\cos(20°)$ is the root of the polynomial[19] $4x^3 - 3x - \cos(60°) = 4x^3 - 3x - 1/2$, or equivalently, a root of $8x^3 - 6x - 1$. We have encountered this polynomial already and proved that it is irreducible. Because the degree is not a power of 2, $\cos(20°)$ is not constructible. And hence, it is impossible to trisect a 60° angle.

We should pause and ask what happens when θ is an angle we *can* trisect—such as 90° or 45°. Where does this argument break down? When $\theta = 90°$, we obtain the polynomial $4x^3 - 3x - \cos(90°) = 4x^3 - 3x = x(4x^2 - 3)$. In this case, the polynomial is reducible: $\cos(30°) = \sqrt{3}/2$ is the root of a quadratic equation. So it is constructible.

Something different—and more interesting—happens when $\theta = 45°$. Now we obtain the polynomial $4x^3 - 3x - \cos(45°) = 4x^3 - 3x - \frac{1}{\sqrt{2}}$, which does not have rational coefficients! So it is not the minimal polynomial for $\cos(15°)$. In this case, $\cos(15°) = (\sqrt{2} + \sqrt{6})/4$ is a root of the irreducible 4th-degree polynomial $16x^4 - 16x^2 + 1$.

The final theorem on Wantzel's page was the impossibility of constructing certain regular polygons. Gauss proved that if $n = 2^k p_1 \cdots p_r$ where the p_i are distinct Fermat primes, then it is possible to construct a regular n-gon. He stated that these were the only constructible regular polygons. Wantzel proved that this was true.

Suppose we are able to construct a regular n-gon, and p is a prime divisor of n. In fact, suppose that p^k is the largest power of p that divides n; say that $n = mp^k$. If we begin at any vertex of the polygon and go around it connecting every mth vertex, we obtain a regular p^k-gon; so, because the n-gon is constructible, the regular p^k-gon is too. Consequently, rather than focusing on the regular n-gon, we can restrict our attention to the regular p^k-gon. We must prove that either $k = 1$ and p is a Fermat prime or that $p = 2$.

Suppose the regular p^k-gon is constructible and that its vertices are the roots of the polynomial $z^{p^k} - 1$ in the complex plane. Gauss proved that when p is prime, $(z^p - 1)/(z - 1) = z^{p-1} + z^{p-2} + \cdots + z + 1$ is irreducible.[20] When we divide $z^{p^k} - 1$ by $z - 1$ we obtain a polynomial, but it may be reducible. However, Wantzel performed a clever trick to obtain an irreducible polynomial: replace z in the equation $\frac{z^p-1}{z-1} = z^{p-1} + z^{p-2} + \cdots + z + 1$ with $z^{p^{k-1}}$. This gives

$$\frac{(z^{p^{k-1}})^p - 1}{(z^{p^{k-1}}) - 1} = (z^{p^{k-1}})^{p-1} + (z^{p^{k-1}})^{p-2} + \cdots + (z^{p^{k-1}}) + 1,$$

which simplifies to

$$\frac{z^{p^k} - 1}{z^{p^{k-1}} - 1} = z^{(p-1)p^{k-1}} + z^{(p-2)p^{k-1}} + \cdots + z^{p^{k-1}} + 1.$$

This looks messy and confusing, but here's a concrete example to illustrate the equality. When $p = 5$ and $k = 3$ the equation is simply

$$\frac{z^{125} - 1}{z^{25} - 1} = z^{100} + z^{75} + z^{50} + z^{25} + 1.$$

The upshot of all this work is that the degree of this polynomial is $(p-1)p^{k-1}$, and because the polynomial was irreducible before we replaced z with $z^{p^{k-1}}$, it is irreducible afterward. Because the regular p^k-gon is constructible, the degree is a power of 2. That is, $(p-1)p^{k-1} = 2^l$ for some l. The only way this can happen is if $p = 2$ or if $k = 1$ and $p - 1 = 2^l$, in which case p is a Fermat prime.[21]

And with that, we reach the end of Wantzel's page, and the end of three of the four problems of antiquity: it is impossible to double the cube, to trisect all angles, and to construct every regular polygon.

Another Proof for Polygons

Before moving on, let's give another proof of the theorem for polygons—one that uses Wantzel's theorem and in which it is a little easier to see what is going on. We saw that if we could construct $\cos(2\pi/n) + i\sin(2\pi/n)$, which is the vertex adjacent to the vertex corresponding to 1, then we could construct the entire n-gon. It turns out that this vertex is not the only vertex we could use to construct the polygon.

The vertices of a regular n-gon are all nth roots of unity, but not all nth roots of unity are equivalent. Consider the 6th roots of unity,

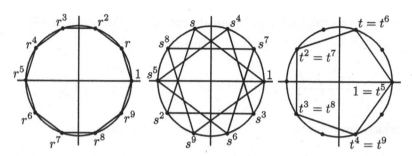

FIGURE 20.2. The primitive 10th roots of unity: r and s generate all vertices of the regular 10-gon.

± 1 and $(\pm 1 \pm i\sqrt{3})/2$. Some of these values are kth roots of unity for k smaller than 6. The value -1 is a square root of unity and a 4th root of unity. And $(-1 \pm i\sqrt{3})/2$ are 3rd roots of unity (they, together with 1, form an equilateral triangle). However, $(1 \pm i\sqrt{3})/2$ are 6th roots of unity but not kth roots of unity for any k smaller than 6.

A complex number z is a *primitive nth root of unity* if $z^n = 1$, but $z^m \neq 1$ for any m smaller than n. For example, $\pm i$ are primitive 4th roots of unity, and $(1 \pm i\sqrt{3})/2$ are primitive 6th roots of unity. A primitive nth root of unity z can generate all the nth roots of unity by taking powers: $z, z^2, \ldots, z^{n-1}, z^n = 1$. The primitive 4th root of unity i generates $i^2 = -1$, $i^3 = -i$, and $i^4 = 1$. In figure 20.2 we consider three 10th roots of unity, r, s, and t, the first two of which are primitive. We see that r and s generate all vertices of the regular 10-gon, but t generates only a regular pentagon. The takeaway is that if we can construct a primitive nth root of unity, then by using it, 1, and a compass, we can construct all vertices of the n-gon.

For a given n, the number of primitive nth roots of unity is precisely the number of integers between 1 and n that are relatively prime to n. This value is denoted $\varphi(n)$, and φ is known as *Euler's phi function* or *Euler's totient function*. For $n = 4$, the numbers 1 and 3 are relatively prime to 4, but 2 is not, so $\varphi(4) = 2$. If p is a prime number, then $1, \ldots, p-1$ are all relatively prime to p, so $\varphi(p) = p - 1$. Table 20.1 shows $\varphi(n)$ for $n = 1, \ldots, 13$.

The following properties of this function make it easy to evaluate:

(1) If p is prime, then $\varphi(p^k) = (p-1)p^{k-1}$.
(2) If m and n are relatively prime, then $\varphi(mn) = \varphi(m)\varphi(n)$.

TABLE 20.1. Euler's phi function

n	1	2	3	4	5	6	7	8	9	10	11	12	13
$\varphi(n)$	1	1	2	2	4	2	6	4	6	4	10	4	12

TABLE 20.2. Cyclotomic polynomials

n	$\varphi(n)$	$\Phi_n(z)$
1	1	$z-1$
2	1	$z+1$
3	2	z^2+z+1
4	2	z^2+1
5	4	$z^4+z^3+z^2+z+1$
6	2	z^2-z+1

As a concrete example,

$$\varphi(5544) = \varphi(2^3 \cdot 3^2 \cdot 7 \cdot 11)$$
$$= \varphi(2^3)\varphi(3^2)\varphi(7)\varphi(11)$$
$$= (2-1)2^2(3-1)3^1(7-1)(11-1)$$
$$= 1440.$$

So 1440 numbers between 1 and 5544 are relatively prime to 5544.

Consider the polynomial that has as its roots all the primitive nth roots of unity. This is called the nth *cyclotomic polynomial*, denoted Φ_n, and it has degree $\varphi(n)$. For example, the two primitive 4th roots of unity are $\pm i$. Thus $\Phi_4(z) = (z+i)(z-i) = z^2+1$. Table 20.2 shows the first six cyclotomic polynomials.

Remarkably, not only are the coefficients of $\Phi_n(z)$ real numbers, they are integers.[22] Moreover, Gauss proved that $\Phi_n(z)$ is irreducible for prime n, and in his diary on June 12, 1808, he claimed he had a proof for composite n: "The equation . . . that contains all primitive roots of the equation $z^n - 1 = 0$ cannot be decomposed into factors with rational coefficients, proved for composite values of n." His proof is lost.[23] The first proof to appear in print is Leopold Kronecker's (1823–1891) from 1854.

We are now ready to give a one-paragraph proof of the converse of Gauss's theorem on constructible polygons: Suppose we can construct a regular n-gon where n has prime factorization $n = p_1^{a_1} p_2^{a_2} \cdots p_k^{a_k}$. Then we can construct a primitive nth root of unity. Its minimal polynomial is the nth cyclotomic polynomial $\Phi_n(z)$, which is irreducible, has integer coefficients, and has degree $\varphi(n)$. By Wantzel's theorem,

$$
\begin{aligned}
\varphi(n) &= \varphi(p_1^{a_1} p_2^{a_2} \cdots p_k^{a_k}) \\
&= \varphi(p_1^{a_1})\varphi(p_2^{a_2}) \cdots \varphi(p_k^{a_k}) \\
&= (p_1 - 1)p_1^{a_1-1}(p_2 - 1)p_2^{a_2-1} \cdots (p_k - 1)p_k^{a_k-1} \\
&= 2^m
\end{aligned}
$$

for some m. One of the primes, p_1 say, could be 2, in which case a_1 can have any value. But for the rest, $p_i - 1$ is a power of 2 and $a_i = 1$; so every prime other than 2 is a Fermat prime appearing at most once.

Some Interesting Consequences

Degree measure for angles is arbitrary, but it's what we are used to. So we may as well ask: Which angles (in integer-degree measure) are constructible? It is impossible to construct a 1° angle. If it were possible, we could construct a regular 360-gon. But $360 = 2^3 \cdot 3^2 \cdot 5$ does not have the form required by the Gauss–Wantzel theorem for constructibility. Similarly, a 2° angle is not constructible, but a 3° angle is, because it is the central angle of a regular polygon with $120 = 2^3 \cdot 3 \cdot 5$ sides.[24] It follows that we can construct any angle of the form $(3k)°$, where k is an integer. However, we cannot construct any angle of the form $(3k + 1)°$ or $(3k + 2)°$; if we were able to, then we could construct a $(3k + 1)° - (3k)° = 1°$ or $(3k + 2)° - (3k)° = 2°$ angle.

We have examples of constructible angles that can be trisected (like a 90° angle) and constructible angles that cannot be trisected (like 60°). It turns out that it is possible to find angles that cannot be constructed, but that if one such angle is given, can be trisected![25] The angle $\theta = (360/7)°$ cannot be constructed because it is the central angle of a

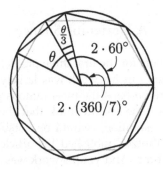

FIGURE 20.3. The angle θ cannot be constructed, but it can be trisected.

regular heptagon. But if we are given θ, we can construct the heptagon in a circle. We can also construct an inscribed regular hexagon so that it shares a vertex with the heptagon (in gray in figure 20.3). Because

$$2 \cdot 60° - 2 \cdot \left(\frac{360}{7}\right)° = \left(\frac{120}{7}\right)° = \frac{1}{3} \cdot \left(\frac{360}{7}\right)° = \frac{\theta}{3},$$

the central angle between the two neighboring vertices of the heptagon and the hexagon shown in the figure trisects the angle θ.

What about an angle that can't be constructed and can't be trisected? Yes, a $(360/42)°$ angle has this property. It can't be constructed because it is the central angle of a $42 = 2 \cdot 3 \cdot 7$-gon, and 7 is not a Fermat prime. The argument that it is not trisectable—even if we start with a $(360/42)°$ angle—is beyond the scope of this book.[26]

It is possible to bisect any angle. It is not possible to trisect every angle. It is possible to divide any angle into four equal parts—just bisect it twice. This prompts us to pose the multisection question: For which n can we divide an arbitrary angle into n equal parts?

By the process of repeated bisecting, we can divide any angle into $n = 2^k$ parts for any k. It turns out that these are the only n that work. Here's why. Suppose we can divide any angle into n equal parts.[27] Then we can divide $360°$ into n equal parts. Now divide these angles into n equal parts. We can use these angles to construct a regular n^2-gon. By the Gauss–Wantzel theorem, we can do this only if n is a power of 2.

A Deafening Silence

The impossibility proofs for the first three problems of antiquity were due to Pierre Wantzel, but his work was largely ignored for almost a century. These were perhaps the most famous problems in the history of mathematics. One would think that their resolution—even if it was a negative resolution—would be notable and newsworthy, especially because Wantzel published his article in one of the premier journals of the time. And yet his work was almost immediately forgotten.

A decade and a half after the proof, prominent mathematicians were unaware of the result. On December 18, 1852, Sir William Rowan Hamilton (1805–1865) wrote to De Morgan:[28]

> Are you *sure* that it is impossible to trisect the angle by Euclid? I have not to lament a single hour thrown away on the attempt, but fancy that it is rather a tact, a feeling, than a proof, which makes us think that the thing cannot be done. No doubt we are influenced by the cubic form of the algebraic equation. But would Gauss's inscription of the regular polygon of seventeen sides have seemed, a century ago, much less an impossible thing, by line and circle?

De Morgan replied on Christmas Eve:

> As to the trisection of the angle, Gauss's discovery increases my disbelief in its possibility. When $x^{17} - 1$ is separated into quadratic factors, we see how a construction by circles may tell. But, it being granted $ax^3 + bx^2 + cx + d$ is *not* separable into a real quadratic and a linear factor, I cannot imagine how a set of intersections of circles can possibly give no more or less than three distinct points.

Complicating matters, the Danish mathematician Julius Petersen (1839–1910) proved Wantzel's theorem in an algebra textbook in 1877,[29] but he did not mention Wantzel. However, he knew about Wantzel's work, because he mentioned it in his PhD thesis.

In 1897 Felix Klein (1849–1925) wrote a book called *Famous Problems of Elementary Geometry: The Duplication of the Cube, the Trisection of an Angle, the Quadrature of the Circle*. In the introduction he wrote,[30]

> [The proof of the impossibility of the duplication of the cube and the trisection of an arbitrary angle] is implicitly involved in the Galois theory as presented to-day in treatises on higher algebra. On the other hand, we find no explicit demonstration in elementary form unless it be in Petersen's textbooks.

Klein did not mention Wantzel. Moreover, further muddying the water he incorrectly credited Gauss with the proof of the impossibility of constructing all regular polygons:[31]

> In his *Disquisitiones Arithmeticae*, Gauss extended this series of numbers $[2^h, 3, \text{and } 5]$ by showing that the division is possible for every prime number of the form $p = 2^{2^\mu} + 1$ but impossible for all other prime numbers and their powers.

In 1914 Raymond Archibald (1875–1955) reviewed two books on this topic. In the review of Hobson's *"Squaring the Circle": A History of the Problem*[32] (which stated Wantzel's theorem, but did not mention Wantzel by name) he wrote, "Who first proved the impossibility of the classic problem of trisection of an angle? I have not met with a statement of this fact in any of the mathematical histories, but surely it was before 1852, when Sir William Rowan Hamilton wrote to De Morgan."[33]

In his review of Klein's book, Archibald wrote,[34]

> Now the implication referred to above [that Gauss proved the converse] is not correct, as Professor Pierpont interestingly sets forth in his [1895] paper "On an undemonstrated theorem of the *Disquisitiones Arithmeticæ*."

James Pierpont (1822–1893) did squash the Gauss misinformation, but he did not give Wantzel credit. Instead, he gave his own proof. He wrote,[35]

> It is, however, vastly more important to know that *only* these polygons can be geometrically constructed as thereby the theory of regular polygons, as far as their construction by straight-edge and compasses is concerned, is complete.... [Our proof] will be of interest as filling up a *lacuna* which all readers of the Disquisitiones must have felt.

Many mathematics books (even books devoted to the history of mathematics) in the late nineteenth and early twentieth centuries discussed the classical problems but did not include their eventual solutions. When they did, they often referred to the eventual solution of the problem of circle squaring. Often they misattributed the polygon proof to Gauss. As for the proofs of impossibility for angle trisection and the doubling of the cube—they either didn't know whether it had been proved, didn't know who gave the first proof, or misattributed the proof. Many authors simply cited earlier texts that gave the proof, but with no explicit claim that it was the first.

Finally, in 1913 Florian Cajori set the record straight:[36]

> Quite forgotten are the proofs given by Wantzel of three other theorems of note, viz., the impossibility of trisecting angles, of duplicating cubes, and of avoiding the "irreducible case" in the algebraic solution of irreducible cubics. For these theorems Wantzel appears to have been the first to advance rigorous proofs.... So far as now known, Wantzel's priority in publishing detailed, explicit and full proofs... is incontested.

Nevertheless, the damage was done. The misinformation was already widely spread by that point. Well-meaning authors could easily miss Cajori's piece and instead get their information from one of the many incorrect sources. They could then pass on the misinformation.

For instance, in 1937, 100 years after Wantzel's proof, E. T. Bell published his popular book *Men of Mathematics*. It is an entertaining and inspiring tale of some of history's greatest mathematicians—although the book has been widely criticized for its focus only on male mathematicians and for its overly dramatic, and sometimes faulty, accounts of the history of mathematics. In it, Bell wrote,[37]

> The young man proved that a straightedge and compass construction of a regular polygon having an odd number of sides is possible when, and only when, that number is either a prime Fermat number or is made up by multiplying together different Fermat primes.... His name was Gauss.

No doubt this assertion in this influential book reinvigorated the belief that Gauss proved these theorems.[38]

Repeatedly throughout the twentieth century—even as late as 1990—mathematicians and historians of mathematics overlooked Wantzel and his contributions.[39] In 1986 Richard Francis wrote the following about the polygon theorem, but it could apply to all of Wantzel's theorems:[40]

In this day of swift communication, a worldwide mathematical community, and an abundance of research journals, such confusion about a current problem's status is hard to understand. However, rumor, reinforced by many years of acceptance and enthusiastic exaggeration, dies hard.

In short, for over a century and a half, there was widespread confusion about who proved what, when, or if at all. Some mathematicians did not know whether there was a proof, some thought we'd known it for many years, others attributed all the results to Gauss, and others attributed the polygon result to Gauss. But almost no one gave Wantzel credit.

One can imagine going back and asking mathematicians about the impossibility theorems. These might be some of their reactions:

Haven't we always known this? The impossibility of these constructions had been confidently asserted for 2000 years. Ever since the Greeks there was a widespread belief that they were impossible. When Pappus classified problems as plane, solid, and linear, he was asserting that solid and linear problems could not be solved with a compass and straightedge. It was ingrained in the fabric of every mathematician's belief system.

Didn't Descartes prove this? Descartes took it up a notch. He extended Pappus's classification system and gave sophisticated-sounding arguments about what was possible and impossible. But his mathematics was confusing, convoluted, and in the end, far from rigorous.

Didn't Gauss prove this? Gauss did indeed discover the necessary and sufficient conditions for the constructibility of regular polygons. But he did not prove both directions. There's a well-established history of Gauss making bold assertions about ideas he knew but hadn't published, and the details would later be discovered among his notes. This, it seems, was not one of those cases—we have found no evidence that he wrote down a proof of the converse of his polygon theorem.

Gauss did not weigh in at all about the angle trisection problem or the problem of doubling the cube. However, if Gauss had proved the converse of his polygon theorem, then it would have implied that it is impossible to construct the regular nonagon. Because the nonagon is not constructible, neither is a 40° angle. But a 120° angle is constructible, so he would also have been only a few steps away from proving the impossibility of angle trisection. But he did not do so.

Wasn't it a trivial extension of Gauss's work? Wasn't it merely a formality to write it all down? Jespar Lützen, who wrote an excellent article called "Why was Wantzel overlooked for a century? The changing importance of an impossibility result," argued that this was not the case. Wantzel's "algebraic characterization of the problems constructible with ruler and compass were lifted directly from Gauss, and the discussion of irreducibility followed ideas in Abel's work. However, these novel algebraic techniques were probably not so well assimilated in Wantzel's day that his readers would consider his proof as trivial."[41]

Although Descartes and others who came after him were able to translate the trisection problem and the problem of doubling the cube into algebraic equations, it was not until Wantzel that complete end-to-end translation of the geometric problems into an algebraic problem happened.

There is no question that Wantzel—as Newton famously said of himself—stood on the shoulders of giants. But it is not clear that those giants could have (or at least there is no evidence that they did) reach the heights that Wantzel did.

Who cares? Lützen suggested that Wantzel may have been ahead of his time. Wantzel proved his theorem before the heyday of impossibility results. At the time he proved his theorem, many mathematicians were still working under a "constructive paradigm"—the emphasis was on solving problems. Moreover, Lützen wrote that "the method of using algebra to *prove* a theorem in geometry would certainly appear as a very unnatural and backward idea to most early modern mathematicians. The fact that the idea did occur must therefore be valued as a great step forward, not just as a trivial matter of course."[42]

A meta-mathematical impossibility theorem like Wantzel's was not in vogue. As Lützen pointed out, Gauss did not want to spend a few extra pages in his book justifying the impossibility of constructing certain regular polygons. Instead, he simply warned the readers not to waste their time trying. He argued that Wantzel's style of mathematics, which would be so popular in the second half of the nineteenth century, was, in the 1930s and 1940s, "a young man's game."[43]

Pierre who? Unfortunately, because of Wantzel's career choices and because of his untimely death, he was not widely known in mathematical circles. He never became part of the mathematical establishment. For instance, he was not elected to the French Académie des Sciences.

Wantzel was a busy engineering student when he proved his theorem. And later he threw himself into his work at the École Polytechnique. Perhaps he did not "play the game" with the other mathematicians of the era—meeting and corresponding with other academics, sharing copies of his publication, speaking about his work, and so on.

We don't know why Wantzel and his work went unnoticed for so long. Perhaps it was a perfect storm of all these reasons and they all played together to enable his work to disappear from public view. But the theorems were not proved by Pappus, Descartes, Gauss, or anyone else before Wantzel—as Wantzel knew. He wrote, "It seems to us that it has not been demonstrated rigorously until now that these problems, so famous among the ancients, are not capable of a solution by the geometric constructions they valued particularly."[44]

TANGENT
What Can We Construct with Other Tools?

> If they be two, they are two so
> As stiff twin compasses are two,
> Thy soul the fixed foot, makes no show
> To move, but doth, if the other do.
>
> And though it in the centre sit,
> Yet when the other far doth roam,
> It leans, and hearkens after it,
> And grows erect, as that comes home.
>
> Such wilt thou be to me, who must
> Like th' other foot, obliquely run.
> Thy firmness makes my circle just,
> And makes me end, where I begun.
> —From John Donne's ca. 1609 poem "A Valediction:
> Forbidding Mourning"[1]

PAPPUS CLASSIFIED GEOMETRIC problems as plane (those that can be solved with a compass and straightedge), solid (can be solved using conic sections), and linear (all else). This book has largely focused on understanding the plane problems. But we have seen a variety of other geometric construction tools. What can we construct using those?

Compass and Straightedge

The set of numbers constructible with a compass and straightedge consists of those that can be written using the integers, the four arithmetic operations, and square roots. As a consequence, it is impossible, using only a compass and straightedge, to trisect every angle and to find

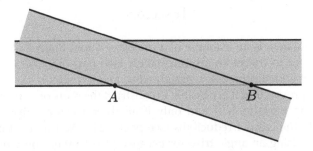

FIGURE T.36. Four of the seven lines we can draw from these two points using a double-edged straightedge.

two mean proportionals. Moreover, it is possible to construct a regular n-gon if and only if $n = 2^i p_1 \cdots p_r$, where the p_k are distinct Fermat primes (prime numbers of the form $2^{2^k} + 1$).

We've encountered some tools that are equivalent to the compass and straightedge. We can construct every Euclidean point if we have a collection of toothpicks (see page 67); a rusty compass and a straightedge (page 178); only a compass (page 183); or one circle and a straightedge (page 188).

As another construction tool, consider an ordinary straightedge that has two parallel edges. What if we were free to use *both* edges during a construction? There are as many as four ways to align such a straightedge along two points. We can always align one edge with both points, and there are two ways to do this. If the distance between the points is greater than the width of the straightedge, we can align the straightedge so that one point is on each edge. There are two ways to do this (or one if the distance between the points is equal to the width of the straightedge). Figure T.36 shows two configurations. We can draw a line along either edge of the straightedge, so we can draw as many as seven lines from these two points. The reader is encouraged to try performing various geometric constructions using this tool.[2] In 1822 Poncelet proved that if we have two points farther apart than the width of a double-edged straightedge, then a point is constructible if and only if it is Euclidean constructible.[3]

Trisectors

The tomahawk is an example of a *trisector*—a tool that would enable a geometer to trisect an arbitrary angle (see page 55). What can we construct with a compass, a straightedge, and a trisector?

Viète proved that it is possible to construct the roots of a cubic equation using these tools if and only if all three roots are real; this is precisely the casus irreducibilis (see page 224). We cannot duplicate the cube using an angle trisector because $x^3 - 2 = 0$ has only one real root. In fact, a further consequence of Viète's work is that the numbers that require a trisector and those that require a "cube rooter" (that is, that require being able to construct two mean proportionals) are disjoint![4] That is, if a number is constructible with a trisector and is constructible with a cube rooter, then it must be constructible without either.

A trisector gives us a collection of newly constructible regular polygons. It allows us to triple the number of sides of an existing polygon, so we can turn an equilateral triangle into a nonagon. But we can do more. Euclidean constructions are tied to Fermat primes, but with a trisector, we are introduced to a new class of primes: *Pierpont primes*,[5] prime numbers of the form $2^s 3^t + 1$. For instance, 7, 13, and 19 are Pierpont primes but not Fermat primes. With a compass, a straightedge, and an angle trisector we can construct a regular n-gon if and only if $n = 2^i 3^j p_1 \cdots p_r$ where the p_k are distinct Pierpont primes.[6]

While there are only five known Fermat primes, there are many known Pierpont primes. There are 18 less than 1000: 2, 3, 5, 7, 13, 17, 19, 37, 73, 97, 109, 163, 193, 257, 433, 487, 577, and 769. We know some that are quite large, such as the 1,529,928-digit[7] $3 \cdot 2^{5,082,306} + 1$. We conjecture that there are infinitely many.

Andrew Gleason observed that if, in addition to a compass and straightedge, we have a tool that can divide any angle into p equal parts for each prime that divides $\varphi(n)$ (where φ is Euler's phi function, which we saw on page 332), then it is possible to construct a regular n-gon. So we could use a quintisector, which can divide an angle into five equal parts, to construct regular 11-, 41-, and 101-gons.[8]

Conic Sections

What if, in addition to lines and circles, we include conic sections?[9] Pappus gave a proof, which likely dates back to the time of Euclid, that it is possible to trisect an angle using a hyperbola,[10] and Menaechmus proved that we can construct two mean proportionals using a parabola and a hyperbola or with two parabolas (see page 75). So being able to use conics is at least as good as having a compass, a straightedge, a trisector, and a cube rooter. Conversely, Viète proved that we can use these tools to construct the real roots of any polynomial of degree 4 or less. So we can solve any problem that is solvable with conics. Thus, Pappus's solid problems are precisely those that can be solved with a trisector and a cube rooter.

Carlos Videla determined exactly the set of conic-constructible points viewed as points in the complex plane.[11] He gave several characterizations, one of them being that it is the smallest field of complex numbers that is closed under square roots, cube roots, and complex conjugation. As an interesting side note, observe that taking the cube root of a complex number is a more complicated process than taking the cube root of a real number. If $z = r(\cos\theta + i\sin\theta)$ is a complex number, then it has three cube roots, one of which is $\sqrt[3]{r}(\cos(\theta/3) + i\sin(\theta/3))$. In particular, taking the cube root of a complex number requires taking the cube root of a real number and trisecting an angle.

One would suspect that gaining the ability to take cube roots in addition to the ability to trisect angles would add a host of new constructible polygons. However, that isn't the case! In 1895 Pierpont determined exactly which regular n-gons can be constructed using a compass, a straightedge, and conic sections, and they are the same ones that are constructible with a trisector, which we described in the previous section.[12] Of course, this is chronologically backward; Pierpont gave his proof for conic-constructible polygons, and almost 100 years later Gleason proved that these could be constructed using a trisector.

It turns out that geometric constructions using conic sections are equivalent to other seemingly different construction methods. They are equivalent to constructions using the Mira (see page 316),[13] to origami constructions (see page 192),[14] and, as we shall see in the next section, to a certain class of neusis constructions.

Marked Straightedge

A marked straightedge allows us to construct more than we could with an unmarked straightedge. Archimedes gave a method of trisecting an angle (see page 149), and Nicomedes showed us that we can find

FIGURE T.37. Each letter n represents a regular n-gon. The X-ed cells are not neusis constructible, the striped cells may be, and the rest are. The black cells are Euclidean constructible, and the gray cells require conic sections.

two mean proportionals (page 151). Thus, we can solve every solid problem using a neusis construction. But can we solve any of Pappus's linear problems—problems that cannot be solved using conic sections? To answer that, we need to take a closer look at our rules.

The neusis technique allows us to construct a line through a point along which the distance between two curves is some fixed distance. If our constructions were made with a compass and straightedge, then these two curves could be two lines, a line and a circle, or two circles.

If we are allowed to perform the neusis construction only between two lines, then we can still trisect any angle and find two mean proportionals. (Archimedes's angle trisection was a line–circle neusis construction, but recall that Pappus gave a line–line neusis trisection [see page 149].) So we can solve any solid problem using line–line neusis constructions. And as it turns out, that is all we are able to do.[15]

However, the standard assumption is that we allow line–circle neusis constructions, such as Archimedes's trisection, and circle–circle neusis constructions. And in this case, we can construct more than we can with conics. Elliot Benjamin and Chip Snyder gave an exact characterization of the set of constructible numbers if we allow line–line and line–circle neusis constructions (they call these *restricted marked ruler and compass constructible numbers*) and showed that it is strictly larger than the set of conic-constructible numbers.[16] We do not have a full characterization of neusis-constructible numbers, although Arthur Baragar gives necessary conditions for a number to be neusis constructible.[17] It is unknown whether circle–circle constructions allow us to construct more than line–circle constructions do. Nevertheless, we can use neusis constructions to solve some of Pappus's linear problems.

As a concrete example, consider the polynomial

$$x^5 - 4x^4 + 2x^3 + 4x^2 + 2x - 6,$$

which has three real roots. Baragar proved that the roots are neusis constructible but they are not expressible by radicals and thus are not conic constructible.

We can also construct polygons using neusis constructions that we can't construct using conics. Benjamin and Snyder showed that a regular 11-gon (hendecagon) is neusis constructible, but because it is not a Pierpont prime, it is not constructible using conics.[18] Baragar proved that there are some regular n-gons that cannot be constructed using neusis, and he gave others that were possibly constructible. Figure T.37 collects the information on constructible polygons.

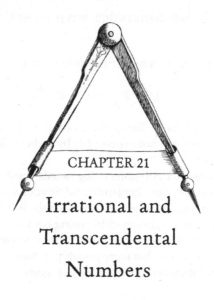

CHAPTER 21

Irrational and Transcendental Numbers

You intended to devote an entire year to the study of the religious problem
and the summer months of 1882 to square the circle and win that million.
Pomegranate! From the sublime to the ridiculous is but a step.
—James Joyce, *Ulysses*, Virag to Bloom[1]

TO PROVE THAT it is impossible to square the circle, we need to
show that π is not a constructible number. By Wantzel's theorem,
we must prove that it is not the root of an irreducible polynomial of
degree 2^k for any k. For the other three problems, we showed that the
associated number was the root of an irreducible polynomial of some
other degree. In this case, we must do something different: prove that
π is not the root of *any* polynomial with integer coefficients. Such a
number is called *transcendental*.

Transcendental numbers are extremely abundant on the real number line. In a sense we can make precise, more real numbers are
transcendental than are not. Nevertheless, it is very difficult to prove
that any given number—like π—is transcendental. This is why the
problem of squaring the circle was the last to fall.

However, in order to understand transcendental numbers, we must first understand irrational numbers. So we must finish our conversation about these numbers that began with Hippasus of Metapontum and the Pythagoreans in chapter 4.

Irrational Numbers

It is very common to come across books asserting that the Greeks—Hippasus of Metapontum in particular—*discovered* irrational numbers. Recall that Hippasus showed that the side and diagonal of a square (or a regular pentagon) are incommensurable magnitudes. Today we see this as proof that there exist irrational numbers. But the Greeks and many future generations of mathematicians did not view the discovery that way. His discovery was a fact about magnitudes, not numbers. It took centuries for irrational numbers to be called numbers.

Irrational numbers arise naturally when solving quadratic equations. The roots of the equation $x^2 - 6x + 4$ are the two irrational numbers $3 \pm \sqrt{5}$. Al-Khwārizmī referred to irrational numbers as "inaudible" numbers, which was later translated to the Latin *surdus*, which means "deaf" or "mute." From this we obtained the term surd, which is still used occasionally, although typically only for the square roots of nonsquare integers. He also treated irrationals as number-like quantities and performed computations like

$$\left(20 - \sqrt{200}\right) + \left(\sqrt{200} - 10\right) = 10.$$

The algebraic techniques of Arabic mathematicians were adopted by the Europeans, although there was general unease about how they fit into the concept of number—accepting negative numbers was enough of a struggle. In the early sixteenth century, mathematicians like Luca Pacioli (1445–1517), Michael Stifel (ca. 1486–1567), Simon Stevin, and Cardano used irrational numbers freely. They helped solve problems that were otherwise unsolvable. However, mathematicians were still not ready to accept them as bona fide numbers. Stifel summed it up well in his influential algebra text, *Arithmetica integra* (1544):[2]

Since, in proving geometrical figures, when rational numbers fail us irrational numbers take their place and prove exactly those things which rational numbers could not prove . . . we are moved and compelled to assert that they truly are numbers, compelled that is, by the results which follow from their use—results which we perceive to be real, certain, and constant. On the other hand, other considerations compel us to deny that irrational numbers are numbers at all. To wit, when we seek to subject them to numeration [decimal representation] . . . we find that they flee away perpetually, so that not one of them can be apprehended precisely in itself. . . . Now that cannot be called a true number which is of such a nature that it lacks precision. . . . Therefore, just as an infinite number is not a number, so an irrational number is not a true number, but lies hidden in a kind of cloud of infinity.

But in the words of John von Neumann, "In mathematics, you don't understand things. You just get used to them."[3] By the end of the sixteenth century, mathematicians were routinely taking square roots of nonsquare numbers, and many were treating them as legitimate numbers. They ignored the stubborn subtleties that were showcased in book X of *Elements*. Bos wrote that proclaiming "the uselessness of *Elements* X became a kind of partisan slogan of those who favored the use of irrational numbers to simplify matters in geometry."[4]

In his 1585 book *Arithmetic*, Stevin forcefully argued that the system of numbers, like geometric magnitudes, were continuous quantities. He argued that values which could (such as $\sqrt{2}$) and could not (such as $\sqrt[3]{2}$) be constructed geometrically are valid numbers. He wrote, "There are no absurd, irrational, irregular, inexplicable, or surd numbers."[5]

Nevertheless, even in the seventeenth century, mathematicians like Pascal and Barrow asserted that numbers such as $\sqrt{2}$ can be understood only geometrically—$\sqrt{2}$ is merely a collection of symbols to represent a magnitude, such as the diagonal of a square with unit-length sides. This was also the view Newton espoused in *Universal Arithmetick*. He presented numbers as abstract quantities, yet tied them to magnitudes:[6]

By a "number" we understand not so much a multitude of units as the abstract ratio of any quantity to another quantity of the same kind which is considered to be unity. It is threefold: integral, fractional, and surd. An integer is measured by unity, a fraction by a multiple part of unity, while a surd is incommensurable with unity.

A correct understanding of the real number line was a necessary step in making calculus fully rigorous. Newton, Leibniz, and their seventeenth-century contemporaries deduced the basic theorems of calculus, but the proofs were built upon a shaky foundation of undefined concepts such as "infinitesimals." In his 1734 book *The Analyst*, Bishop George Berkeley (1685–1753) famously pointed out the lack of rigor in Newton's calculus:[7]

> What are these . . . evanescent increments? They are neither finite quantities nor quantities infinitely small, nor yet nothing. May we not call them the ghosts of departed quantities?

Mathematicians in the eighteenth century—most notably Euler— pushed calculus much further ahead. But there were still unresolved questions of the rigorous underpinnings. It was in the nineteenth century that mathematicians tied up the loose ends and came up with the definitions and theorems that can be found in today's textbooks. As math historian Judith Grabiner so succinctly stated about the derivative from calculus, it "was first used, then discovered, explored and developed, and only then, defined."[8]

Georg Cantor and Richard Dedekind (1831–1916) independently devised the first rigorous constructions of the real numbers. Their first articles on this topic were both published in 1872. The main ingredient that was previously missing was the notion of *completeness*, which says roughly that there are no "holes" in the real number line. Or, stated another way, the set of rational numbers is full of holes, and Cantor and Dedekind described how to fill the holes.

e Is Irrational

By the eighteenth century, the only known irrational numbers were those expressed using nth roots. The number e was the next number proved to be irrational.

In 1737 Euler gave the proof in an important and extensive article on continued fractions.[9] We will mention only one fact about this rich and beautiful method of expressing numbers: the simple continued fraction expansion of a number terminates if and only if it is a rational number. For example, the continued fraction for the rational number $43/30$ terminates and for $\sqrt{2}$ does not:

$$\frac{43}{30} = 1 + \cfrac{1}{2 + \cfrac{1}{3 + \cfrac{1}{4}}} \quad \text{and} \quad \sqrt{2} = 1 + \cfrac{1}{2 + \cfrac{1}{2 + \cfrac{1}{2 + \cdots}}}.$$

(This result is similar to the familiar fact that a number is rational if and only if its decimal expansion terminates or eventually repeats.)

So one way to prove that a number is irrational is to show that its continued fraction expansion is infinite. Euler did just that, showing that e has a stunningly beautiful form that is not repeating, but has a clear pattern with infinitely many terms. It begins

$$e = 2 + \cfrac{1}{1 + \cfrac{1}{2 + \cfrac{1}{1 + \cdots}}}$$

and then continues with 1, 4, 1, 1, 6, 1, 1, 8, 1, 1, 10, and so on.

As wonderful as this continued fraction is, the justification is complicated. Today it is more common to present the proof discovered by Joseph Fourier (1768–1830) in 1815. The key ingredient in his proof is the infinite series $e = 1 + 1/1! + 1/2! + 1/3! + \cdots$. Apparently, Fourier told the proof to Louis Poinsot (1777–1859), and Poinsot told Janot de Stainville (1783–1828), who put it in his book *Mélanges d'analyse algébrique et de géométrie*.[10] We will give the proof because the method of proof is similar to the others we will mention in this chapter.

It is a proof by contradiction. Assume that e is rational; that is, we can write it as h/k where h and k are positive integers. In fact, because e is not an integer, $k > 1$. Now define a new number x as

$$x = k! \left(e - 1 - \frac{1}{1!} - \frac{1}{2!} - \cdots - \frac{1}{k!} \right).$$

To obtain a contradiction, we will show that x is an integer strictly between 0 and 1. Because this is impossible, we will obtain our contradiction.

First, replace e with h/k, then multiply through by $k!$:

$$x = k! \left(e - 1 - \frac{1}{1!} - \frac{1}{2!} - \cdots - \frac{1}{k!} \right)$$

$$= k! \left(\frac{h}{k} - 1 - \frac{1}{1!} - \frac{1}{2!} - \cdots - \frac{1}{k!} \right)$$

$$= \frac{k! \cdot h}{k} - k! - \frac{k!}{1!} - \frac{k!}{2!} - \frac{k!}{3!} - \cdots - \frac{k!}{k!}.$$

The denominator in each fraction cancels with terms in the numerator. For instance,

$$\frac{k!}{3!} = \frac{k \cdot (k-1) \cdot \ldots \cdot 5 \cdot 4 \cdot \cancel{3} \cdot \cancel{2} \cdot \cancel{1}}{\cancel{3} \cdot \cancel{2} \cdot \cancel{1}} = k \cdot (k-1) \cdot \ldots \cdot 5 \cdot 4.$$

In particular, each term is an integer, so x is an integer.

Next we return to our original expression for x, replace e with its series representation and do some canceling:

$$x = k! \left(e - 1 - \frac{1}{1!} - \frac{1}{2!} - \cdots - \frac{1}{k!} \right)$$

$$= k! \left(\left(1 + \frac{1}{1!} + \frac{1}{2!} + \cdots \right) - 1 - \frac{1}{1!} - \frac{1}{2!} - \cdots - \frac{1}{k!} \right)$$

$$= k! \left(\frac{1}{(k+1)!} + \frac{1}{(k+2)!} + \frac{1}{(k+3)!} + \cdots \right).$$

Because each term is positive, x is positive.

Next, multiply through by $k!$, and cancel terms:

$$x = k! \left(\frac{1}{(k+1)!} + \frac{1}{(k+2)!} + \frac{1}{(k+3)!} + \cdots \right)$$

$$= \frac{k!}{(k+1)!} + \frac{k!}{(k+2)!} + \frac{k!}{(k+3)!} + \cdots$$

$$= \frac{1}{k+1} + \frac{1}{(k+1)(k+2)} + \frac{1}{(k+1)(k+2)(k+3)} + \cdots.$$

Notice that $k + n > k + 1$ for all $n > 1$, so $\frac{1}{k+n} < \frac{1}{k+1}$ for all $n > 1$. Hence,

$$x = \frac{1}{k+1} + \frac{1}{(k+1)(k+2)} + \frac{1}{(k+1)(k+2)(k+3)} + \cdots$$

$$< \frac{1}{k+1} + \frac{1}{(k+1)(k+1)} + \frac{1}{(k+1)(k+1)(k+1)} + \cdots$$

$$= \frac{1}{k+1} + \left(\frac{1}{k+1}\right)^2 + \left(\frac{1}{k+1}\right)^3 + \cdots$$

$$= \frac{\left(\frac{1}{k+1}\right)}{1 - \left(\frac{1}{k+1}\right)} \qquad \text{by the geometric series formula}$$

$$= \frac{1}{k}$$

$$< 1.$$

Thus $0 < x < 1$, and we obtain our desired contradiction. We conclude that e is irrational.

π Is Irrational

Edward Titchmarsh (1899–1963) famously wrote, "It can be of no practical use to know that π is irrational, but if we can know, it surely would be intolerable not to know."[11]

Did the Greeks suspect that π is irrational? Eutocius wrote of Archimedes's computation of the value of π, "As I have said many times . . . it is not possible exactly to find the straight line equal to the circumference of the circle by the things said here."[12] It seems like he is asserting the irrationality of π, but Knorr argued that he may be saying something weaker—that Archimedes's technique, which requires repeated approximations, will not yield the value of π.

In this excerpt, the twelfth-century Rabbi Maimonides (1135–1204) seems to imply that π is irrational:[13]

> You ought to know that the ratio of the diameter of the circle to its circumference is unknown, nor will it ever be possible to express it precisely. This is not due to any shortcoming of knowledge on our part, as the ignorant think. Rather, this matter is unknown due to its nature, and its discovery will never be attained.

In his *Introductio*, Euler wrote, "Then it is obvious that the circumference of this circle cannot be exactly expressed in rational numbers."[14] Although Euler was able to find the continued fraction for e, he was not able to do the same for π—and for good reason. Unlike for e, the simple continued fraction for π does not have any apparent pattern.

Nevertheless, continued fractions were the key to the first proof of the irrationality of π. It came from Johann Heinrich Lambert in 1761.[15] Lambert found the following (nonsimple) continued fraction for the tangent function:

$$\tan x = \cfrac{x}{1 - \cfrac{x^2}{3 - \cfrac{x^2}{5 - \cdots}}} = \cfrac{1}{\frac{1}{x} - \cfrac{1}{\frac{3}{x} - \cfrac{1}{\frac{5}{x} - \cdots}}}.$$

He used this expression to prove that if x is a nonzero rational number, then $\tan x$ is irrational. (He proved analogous theorems for e^x and $\ln x$ as well.) Said another way, if $\tan x$ is a rational number, then x is either 0 or irrational. In particular, $\tan(\pi/4) = \tan(45°) = 1$ is rational, so $\pi/4$, and hence π, is irrational. Lambert wrote,[16]

> Hence *the circumference of the circle does not stand to the diameter as an integer to an integer.* Thus we have here this theorem in the form of a corollary to another theorem that is infinitely more universal. Indeed, it is precisely this absolute universality that may well surprise us.

In his popular 1794 textbook *Éléments de géométrie*, Adrien-Marie Legendre (1752–1833) used continued fractions to prove that π^2 is irrational. Note that π^2 being irrational implies that π is irrational, but not vice versa. (Recall that $\sqrt{2}$ is irrational, but $(\sqrt{2})^2$ is not!)[17]

In 1947 Ivan Niven (1915–1999) published a short, elementary proof that π is irrational.[18] When we say that the proof is elementary, we do not mean that it is easy. Niven packed a lot of mathematics into his one paragraph proof that fills two-thirds of a journal page. But the mathematical tools he used are available to a student who has completed a first-year calculus class. We will give a brief sketch here.[19]

For the sake of contradiction, assume that $\pi = a/b$ is a reduced fraction. Choose an integer n large enough that $\pi^{n+1}a^n/n! < 1$. We are guaranteed to find such an n because in the long run, factorials grow faster than any exponential function. Now define a new function $f(x) = x^n(a - bx)^n/n!$. Because of the way we defined f, it is possible to show that

$$0 < \int_0^\pi f(x) \sin x \, dx < 1.$$

Next define another new function,

$$F(x) = f(x) - f^{(2)}(x) + f^{(4)}(x) - \cdots,$$

where $f^{(j)}$ is the jth derivative of f. Because of all the ingredients that went into making F, then $F(\pi)$ and $F(0)$ are integers, and

$$\int_0^\pi f(x) \sin x \, dx = F(\pi) - F(0),$$

which is the difference of integers, and is hence an integer. Just as with Fourier's proof of the irrationality of e, we end up with an integer strictly between 0 and 1, which is impossible. So π is irrational.

Niven's idea of finding these functions f and F and the integral relating them was inspired by Charles Hermite's (1822–1901) 1873 proof that e is transcendental, which we will discuss shortly.[20] Hermite's technique is very useful in proving numbers are irrational. For example, in Niven's text *Irrational Numbers* he used this technique to show that π, π^2, e^r, and $\cos r$ (for any nonzero rational number r) are irrational. Once we have these results, it is not difficult to show that any trigonometric function with nonzero rational arguments is irrational (for instance, $\sin(1)$, $\cos(3/4)$, and so on), and if r is a positive rational number (not equal to 1), then $\ln r$ is irrational.[21]

Liouville Numbers

Just because a number is irrational does not mean it is not constructible; after all, $\sqrt{2}$ is irrational and constructible. By Wantzel's theorem, we need to show that the number is not the root of a polynomial with integer coefficients of degree 2^k for any k. As we shall see, some numbers are not the root of any polynomial with integer coefficients.

A real number is called *algebraic* if it is the root of a polynomial with integer coefficients. Examples of algebraic numbers are $1/2$, which is the root of $2x - 1$; $\sqrt{2}$, which is a root of $x^2 - 2$; the golden ratio, a root of $x^2 - x - 1$; and the single real root of the quintic polynomial $x^5 - x + 1$, which cannot be expressed with radicals. A real number that is not algebraic is *transcendental*. Every transcendental number is irrational, but not every irrational number is transcendental.

Euler may have been the first mathematician to define transcendental numbers,[22] but his definition is not clearly articulated, and it is likely not equivalent to the modern definition.[23]

The French mathematician Joseph Liouville (1809–1882) was the true pioneer of the realm of transcendental numbers. In an important 1844 theorem, Liouville proved that irrational algebraic numbers are not well approximated by rational numbers.[24] Roughly speaking, the only way a rational number can be close to such a number is by making the denominator very large.

This theorem gives a way to show that *some* numbers are transcendental. If an irrational number can be well approximated by rational numbers, then it must be transcendental. We call such numbers *Liouville numbers*. In 1844 Liouville used this theorem to produce the first transcendental number, which he expressed as a continued fraction.[25]

In 1851 Liouville observed that long strings of zeros make the number easy to approximate with rationals.[26] For instance, $1/2$ is a very good approximation of $a = 0.500000000000001\ldots$ ($a - 1/2$ is approximately 10^{-15}). He used this observation to produce the first decimal description of a transcendental number. His number has all zeros except in the $n!$ places, which are 1s. So there would be a 1 in the places $1! = 1$, $2! = 2$, $3! = 6$, $4! = 24$, and so on: $0.110001000000000000000000010\ldots$.

Here's an intuitive way to see that a number like this one is transcendental. Not only does this number have long strings of zeros, the farther out we look, the longer the strings get. When we square, cube, and raise the number to the nth power, there are still many long strings of zeros. But we want to plug the number into a polynomial and have everything cancel out to get zero. The long strings of zeros, which get longer and longer as we look farther to the right, prevent this from happening for any polynomial, regardless of how large the coefficients and how high the degree. Thus, it is transcendental.[27]

Squaring the Circle

After Lambert proved that π is irrational, he conjectured that it could not be expressed in terms of radicals. A few years later, Euler made the conjecture that π is transcendental (although he did not use the modern definition of transcendental[28]). He wrote, "It appears to be fairly certain that the periphery of a circle constitutes such a peculiar kind of transcendental quantities that it can in no way be compared with other quantities, either roots or other transcendentals."[29]

Legendre wrote, "It is probable that this number π is not even included among algebraical irrational quantities, . . . but a rigorous demonstration of this seems very difficult to find; we can only show that the square of π is also an irrational number."[30]

Then, in the middle of the nineteenth century, Liouville gave his criteria that could be used to show a number is transcendental. Unfortunately, not all transcendental numbers satisfy Liouville's condition.[31] In particular, π does not; it is not a Liouville number. So we need another method to prove that it is transcendental. The difficulty is that the definition of transcendental is a negative statement, just like the definition of irrational—*there does not exist* a polynomial having the number as a root. So, just like a proof of irrationality, a proof that a number is transcendental is difficult and often done by contradiction.

Like π, e is not a Liouville number. Thus, we need to be clever to prove it is transcendental. Charles Hermite was the clever mathematician we were waiting for. Hermite was a talented French mathematician whose name is attached to many mathematical theorems and objects. He published widely in areas such as analysis, algebra, and number theory. He was the author of textbooks that were popular and highly acclaimed. His contemporary Paul Mansion (1844–1919) wrote, "Uncontested, the scepter of higher arithmetic and analysis passed from Gauss and Cauchy to Hermite who wielded it until his death."[32]

In 1873 he published an "epoch-making memoir . . . [that] began a new era."[33] In it, he gave two proofs that e is a transcendental.[34] Although Liouville had discovered infinitely many transcendental numbers, e was the first noncontrived number—the first number that we cared about—to be proved transcendental. We will not give Hermite's proof, but it is similar to Niven's proof that π is irrational (Niven's proof was inspired by Hermite's).

In 1873, the same year that Hermite proved the transcendence of e, a German mathematics student named Ferdinand von Lindemann earned his PhD from Felix Klein in Erlangen. Shortly afterward, he headed abroad to visit mathematicians in England and France. It was during his stop in Paris that Lindemann met Hermite and was able to discuss with him his methods.

Nine years later, in 1882, Lindemann proved that π is transcendental. Some mathematicians have expressed disappointment that

Hermite, who had devised the key ideas for what would eventually be Lindemann's proof, did not prove it. Although Lindemann had a good career, producing 60 PhD students, he was not seen as Hermite's equal. But Hermite was generous, inspiring, and shared his knowledge far and wide through his correspondences. Freudenthal pointed out that this was often to Hermite's detriment, because it allowed others to achieve the important results—such as Lindemann's proof that it was impossible to square the circle—that were "rightly" Hermite's:[35]

> In a sense [his proof that e is transcendental] is paradigmatic of all of Hermite's discoveries. By a slight adaptation of Hermite's proof, Felix Lindemann, in 1882, obtained the much more exciting transcendence of π. Thus, Lindemann, a mediocre mathematician, became even more famous than Hermite for a discovery for which Hermite had laid all the groundwork and that he had come within a gnat's eye of making.

That said, Hermite chose not to pursue the proof. In a letter to Carl Wilhelm Borchardt (1817–1880), Hermite wrote, "I shall risk nothing on an attempt to prove the transcendence of π. If others undertake this enterprise, no one will be happier than I in their success. But believe me, it will not fail to cost them some effort."[36] Thus, he left it open for another mathematician to prove—and that mathematician was Lindemann.

In the introduction to Lindemann's paper, he wrote, "The impossibility of the quadrature of the circle will thus have been established when one has proved that *the number π can never be the root of any algebraic equation of any degree with rational coefficients*. We seek to prove this in the following pages."[37] Thus, Lindemann knew that his proof would put to rest the notorious problem of squaring of the circle.

Lindemann actually proved something stronger.

Lindemann's theorem. If a is a nonzero algebraic number, then e^a is a transcendental number.

For instance, taking $a = 1$, we conclude that e is transcendental. Likewise, e^2, e^3, $1/e$, $e^{\sqrt{2}}$, and e^{ϕ} (where ϕ is the golden ratio) are transcendental. We can also state Lindemann's theorem using logarithms.[38] It says that if b is algebraic and $b \neq 1$, then $\ln b$ is transcendental. So $\ln 2$ and $\ln \phi$ are transcendental. To prove that π is transcendental we need the logically equivalent contrapositive of Lindemann's theorem: if e^a is an algebraic number, then a is either zero or transcendental.

It is important to point out that Lindemann's theorem also applies to complex numbers. In this context "rational numbers" means the same thing as always—they are numbers that can be expressed as quotient of integers. Rather than risking confusion by calling all other complex numbers irrational, we will call those numbers "not rational." The algebraic numbers are still the roots of polynomials with integer coefficients, but they now include numbers like i and $-i$, which are the roots of $x^2 + 1$. All other complex numbers are transcendental.

Now, after thousands of years of hard work by countless mathematicians, we are able to prove that π is transcendental, and hence that it is impossible to square the circle. It is fitting that the key to proving the most famous and most difficult problem of antiquity is the most beloved expression in all of mathematics—Euler's identity: $e^{\pi i} = -1$. Because -1, and hence $e^{\pi i}$, is an algebraic number, Lindemann's theorem tells us that πi is transcendental. However, because i is algebraic, π is transcendental, and it is impossible to square the circle![39] And with that result, the final problem of antiquity was proved impossible.

But this was just the beginning of mathematicians' work on transcendental numbers. In an 1855 article, Karl Weierstrass (1815–1897) praised Lindemann's result as "one of the most beautiful theorems of arithmetic,"[40] and then proved the following generalization of Lindemann's theorem that Lindemann stated and whose proof he sketched.

Lindemann–Weierstrass theorem.[41] If $a_1, a_2, \ldots, a_m$ are distinct algebraic numbers and $b_1, \ldots, b_m$ are nonzero algebraic numbers, then $b_1 e^{a_1} + b_2 e^{a_2} + \cdots + b_m e^{a_m} \neq 0$.

Again, the contrapositive is useful for proving that a number is transcendental: if $b_1 e^{a_1} + b_2 e^{a_2} + \cdots + b_m e^{a_m} = 0$, then either one of the a_i or b_i is transcendental or all the $b_i = 0$. We can use this theorem to prove that if a is a nonzero algebraic number, then $\sin(a)$, $\cos(a)$, and $\tan(a)$ are transcendental. To prove these results we must use the exponential form of the trigonometric functions. Say a is a nonzero algebraic number and $\sin a = b$. Then, rewriting this expression using exponentials, $\sin(a) = \frac{1}{2i}(e^{ia} - e^{-ia}) = b$, or equivalently, $e^{ia} + (-1)e^{-ia} + (-2ib)e^0 = 0$. By the Lindemann–Weierstrass theorem, $-2ib$ must be transcendental, which implies that b is transcendental.

Hilbert's Seventh Problem

In 1900 David Hilbert—a student of Lindemann's and one of the most eminent mathematicians of the era—delivered a lecture before the Second International Congress of Mathematicians in Paris. It began,[42]

> Who of us would not be glad to lift the veil behind which the future lies hidden; to cast a glance at the next advances of our science and at the secrets of its development during future centuries? What particular goals will there be toward which the leading mathematical spirits of coming generations will strive? What new methods and new facts in the wide and rich field of mathematical thought will the new centuries disclose?... The close of a great epoch not only invites us to look back into the past but also directs our thoughts to the unknown future.

Then Hilbert gave what he viewed as the 10 most important unsolved mathematical problems of the day. The written version of his speech contained 23 problems. Ever since, mathematical explorers have set out to conquer these mathematical summits. The problems became so famous that it was often enough to refer to them by number.

Hilbert's seventh problem was to generalize the work of Hermite, Lindemann, and Weierstrass. He wrote,

> Hermite's arithmetical theorems on the exponential function and their extension by Lindemann are certain of the admiration of all generations of mathematicians.... I should like, therefore, to sketch a class of problems which, in my opinion, should be attacked as here next in order....
>
> *The expression α^β, for an algebraic base and an irrational algebraic exponent, e.g., the number $2^{\sqrt{2}}$ or $e^\pi = i^{-2i}$, always represents a transcendental or at least an irrational number.*
>
> It is certain that the solution of these and similar problems must lead us to entirely new methods and to a new insight into the nature of special irrational and transcendental numbers.

Hilbert's conjecture is a stronger version of one given by Euler in 1748 in his *Introductio*,[43] which we may phrase as follows: if a is a nonzero rational number and b is an algebraic number, then a^b is irrational. Euler conjectured that the number $2^{\sqrt{2}}$ is irrational, whereas Hilbert contended that it is transcendental.

Hilbert thought that his seventh problem was a real stumper—one of the toughest in his list. In a 1920 lecture, Hilbert told his audience that he believed he would see a proof of the Riemann hypothesis (his eighth problem) before he died, and the youngest members of the gathered crowd might see a proof of Fermat's last theorem (which was surprisingly missing from his list, but is related to the tenth problem). He did not think anyone in the room would see a proof that $2^{\sqrt{2}}$ is transcendental. His order was completely backward. The Riemann hypothesis is still one of the most infamous unsolved problems in mathematics (there is even a \$1 million bounty for its proof). Fermat's last theorem was proved by Andrew Wiles in 1994. And the audience had to wait only 10 years to learn the true nature of $2^{\sqrt{2}}$.

In 1929 the 23-year-old Russian Aleksander Gelfond (1906–1968) made the first crack in the armor of this challenging problem. He proved that if a is algebraic and not 0 or 1, and $b = \pm i\sqrt{d}$, where d a positive rational number, then a^b is transcendental. In particular, he concluded that $2^{\sqrt{2}i}$ and e^{π} are transcendental. The latter required some trickery using Euler's identity: $e^{\pi} = e^{-i\cdot\pi i} = (e^{\pi i})^{-i} = (-1)^{-i}$. Thus, he had taken down one of Hilbert's two values and almost the second.[44] As a bonus, we see that the real number $i^i = 0.2078795763\ldots$, which we encountered on page 298, is transcendental.

When Gelfond's result was published, Carl Ludwig Siegel (1896–1981) saw how the proof could be extended to b a real quadratic irrational number, thus showing that $2^{\sqrt{2}}$ is irrational. Siegel wrote to Hilbert announcing Gelfond's discovery and his extension. Hilbert was interested only in Siegel's result and wanted him to publish it. Siegel thought Gelfond had done all the hard work, so he did not publish.[45] Rodion Kuzmin (1891–1949) published the first proof in 1930.[46]

An interesting consequence of this result is that it is possible to raise a transcendental number to an irrational power (or an irrational number to a transcendental power) and obtain a rational number:[47]

$$\left(\sqrt{2}^{\sqrt{2}}\right)^{\sqrt{2}} = \sqrt{2}^{\sqrt{2}\cdot\sqrt{2}} = \sqrt{2}^2 = 2.$$

Finally, in 1934, within a few weeks of each other, Gelfond and Theodor Schneider (1911–1988) proved the following general theorem, which implies, for instance, that $2^{\sqrt[3]{2}}$ is transcendental.

Gelfond–Schneider theorem.[48] If a and b are algebraic numbers with $a \neq 0, 1$ and b not rational, then a^b is transcendental.

We know more about transcendental numbers now than we did when Hilbert's problem was solved. In a sequence of articles in 1966 and 1967, Alan Baker (1939–2018) extended this work even further, and for this work he won the Fields Medal in 1970. For instance, he proved that certain products of the form $a_1^{b_1} \cdots a_k^{b_k}$, like $2^{\sqrt{2}} 2^{\sqrt[3]{2}}$ and $2^{\sqrt{2}} 3^{\sqrt{3}}$, are transcendental.[49]

However, before we start feeling too confident in our ability to identify transcendental numbers, let's take a look at a few numbers that are not yet known to be transcendental. Because the sums and products of algebraic numbers are algebraic and $\frac{1}{2}((e - \pi) + (e + \pi)) = e$ is transcendental, we know $e - \pi$ or $e + \pi$ (or both) must be transcendental. But we do not know which one. (Seriously, does anyone believe that either one is algebraic?) Similarly, we don't know whether π^e, e^e, or π^π are transcendental. We have a long way to go.

The Ubiquitous Transcendental Numbers

There are infinitely many whole numbers. There are infinitely many even, odd, and prime numbers. There are infinitely many rational, irrational, transcendental, and algebraic numbers. One might think that there is little more to say about the sizes of these sets. Surprisingly, as we mentioned on page 35, we can say more. In 1874 Georg Cantor proved that infinity comes in different sizes—infinitely many different sizes.[50] This revelation shocked the mathematical community.

The smallest infinite set is called *countably infinite*. We count finite sets by forming a one-to-one relationship between the set and the numbers 1 through n for some n. Children exhibit this every time they count anything. They point at their uneaten carrots one at a time: "One, two, three, four, ..." This is exactly the idea behind countably infinite sets, except that we form a one-to-one relationship between our set and the counting numbers—the positive integers.

Figure 21.1 shows that there are countably many positive even numbers—every counting number is matched up with its double. Figure 21.1 also shows that the nonnegative rational numbers are countably infinite. In this case we list all of the reduced fractions whose numerator and denominator sum to 1 (only 0/1), then those that sum

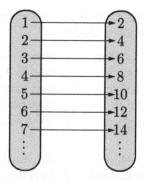 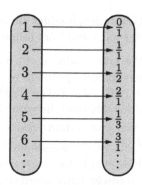

FIGURE 21.1. The positive even numbers and the nonnegative rational numbers are countably infinite.

to 2 (only 1/1), and so on. We skip unreduced fractions like 2/2 and 3/6 because they will have already appeared in the list. Other countably infinite sets include the integers, the odd numbers, the prime numbers, and the rational numbers (positive, negative, and zero). Thus, all countably infinite sets have the same cardinality—they represent the same size of infinity.

Perhaps this conversation is *already* shocking. The integers are very spaced out along the number line whereas the rational numbers are very tightly packed. Between any two real numbers—regardless of how close together—there are infinitely many rational numbers! It is difficult to envision these sets having the same size.

However, once one becomes accustomed to this fact, we face Cantor's next surprise: the set of real numbers is not countably infinite; we say they are *uncountable*.[51] That is, there are so many real numbers, it is impossible to find a one-to-one relationship like those in figure 21.1. Because the real numbers are composed of rational and irrational numbers, this implies that the set of irrational numbers is uncountably infinite. Thus, there are more irrational numbers than there are rational numbers! Cantor went on to prove that there are infinitely many different sizes of infinity, although we will not discuss that here.

So what does this have to do with transcendental numbers? In the same 1874 article, Cantor proved that there are countably many algebraic numbers, and hence uncountably many transcendental numbers. So the overwhelming majority of real numbers are not the root of any polynomial with integer coefficients.

Here's a way to think about this. There are countably many degree-1 polynomials with integer coefficients; that is, we have countably many choices for a_0 and a_1 in $a_0 + a_1 x$. There are countably many polynomials of degree 2 (of the form $a_0 + a_1 x + a_2 x^2$). And so on. Because there are countably many possible degrees, there are countably many polynomials with integer coefficients. Moreover, each polynomial has a finite number of roots; a polynomial of degree n has at most n roots. These roots are precisely the algebraic numbers. Thus, there are countably many algebraic numbers. Because a real number is either algebraic or transcendental, there must be uncountably many transcendental numbers. The vast, vast majority of real numbers are transcendental.

TANGENT
Top 10 Transcendental Numbers

CYRIL: As for fashion, they forswear it,
So they say (so they say);
And the circle—they will square it
Some fine day (some fine day);

—Gilbert and Sullivan, *Princess Ida*[1] (the musical debuted on January 5, 1884, less than two years after squaring of the circle was proved impossible)

EVERYONE LOVES TOP-10 lists. What could be better than a numbered list about numbers? Here are the top 10 transcendental numbers (with a bonus of two that we did not discuss).

(1) $\sum_{k=1}^{\infty} \frac{1}{10^{k!}} = 0.110001000000000000000000010\ldots$ (Liouville, 1851)

(2) e (Hermite, 1873)

(3) π (Lindemann, 1882)

(4) $\ln(2)$ (Lindemann, 1882)

(5) $\sin(1)$ (Weierstrass, 1855)

(6) e^{π} (Gelfond, 1929)

(7) i^i (Gelfond, 1929)

(8) $2^{\sqrt{2}}$ (Siegel and Kuzmin, 1930)

(9) $2^{\sqrt[3]{2}}$ (Gelfond and Schneider, 1934)

(10) $2^{\sqrt{2}} 2^{\sqrt[3]{2}}$ (Baker, 1966)

(11) $\cfrac{1}{1 + \cfrac{1}{2 + \cfrac{1}{3 + \cfrac{1}{\ldots}}}} = 0.6977746579\ldots$ (Siegel, 1929[2])

(12) $0.12345678910111213141516171819120\ldots$ (Mahler, 1937[3]). This is the *Champernowne constant*. Do you see the pattern in the digits?

Sirens or Muses?

The quality of the human mind, considered in
its collective aspect, which most strikes us, in surveying
this record, is its colossal patience.
—Ernest Hobson[1]

WERE THE PROBLEMS of antiquity Sirens or Muses?

In Greek mythology the Sirens were beautiful creatures that played seductive music and sang enchanting songs. Their irresistible melodies drew in sailors, but rather than meeting these alluring Sirens, the sailors' vessels would crash upon the rocky shore.

These alluring problems sang to many mathematicians. They were irresistible to generation after generation of thinkers—from the greatest minds in history to mathematical hobbyists—from Archimedes to Leonardo da Vinci, from Isaac Newton to Thomas Hobbes, from René Descartes to Galileo. Yet all attempts to solve them were destined to fail. The investigators crashed on the shores of impossibility. Even today, after many decades of unambiguous evidence that the problems are impossible, mathematical cranks fail to heed the warnings, continue to become entranced with the Sirens' songs, and crash into the problems' rocky shores.

I argue, however, that the problems were the Muses. The Greeks believed that these goddesses were the sources of inspiration for science, literature, and the arts. Although these classical problems were impossible, they inspired many of the greatest thinkers in history, and from these problems came many other discoveries.

Archimedes, who was likely working on the problem of squaring the circle, made important discoveries about π—relating it to the circumference and area of the circle and to the volume and surface area of the sphere. Descartes built his *Geometry* around the study of such problems. The problem of constructing regular polygons kicked off the mathematical career of the teenage Gauss. The fact that the problems are impossible is not the right way to view them. They were the sources of inspiration that never stopped giving.

They aroused the curiosity of centuries of geometers. What exactly can be accomplished with a compass and straightedge? What can be accomplished if we add more to this geometric toolkit? What happens if we have less? What happens if we have completely different tools?

Mathematicians who were interested in tackling these problems invented new areas of mathematics or brought in seemingly unrelated tools to help them understand the problems—classical geometry, analytic geometry, algebra, complex analysis, and number theory. The problems forced mathematicians to face the difficult question of "What is a number?"—from whole numbers to zero to rational numbers to irrational numbers to negative numbers to constructible numbers to algebraic and transcendental numbers. The acceptance of each type of number was hard fought, but an understanding of these numbers was crucial to cracking the problems.

Even after the proofs of impossibility, the problems lived on, continuing to inspire. Digit hunters devised ever more efficient means of computing digits. After Lindemann's proof that π is transcendental and David Hilbert's list of the greatest mathematical challenges of the twentieth century, the study of transcendental number theory became an active and fruitful area of study.

The study of algebraic numbers was also a high priority for mathematicians. Today the problems of antiquity, together with the study of polynomials of degree 5 and higher, reside in an area of abstract algebra known as *Galois theory*, named for the tragic figure Évariste Galois, the young mathematician who conducted pioneering work in the field. Although it is a cliché to say so, he was "ahead of his time." Galois theory is viewed by many mathematicians as one of the deepest, most beautiful fields of mathematics. It brings together and shows the interconnections between seemingly different areas of mathematics—geometry, field theory, group theory, and algebra.

Who could have known that the compass and the straightedge, two tools that are prized for their simplicity, could be the source of so much majesty. As Robert Yates wrote, "Indeed, it seems that a game built around such scanty outlay would be a disappointing affair. Nothing, however, could be further from the truth. Probably the most fascinating game ever invented, it is awe-inspiring in its extent to the novice, and a thoroughly absorbing occupation to the expert."[2]

Notes

1. Westlake (1990, p. 41)

Introduction

1. Carroll (1917, p. 81)
2. As quoted in Clary (2004, p. 19).
3. The other two postulates are "4. That all right angles are equal to each other. 5. That, if a straight line falling on two straight lines makes the interior angles on the same side less than two right angles, the two straight lines, if produced indefinitely, meet on that side on which are the angles less than two right angles." The fourth postulate gave a uniform standard ("the right angle") throughout the plane, and the fifth one is the famous "parallel postulate" (Heath, 1908a, p. 154).
4. Heath (1931, p. 137)
5. Hobson (1913, p. 12)

Chapter 1. The Four Problems

1. This biblical phrase (Daniel 12:4) is also the inscription on the frontispiece of Francis Bacon's *Instauratio magna*, which features a ship sailing through the pillars of Hercules, the classical boundaries of knowledge.
2. Aristophanes (2000, p. 155). See the explanation in Heath (1921a, pp. 220–21).
3. As quoted in Weik (1922, pp. 140–41).
4. Plato wrote about the problem of doubling the square in *Meno* (82b–85b). He presents a dialogue between Socrates and one of Meno's slave boys. Socrates asks the slave how to find the length of the side of a square that has double the area of a 2-by-2 square. The slave first says 4 units. But Socrates points out that it would have area 16, not 8. Then the boy suggests 3 units. When Socrates points out that this too is incorrect, the slave has no more guesses. Socrates concludes by showing the slave that the diagonal of the original square is the desired segment.
5. There's debate about whether this letter, which is in Eutocius's commentary of Archimedes's *On the Sphere and Cylinder*, is legitimate. See Knorr (1993, pp. 17–24) and Heath (1921a, pp 244–45). This excerpt is in Seidenberg (1961).
6. Heath (1921a, p. 245). Knorr (1993, p. 23) gave a translation in which it isn't as clear that there was an error. The poet may have intended that each dimension be doubled—not the volume. He wrote, "Evidently, Eratosthenes has misconstrued the passage in his desire to find precedents for an interest in this problem."
7. As quoted in Heath (1921a, pp. 245–46).
8. Kazarinoff (1970, p. 28)

9. This argument can be found in Knorr (1993, pp. 22–23).

10. Theon of Smyrna wrote that this story appears in Eratosthenes's work *Platonicus*. This was van der Waerden's conclusion in van der Waerden (1954, p. 161).

11. Seidenberg (1961). See also Seidenberg (1981).

TANGENT: CRANKS

1. From the December 6, 1889, entry of Bassetto (1937, p. 264).
2. De Morgan (1915a, p. 2)
3. De Morgan (1915b, p. 210)
4. P. (1906)
5. National Security Agency (2012)
6. *Time* (1931)
7. *Pittsburgh Press* (1931)
8. See Sanders (1931) for details of Callahan's argument.
9. Here's one way to divide a segment AB into n equal parts (we will show the case $n = 3$). Draw circles with centers A and B and radii AB. They meet at C. Draw the line BC. Use the compass to mark three equal segments on the line BC, each with length BC. In the figure this yields points D and E. Draw AE. Then draw lines through C and D parallel to AE. They meet AB at F and G respectively, which trisect the segment.

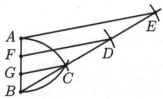

10. Jacob (2005)
11. "Si les Géomètres osaient se prononcer sans des démonstrations absolues, et qu'ils se contentassent de vraisemblances les plus fortes, il y a longtemps qu'ils auraient décidé tout d'une voix que la quadrature du cercle est impossible." (Jacob, 2005)
12. "Ce ne sont pas les Géomètres fameux, les vrais Géomètres qui cherchent la quadrature du cercle: Ils savent trop de quoi il s'agit. Ce sont les demi-Géomètres qui savent à peine Euclide." (Jacob, 2005)
13. Académie des Sciences, France (1778, p. 61)
14. Baez (1998)
15. Caldwell (2017)
16. See Yates (1942, ch. 5) for other profiles of mathematical cranks.
17. Dudley (1962)
18. Dudley (1994, pp. 20–33), Dudley (1983)
19. Dudley (1994, p. 33)
20. Dudley (1992)
21. Dudley et al. (2008)

CHAPTER 2. PROVING THE IMPOSSIBLE

1. Shakespeare (1966, p. 29)

2. Merriam-Webster.com (2017)

3. In 10-pin bowling, a player who throws 12 consecutive strikes (knocking down all 10 pins in one roll) scores a perfect 300 game.

4. In 1980 researchers used a particle accelerator to turn bismuth, which is next to lead in the periodic table, into gold (Aleklett et al., 1981).

5. de Lavoisier (1777)

6. Farrington (1900), Marvin (1996)

7. In 1794 Ernst Chladni gave the first correct description of the origin of meteorites, but there was strong resistance to it, and it had to compete with many other explanations (Marvin, 1996). It took a decade for the skeptics to become converted and more than 150 years before the origin of meteorites was fully understood.

8. To make this precise, we must prove that an integer can't be both even and odd. A number n is odd if there is an integer k such that $n = 2k + 1$. Now suppose, for the sake of contradiction, there is a number n that is both even and odd. Then there exist integers j and k such that $n = 2j$ and $n = 2k + 1$. So $2j = 2k + 1$, which implies that $2(j - k) = 1$. So 2 is a factor of 1, which is a contradiction.

9. Sam Loyd took credit for inventing the 15-puzzle, insisting until his death that he was the creator. Recent scholarship shows that the game was invented by Noyes Palmer Chapman in 1874 (Slocum and Sonneveld, 2006).

10. Johnson and Story (1879)

11. Newcomb (1903)

TANGENT: NINE IMPOSSIBILITY THEOREMS

1. For many years it was believed that Fermat was born in 1601, but new evidence shows that he may have been born in 1607. See Barner (2001).

2. This discussion is the modern way of describing the halting problem. Turing's approach was slightly different. He was writing about his Turing machines (as we now call them), and in his case he wanted his machine to continue forever—halting was bad, not good.

CHAPTER 3. COMPASS-AND-STRAIGHTEDGE CONSTRUCTIONS

1. This text was in an interview with Reverend J. P. Gulliver, which appeared in the *Independent* on September 1, 1864. As quoted in Carpenter (1872, p. 315).

2. See Bulmer-Thomas (2008a) for evidence supporting these speculations.

3. As quoted in Bulmer-Thomas (1976).

4. For the history of the copies and translations of *Elements* see Murdoch (2008).

5. Russell (1967, pp. 37–38)

6. In a letter to William Duane, October 1, 1812 (Jefferson, 2008, pp. 366–68).

7. From January 21, 1812 (Cappon, 2012, p. 291).

8. As quoted in Carpenter (1872, pp. 313–14).

9. This reasoning is backward: the existence of these formulas is due to Archimedes's theorem.

10. Knorr (2004, pp. 84–85)

11. "A *point* is that which has no part" and "a *line* is a breadthless length" are the first two definitions in Euclid's *Elements* (Heath, 1908a, p. 153).

12. As quoted in Horadam (1960).

13. Marion Elizabeth Stark's translation of Steiner (1833) (Steiner, 1950, p. 65).

14. If the diameter of the circle that contains the polygon is 8.5 inches, then a side of the polygon would be $8.5 \sin\left(\frac{2\pi}{2 \cdot 65{,}537}\right) = 0.000407457$ inches long, which is approximately 1/10th the thickness of notebook paper.

15. Hobson (1913, p. 6)

16. Plato (1901, p. 224)

17. Book VI, 510d–511a (Plato, 1901, p. 207).

18. Poincaré (1895)

19. Sometimes we can relax our strict requirement of using existing points only. In proposition I.9 (Heath, 1908a, p. 264), Euclid gave the construction for the bisection of an angle $\angle ABC$. He wrote, "Let D be taken at random on AB." Then he constructed a circle with center B and radius BD. Thus, Euclid opened the compass an arbitrary amount, which goes against our rules. But it does not matter because any opening of the compass suffices. If we were to press Euclid on the issue and insist that he use an existing point, he could always take $A = D$.

Here's a more complicated example. In proposition I.12, Euclid constructed a line perpendicular to a line AB through a point C not on AB. One step requires finding two points on AB that are equidistant from C. Euclid wrote, "For let a point D be taken at random on the other side of the straight line AB, and with center C and distance CD let the circle EFG be described" (Heath, 1908a, p. 271). The angle bisector of $\angle ECG$, CH, is the desired line. As with the angle bisection, it was unnecessary for Euclid to choose D at random. He could have used A or B as the point on the circle (as long as the point A or B wasn't the same point as H), or if he truly desired a point on the other side of the line he could have constructed D to be the third point in an equilateral triangle ABD. Although another geometer would not be able to construct Euclid's E or G, the point H is constructible by anyone.

So we could loosen our rules to allow random points, lines, and circles, but we would have to be careful how we did so. We would have to describe when it was allowed and when it was not, and when it was, we would have to classify some of the points, lines, and circles as "intermediate" figures and classify others as constructible. But going to all of that trouble is not worth it. It would not enable us to solve any problems that we couldn't solve by playing by the given rules.

20. As quoted in Heath (1908a, p. 246).

21. The justification is straightforward. Because $EF = DE$ and $BE = CE$, then $CD = BF$, and because $AB = BF$, then $AB = CD$.

22. As quoted in Heath (1908a, p. 124).

23. From the chapter of Plutarch (1917, pp. 471–73) on Marcellus.

24. O'Leary (2010, p. 58)

25. As translated by Bos (2001, p. 38).

26. Heath (1921b, p. 68)

27. Bos (2001, p. 49)

28. Descartes (1954, p. 156)

29. From his 1619 *The Harmony of the World*. Translation: Bos (2001, p. 185).

30. Newton (1769, pp. 469–70)

TANGENT: THE TOMAHAWK

1. Emerson (1893, p. 14)

2. For other devices, see Yates (1942, ch. 3), Bhojane (1987), and Berger (1951).

3. We can't trisect small angles using a tomahawk. In this case, we could make a new tomahawk with a longer handle, or we could repeatedly double the angle until it is large enough for the tomahawk. Then, after trisecting this larger angle, use the Euclidean tools to halve it the same number of times.

CHAPTER 4. THE FIRST MATHEMATICAL CRISIS

1. Ovid (2004, p. 273)

2. The terms rational and irrational appear in *Elements* (see definitions X.3 and X.4), but the meaning is slightly different from ours. According to Cajori (1991, p. 68), the Roman Cassiodorus (ca. 485–ca. 585) was the first to use the terms rational and irrational in the way we do today.

3. Knorr (1975, p. 50) argued that the discovery was *not* a crisis in mathematics. While it showed that one of the foundational beliefs of the Pythagoreans was incorrect, mathematics largely went forward unimpeded. The theoretical work on incommensurable magnitudes developed and filled out in subsequent years.

4. There is no firm evidence of when this discovery was made. For instance, Fritz (1945) argued that it took place in the middle of the fifth century BCE and Knorr (1975, ch. II) argued that it was between 430 and 410 BCE.

5. Hartshorne (2000a, p. 41) discussed the characteristics and properties of this undefined quantity, which he calls "equal content," rather than area.

6. Book VI, 525d–525e (Plato, 1901, p. 222).

7. We follow Grattan-Guinness (1996) and avoid using "equal" when referring to ratios. Euclid viewed ratios as different from numbers and from magnitudes, and he never stated that two ratios were equal; instead, he wrote that one ratio is "in the same ratio as" another. We use :: (not =) to indicate that two ratios are the same. According to Cajori (2007a, pp. 285–86), William Oughtred introduced the :: notation in 1631, although he wrote *a.b* :: *c.d*. The astronomer Vincent Wing modified it to the familiar *A:B* :: *C:D* 20 years later.

8. It is traditional to attribute this theorem to Pythagoras and his followers, but we have no firm evidence that they discovered the theorem or gave the first proof.

9. See Fritz (1945) for more information about his life.

10. Proclus, Pappus, and Iamblichus all wrote about the discovery of incommensurable magnitudes.

11. However, strong evidence points to Theaetetus for this construction. Perhaps Hippasus drew the dodecahedron but did not produce a geometrical construction.

12. Burkert (1972, p. 461)

13. Knorr (1975) gave a detailed history and analysis of incommensurability.

14. See Fritz (1945).

15. Knorr (1975, p. 1)

16. Plato (1992, p. 8)

17. Plato is not clear about whether Theodorus proved that $\sqrt{17}$ is irrational, and because Theodorus's proof is lost we do not know why he stopped at 17. Knorr gives a convincing reconstruction of a possible proof that works for the given values and runs into trouble at 17 (Knorr, 1975, pp. 183–91). See also Bulmer-Thomas (2008g,h) and Artmann (1999, pp. 240–53).

18. *Theaetetus* and *Sophist*.

19. As quoted in van der Waerden (1954, p. 165).

20. This commentary has survived only in Arabic and is generally attributed to Pappus. It is quoted from Bulmer-Thomas (2008f).

21. Using modern terminology, for magnitudes a, b, c, and d, we write $a{:}b :: c{:}d$ if for any rational number n/m, the quantities a/b and c/d are both greater than, both less than, or both equal to m/n.

22. Dedekind (1872)

23. As quoted in Bulmer-Thomas (2008a).

24. As quoted in Bulmer-Thomas (2008a).

25. Aristotle (1869, pp. 350–51)

26. Huxley (2008)

TANGENT: TOOTHPICK CONSTRUCTIONS

1. Clemens (1917, p. 82)

2. Actually, Dawson refers to match sticks, not toothpicks (Dawson, 1939).

CHAPTER 5. DOUBLING THE CUBE

1. Euclid performed this construction in II.4, and it was shown equivalent to the construction for mean proportionals in VI.17.

2. Heath (1908b, p. 188)

3. See Markowsky (1992) for more on misconceptions about the golden ratio.

4. Others argue that he was already well versed in mathematics when he arrived in Athens, having studied under the Pythagoreans in Chios. See Bulmer-Thomas (2008c). Our knowledge of Hippocrates's work comes to us third hand. Aristotle's student, Eudemus of Rhodes, wrote about Hippocrates's work in his now-lost *History of Geometry*. In the sixth century CE, 1000 years later, Simplicius quoted Eudemus in his commentary on Aristotle's *Physics*.

5. As quoted in Dantzig (1955, p. 122).

6. From Eutocius's commentary of Archimedes's *On the Sphere and Cylinder*. As quoted in Knorr (1993, p. 23).

7. See Heath (1921a, pp. 246–55) and Knorr (1993, pp. 51–66).

8. Knorr (1993, p. 50)

9. See Knorr (1993, pp. 50–52), Graesser (1956), or Masià (2016).

10. See Knorr (1993, p. 53) for speculation as to what Eudoxus's proof was.

11. As quoted in Bulmer-Thomas (2008d).

12. As quoted in Bulmer-Thomas (2008d). A similar quote has been attributed to Euclid, writing to Ptolemy I.

13. In 1870 Bretschneider (1870, pp. 157–58) suggested Menaechmus's possible approach to the problem.

14. In 1636 Fermat used his analytic geometry to carry out the parabola–hyperbola construction of two mean proportionals. See Bos (2001, pp. 115–16).

15. Bulmer-Thomas (2008d) gave this as a possibility.

16. Knorr (1993, pp. 57–61) discussed the possible origins of this method.

17. The other two were Heron and Philon of Byzantium; their solutions, while technically different, are only superficially so.

18. Heron's construction did not have a circle. Instead he aimed to construct *FCG* so that *EFG* is an isosceles triangle (Bos, 2001, p. 29).

TANGENT: ERATOSTHENES'S MESOLABE

1. Thoreau (1985, p. 294)

CHAPTER 6. THE EARLY HISTORY OF π

1. Vasari (1998, pp. 22–23)

2. In the United States, March 14 is represented by 3/14, and is thus "pi day."

3. In this formula, μ is the mean and σ the standard deviation.

4. If K is the Gaussian curvature of a Riemann surface S and $\chi(S)$ is the Euler characteristic of S (which rely on the geometry and topology of S, respectively), then $\int_S K \, dA = 2\pi \chi(S)$. See Richeson (2008, ch. 21).

5. Under suitable hypotheses, if f is a function and γ is a closed path in the complex plane enclosing the point a, then $f(a) = \frac{1}{2\pi i} \oint_\gamma \frac{f(z)}{z-a} \, dz$. This remarkable formula says that the value of the function f at $z = a$ can be determined from the values of f along the curve γ.

6. De Morgan (1915b, p. 214)

7. Cajori (2007b, p. 9)

8. Jones (1706, p. 263)

9. Robert Palais pointed out to me that Euler (1736, p. 113) used π for this value in his *Mechanica*. He wrote, "Let $1{:}\pi$ denote the ratio of the diameter to the circumference." Euler (1729) also used π to denote what we would today call 2π, or τ (see page 300): "$1{:}\pi$ is taken for the ratio of the radius to the periphery." He used π for this value again in a letter to Jean d'Alembert (1717–1783) on April 15, 1747: "Let π be the circumference of a circle of which the radius is $= 1$" (Henry, 1886).

10. Translation of Euler (1748) by Struik (1969, p. 347).

11. Richard Green pointed out to me that the 114th digit is incorrect. Euler listed it as 8, but it should be 7.

12. The line of cuneiform running through the tablet—which is item number 7289 from the Yale Babylonian Collection—translates to $1 + 24/60 + 51/60^2 + 10/60^3 = 1.4142129\ldots$, where $\sqrt{2} = 1.4142135\ldots$.

13. This is table 322 from the G. A. Plimpton Collection at Columbia University.

14. In the Babylonian's base-60 notation, $24/25 = 57/60 + 36/60^2$.

15. Seidenberg (1972)

16. This proposed Egyptian derivation was given in Vogel (1958).

17. See Seidenberg (1972) and Smeur (1970) for more information.

18. Seidenberg (1972)

19. Seidenberg (1961)

20. "If you wish to turn a square into a circle stretch a cord of the centre (of the square) towards one of the corners, draw it round the side and describe a circle together with the third part of the piece standing over." Translation: Gurjar (1942).

21. See Seidenberg (1961).

22. Dun (1996)

23. See Lam and Ang (1986) for a discussion of the history of π in China.

24. In fact, Liu used $r = 10$ in his commentary.

25. Let x be the length of segment DE. Then by applying the Pythagorean theorem to OBE we obtain $(r - x)^2 + (s_n/2)^2 = r^2$, and by applying it to BDE we obtain $x^2 + (s_n/2)^2 = s_{2n}^2$. Solving the first of these for x we obtain $x = r - \sqrt{r^2 - s_n^2/4}$. Solving the second for s_{2n} and substituting for x we obtain $s_{2n} = \sqrt{x^2 + s_n^2/4} = \sqrt{2r^2 - r\sqrt{4r^2 - s_n^2}}$.

26. This and all biblical quotes are from the American Standard Version. This description reappears in 2 Chronicles 4:2.

27. This rule appears in the Babylonian Talmud, Eruvin 14a, and elsewhere in the Talmud. See Tsaban and Garber (1998).

28. See Tsaban and Garber (1998), Stern (1985), and Simoson (2009).

29. The cubit is not a well-defined length. It was originally the length of a forearm, or approximately 18 inches.

30. This is the third convergent for π (see page 132), so 333/106 is the best rational approximation for π for any fraction with denominator less than 106.

31. Gardner (1985)

32. In *McLean v. Arkansas Board of Education*, the United States District Court for the Eastern District of Arkansas found that the Arkansas Balanced Treatment Act requiring schools to balance the teaching of creationism and evolution violated the Establishment Clause of the First Amendment to the US Constitution.

33. O'Shaughnessy (1983)

TANGENT: THE GREAT PYRAMID

1. Dudley (1992, p. 234)

2. See Herz-Fischler (2000, chs. 9 and 18) for details.

3. These measurements are from a 1925 survey by J. H. Cole for the Egyptian government (Cole, 1925). I took for the side length the average of the lengths of

the four sides (230.391, 230.253, 230.357, and 230.454 meters). The height given in Cole's report was 481.100 feet, which is 146.639 meters (of course the pyramid is not this tall today, because the top 9.45 meters have disappeared).

4. Using the value for π from the Rhind Papyrus era and the measured value of a would give $b = 231.69$ meters, which is over 1 meter larger than the measured value.

5. Herodotus's visit was over 2000 years after the pyramids were built, so that puts the reliability of the information in doubt. And worse, his writing about the dimensions of the pyramid can be translated in more than one way, and it is not clear that this is what he meant; he may have been referring to lengths and not areas. See for example Verheyen (1992).

6. To make matters worse (or better for the mystical pyramidologists), under these hypotheses the computed value of π happens to be $4\sqrt{1/\varphi} = 3.1446\ldots$, where φ is the golden ratio. It is a nice exercise to derive this relation.

CHAPTER 7. QUADRATURES

1. As quoted in Zubov (1968, p. 175).

2. In Euclid's proposition I.42 he constructed a parallelogram with area equal to the area of the triangle. Because we are free to choose the angle, we may choose a right angle and obtain a rectangle.

3. This is not Euclid's method. We are following Dunham (1990, pp. 11–17). In Elements, Euclid uses proposition I.42 to turn each of the triangles into rectangles. Then he uses a sequence of propositions ending with I.45 to join all of the rectangles into a single rectangle. Then he uses II.14 to square the rectangle.

4. Proclus (1992, p. 335)

5. Lunes have their name because they resemble the crescent moons we see in the sky. However, the true lunar crescent shape is not a mathematical lune—it is bounded on one side by a half circle and on the other by a half ellipse.

6. Heath (1908c, p. 371)

7. See Knorr (1993, pp. 27–29) for a discussion of possible proofs by Hippocrates.

8. To be precise, we have shown that if regions A and B are squarable, so is $A + B$, but we are using that $A - B$ is squarable. This can be shown in a similar way.

9. Bernoulli (1724)

10. Euler wrote about lunes in 1737 (Euler, 1744c) and 1771 (Euler, 1772). The second of these contains the two new squarable lunes. See Sandifer (2007a, pp. 261–68) and Langton (2007) for more information on Euler's articles.

11. Wijnquist (1766)

12. I should say that we conjecture that this was Euler's conjecture. See Langton (2007, p. 59) for a dissection of Euler's poorly worded remarks.

13. Clausen (1840)

14. Many articles assert that the conjecture was fully settled following the work of Edmund Landau in 1902 (Landau, 1903), Liubomir Chakalov in 1929 (Tschakaloff, 1929), Nikolai Chebotarëv in 1935 (Tschebotaröw, 1935), and Anatolii

Vasilievich Dorodnov in 1947 (Dorodnov, 1947). However, they were still looking at a restricted set of lunes, not all possible lunes. In 2003 Kurt Girstmair (Girstmair, 2003) pointed out that by appealing to a 1966 theorem on transcendental numbers by Alan Baker (Baker, 1966), even if we consider all possible lunes, there are still only five that are constructible and squarable.

TANGENT: LEONARDO DA VINCI'S LUNES

1. As quoted in Isaacson (2017, p. 306).
2. Kemp (1981, p. 296)
3. da Vinci (1939, p. 371)
4. See Marinoni (1974, pp. 78–79) for a nice, large, two-page reproduction.
5. Here's a sketch of the proof that the area of the shaded regions is 2 (assuming the outside radius is 1). When two circles of radius r overlap by $1/4$ of their circumference to form a lens (like the four white lenses in the first figure and the removed lens at the top of the second figure) the area is $(\pi/2 - 1)r^2$. The area of the shaded region on the left is the area of the four inner circles minus the area of eight lenses. The radii of the inner circles are $1/2$, so the area of the shaded region is $4\pi(1/2)^2 - 8(\pi/2 - 1)(1/2)^2 = 2$. Now consider the second figure. The medium-sized circle has radius $\sqrt{2}/2$. The area of "Leonardo's claws"—the regions along the left and right sides of the figure—is the area of the large circle minus the area of the medium-sized circle and the lens: that is, $\pi(1)^2 - \pi(\sqrt{2}/2)^2 - (\pi/2 - 1) = 1$. The area of the square is $1/2$ and each lune has area $1/8$. So the shaded region has area $1 + 1/2 + 4(1/8) = 2$. Leonardo's drawings of these figures can be found in Marinoni (1974, p. 203) and Coolidge (1949, p. 47), respectively.

CHAPTER 8. ARCHIMEDES'S NUMBER

1. Mather (1710)
2. In 1963 Karl Popper presented the approximation $\pi \approx \sqrt{2} + \sqrt{3} = 3.14626\ldots$, which he speculated may have been known to Plato. He obtained it by averaging the area of an inscribed octagon and a circumscribed hexagon for the unit circle. However, after presenting his argument he gave the caveat "I must again emphasize that no direct evidence is known to me to show that this was in Plato's mind; but if we consider the indirect evidence here marshaled, then the hypothesis does perhaps not seem too far-fetched" (Popper, 1963, pp. 251–53).
3. Heath (1921a, p. 222)
4. Heath (1921a, p. 221). By "quadrature by means of segments" Aristotle was referring to Hippocrates's attempts to square the circle.
5. See, for instance, Wasserstein (1959).
6. Wasserstein (1959) wondered if, because the theory of proportions was a hot topic at the time, Bryson had tried to connect the problem of squaring the circle to some type of mean, as Hippocrates did with the problem of doubling the cube.
7. Heath (1921a, p. 223)
8. Euclid appealed to proposition X.1, today known as the axiom of Archimedes.

9. Netz and Noel (2007, pp. 35–47)

10. This anecdote is in Vitruvius's architecture manual, which came out 200 years after Archimedes's death. As Netz and Noel (2007, p. 34) pointed out, Archimedes's "method is sound, but it is based on a trivial observation.... This is so trivial that it was not even mentioned in Archimedes's treatise on *Floating Bodies."*

11. This tale was also from the twelfth-century poem by Tzetzes. The television program *MythBusters* tested this myth twice—in January 2006 and in December 2010 (the second time they were urged to do so by then-president Barack Obama). In both cases, the myth was "busted."

12. Plutarch (1917, p. 473)

13. As quoted in Clagett (2008).

14. As quoted in Netz and Noel (2007, p. 38).

15. Plutarch (1917, p. 481)

16. Plutarch (1917, pp. 479–81)

17. As quoted in Heath (1921a, p. 345).

18. The history of arc length is long and fascinating, but a thorough discussion would send us too far afield. See Gilbert Traub's PhD dissertation (Traub, 1984) for a detailed history. We are interested in circles, not general curves, and in fact we are not interested in lengths, but in relative lengths and ratios of lengths ($C:d$).

19. Leibniz (2005, p. 69)

20. Heath (1908a, p. 155)

21. Dijksterhuis (1987, p. 222). We must wait to see if Reviel Netz's fresh translation of *Measurement of a Circle* from the rediscovered Archimedes palimpsest will yield any new information. See Netz and Noel (2007) for the story of the palimpsest and Clagett (1964, pp. 3–6), Dijksterhuis (1987, pp. 33–49), and Knorr (1986) for details on the various Greek and Arabic sources of this treatise.

22. Archimedes (2002, p. 91). Knorr (1986) argued that the original theorem proved by Archimedes didn't create a triangle, but a rectangle. He claimed that the original theorem read something like "The product of the perimeter of the circle and the (line) from its center is double of the area of the circle." He used "the product" as an abbreviation for "the rectangle bounded by."

23. These are propositions I.20 (the triangle inequality) and I.21.

24. We do not know for certain the order in which Archimedes wrote *On the Sphere and Cylinder* and *Measurement of a Circle.* Johan Heiberg (1854–1928) listed *On the Sphere and Cylinder* before *Measurement of a Circle,* but he put a question mark after the latter. This order makes sense logically—postulates first, theorem second. When the influential historian Thomas Heath (1861–1940) reproduced the list, the question mark disappeared. Thus, this order became the standard. However, the historian Wilbur Knorr convincingly argued that *Measurement of a Circle* was written first. He gave specific mathematical evidence showing that Archimedes was a more mature mathematician when he wrote *On the Sphere and Cylinder.* He speculated that when Archimedes wrote *Measurement of a Circle* he accepted the postulates as obvious truths: "The formulation of these principles as explicit axioms in *On the Sphere and Cylinder* would thus result from Archimedes' own later reflections on the formal requirements of such demonstrations" (Knorr, 1993, pp. 153–55).

25. Netz (2004, p. 35). Note that in Netz's translations he used the term "lines" instead of "curves"; when they are straight they are called "straight lines."

26. Netz (2004, p. 36)

27. In fact, Archimedes should have included a third postulate: finite additivity. He used the fact that if a curve is broken into parts, then the total length equals the sum of the individual lengths. Euclid's *Elements* should also have contained this postulate. To make up for this, Christopher Clavius (1538–1612) added the postulate that "the whole is equal to the sum of its parts" to his 1574 version of *Elements* (Traub, 1984, p. 40; Heath, 1908a, p. 323).

28. Netz (2004, p. 36)

29. Netz (2004, p. 41)

30. Archimedes (2002, pp. 91–93)

31. Archimedes (2002, pp. 93–99)

32. If the circle had radius r, then the two polygons could fit inside a circular band of thickness $r \sec(180°/96) - r \cos(180°/96)$. In this book, the radius of the circle would be about 2 centimeters. So the thickness of this band would be approximately 21 microns, which is smaller than the diameter of a human hair.

33. Archimedes may have used the Euclidean algorithm to obtain rational approximations of $\sqrt{3}$. Today we'd describe this by truncating the continued fraction

$$\sqrt{3} = 1 + \cfrac{1}{1 + \cfrac{1}{2 + \cfrac{1}{1 + \cfrac{1}{2 + \cdots}}}}$$

(see page 132). But the truncations required to obtain Archimedes's fractions are after the 9th and 12th terms. Why not use the 11th and 12th to obtain tighter bounds? Another possibility, presented in Acerbi (2008), is to compute the continued fraction for $\sqrt{27}$ and then divide the final fractions by 3. Indeed, truncating it after the 3rd and 4th terms yields the bounds $265/51 < \sqrt{27} < 1351/260$. Dividing by 3 gives Archimedes's values. See Archimedes (2002, pp. lxxiv–xcix) and Dantzig (1955, pp. 152–59) for other possible methods.

34. The bounds in *Measurement of a Circle* are amazing, but according to Heron of Alexandria (ca. 10–ca. 70 CE), Archimedes did better than this (from *Metrica*, quoted in Knorr (75/76)): "Archimedes proves in his work on plinthides and cylinders that of every circle the perimeter has to the diameter a greater ratio than 211,875:67,441, but a lesser ratio than 197,888:62,351. But since these numbers are not well-suited for practical measurements, they are brought down to very small numbers such as 22:7." The problem is that the bounds Heron states are not bounds at all: the lower bound is, in fact, an upper bound, and the upper bound is larger than 22/7. No one dares admit that Archimedes is wrong, thus effort has been made to try to reconstruct Archimedes's better bounds. Knorr wrote about some attempts and gave his own: $3.141528\ldots = 197,888/62,991 < \pi < 211,875/67,441 = 3.141634\ldots$.

35. Archimedes could have used the Euclidean algorithm to obtain these approximations. The convergents for a continued fraction (see page 132) alternate between being less than and greater than the true value, so

$$\frac{29{,}376}{9347} = 3 + \cfrac{1}{7 + \cfrac{1}{667 + \cfrac{1}{2}}} < 3 + \frac{1}{7} = \frac{22}{7}$$

and

$$\frac{25{,}344}{8069} = 3 + \cfrac{1}{7 + \cfrac{1}{10 + \cfrac{1}{2 + \cfrac{1}{1 + \cfrac{1}{36}}}}} > 3 + \cfrac{1}{7 + \cfrac{1}{10}} = \frac{223}{71}.$$

36. Netz (2004, p. 148)

37. Netz (2004, p. 144)

38. Netz (2004, p. 150)

39. Archimedes (2002, pp. 1–2)

40. Plutarch (1917, p. 481)

41. His argument involved finding the balancing point for a triangle and the parabolic segment.

42. If $|r| < 1$, then $1 + r + r^2 + \cdots = 1/(1 - r)$. For Archimedes's sum take $r = 1/4$.

43. Knorr (1993, p. 170)

44. Plutarch (1917, pp. 475–77)

45. Livy (1972, p. 338)

46. As quoted in Jackson (2005).

47. Cicero (1886, pp. 289–90)

48. As quoted in Knorr (1993, p. 364).

TANGENT: COMPUTING π AT HOME

1. De Morgan (1915a, pp. 285–86)

2. A continued fraction is *simple* if all of the numerators are 1s and all of the a_k are positive integers.

3. See Arndt and Haenel (2000, pp. 244–45) for the first 2000 entries for the continued fraction. Note that unlike for e, the values in the continued fraction for π have no apparent pattern.

4. The first completely known continued fraction for π was due to Lord William Brouncker (1620–1684) in 1655. John Wallis (1616–1703) published the formula in his *Arithmetica infinitorum* (Wallis, 1656, p. 182). Recently L. J. Lange found another

elegant continued fraction for π (Lange, 1999). They are, respectively,

$$\frac{4}{\pi} = 1 + \cfrac{1^2}{2 + \cfrac{3^2}{2 + \cfrac{5^2}{2 + \cdots}}} \qquad \text{and} \qquad \pi = 3 + \cfrac{1^2}{6 + \cfrac{3^2}{6 + \cfrac{5^2}{6 + \cdots}}}.$$

5. See Badger (1994) for more about Buffon's needle, including a derivation of the probability and his argument that Lazzarini fabricated his data. Buffon's original work is found in Leclerc (1829) and Lazzarini's in Lazzarini (1901).

6. Shanks (1992). Shanks also gave examples that converge even faster: $g(x) = x + 2\cos(x/2)$ and $h(x) = x + (2\sin x - \tan x)/3$.

7. Theorem 332 in Hardy and Wright (1975, p. 269).

8. Sequence A002088 in the *On-line Encyclopedia of Integer Sequences* gives, for a given n, the number of relatively prime pairs (a, b) with $1 \leq a \leq b \leq n$. The sequence begins 1, 2, 4, 6, 10, 12, 18, ... If the nth element of the sequence is x_n, then the number of relatively prime pairs (a, b) with $a, b \leq n$ is $2x_n - 1$. For instance, $x_{10} = 32$, so there are 63 relatively prime pairs of numbers 1 to 100.

CHAPTER 9. THE HEPTAGON, THE NONAGON, AND THE OTHER REGULAR POLYGONS

1. Abbott (1884, p. 3)

2. *The Seven-Pointed Star* is one of the holy texts and the star is one of the most prominent religious symbols in the fictional Westeros in this book and television series.

3. Weatherall's "Polygon Song" mentions the heptagon and the nonagon, and They Might Be Giants have a song called "Nonagon."

4. Richmond (1893). In Richmond's construction he bisects angle $\angle QRP$ to find a point V on OP. The line through V perpendicular to OP yields the other two vertices of the pentagon.

5. For a discussion of whether this is the work of Archimedes see Hogendijk (1984) and Knorr (1989).

6. For details see Knorr (1989), Heath (1931, pp. 240–42), and Hogendijk (1984).

TANGENT: IT TAKES TIME TO TRISECT AN ANGLE

1. Tolstoy (2002, p. 851)

2. Moser (1947)

3. This angle trisection is inspired by the one in Blonder (2015).

CHAPTER 10. NEUSIS CONSTRUCTIONS

1. Stark (1948). One wonders whether the editor of this column had the details of this anecdote correct. It is not clear why the mathematician said he could square the circle—unless the marks are, say, πr units apart.

2. Some authors refer to a straightedge with one mark on it, but in that case they are viewing the end of the straightedge as being the second mark.

3. Here's a method for scaling a figure. Suppose we have a straightedge with marks a units apart and a segment of length b, and we would like to dilate (or contract) the entire figure by a factor of a/b relative to some point O. It suffices to show how to move one point; so suppose C is c units from O, and we would like it to be $(a/b)c$ units away in the same direction. First, perform a neusis construction somewhere on the figure to obtain a segment of length a. Then construct a point A so that OA has length a and a segment OB of length b along this same line. Draw line BC and a line parallel to it passing through A. This line intersects OC at D. By similar triangles, the length of OD is $(a/b)c$. Thus D is the desired point.

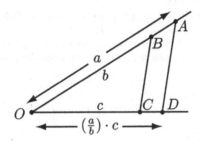

4. Hippocrates may have been aware of this fact, but Simplicius does not mention it (Langton, 2007, p. 55).

5. The reader is encouraged to think about how to trisect an obtuse angle.

6. There are in fact three ways to orient the marked straightedge to carry out the neusis construction. The second one puts D to the right of A. The third way requires the full circle, not the semicircle. In this case, D would lie between A and C. These two additional constructions yield the angles $\theta/3 + 120°$ and $\theta/3 + 240°$, which also trisect the angle. For instance, $3(\theta/3 + 120°) = \theta + 360° = \theta$.

7. Pappus (2010, pp. 148–49). To see that this is a variation of the construction in figure 10.4, rotate figure 10.6 180° and draw a circle with center C and radius BC; the common points in the two diagrams have the same letters. Furthermore, notice that $\theta = \angle ACB = \angle GCB = \angle CBF$ and $\theta/3 = \angle ADB = \angle GDB = \angle DBF$.

8. See Heath (1921a, pp. 260–62) or Bos (2001, pp. 33–34) for a description of Nicomedes's construction and Newton (1972, p. 457) for Newton's.

9. This theorem follows from Euclid's proposition III.36.

10. Viète (1983, pp. 388–417). This translation is from Bos (2001, p. 168).

11. Viète (1646, pp. 245-46); in English, Viète (1983, p. 398).

12. Archimedes (2002, p. xxxii)

13. Proposition 5 of *A Supplement to Geometry* (Viète, 1983, pp. 392–94). See also Bos (2001, pp. 169–71).

14. We will give a very slightly modified construction and justification of Viète's heptagon found in Viète (1983, pp. 408–10), following Hartshorne (2000a, p. 265).

15. We will say more about which polygons are neusis constructible in the Tangent on page 342.

Tangent: Crockett Johnson's Heptagon

1. Keats (1820, p. 91)

2. Johnson (1975)

3. The information about Johnson's life and art was obtained from Stroud (2008) and Cawthorne and Green (2009).

4. The Smithsonian's collection of Johnson's paintings can be seen on their website (Smithsonian, 2018).

5. This zigzag pattern had already been discovered by Finlay (1959).

6. For the full proof, see Johnson (1975).

7. On my blog I posed the question whether there is a geometric, trig-free proof of this result. Dan Lawson sent me such a proof. I reposted it on my blog: Lawson (2016).

Chapter 11. Curves

1. Millay (1923, p. 74)

2. Horadam (1960)

3. Simpson and Weiner (1989)

4. As quoted in O'Grady (2008, p. 12).

5. Some scholars are not convinced that Hippias of Elis discovered the quadratrix because Proclus (1992, pp. 212, 277) attributed the discovery to Hippias, but did not specify which one. However, the circumstantial evidence is convincing. It was common for Proclus to refer to people by their names and surnames initially, then drop the latter when there was no confusion. Proclus mentioned Hippias of Elis on page 52, earlier in the commentary, and no other Hippias by surname. Knorr (1993, pp. 80–82) was unconvinced and argued that the discovery would likely come after Eudoxus and Archimedes, and thus contended that the Hippias mentioned is a second- or third-century BCE scholar who extended the work of Nicomedes.

6. As quoted in Burton (2007, p. 134).

7. See Knorr (1993, pp. 226–33).

8. See Knorr (1993, p. 226) for the proof.

9. See Scott (1976, pp. 102–3) or Heath (1921a, p. 229).

10. Sporus also claimed that we cannot use the quadratrix to square or rectify the circle because to draw it, we must coordinate the linear movement of one segment and the rotational movement of another. But Bos (2001, pp. 42–43, footnote 15) argued that this is not the case. Sporus's criticism assumed we were given the

bounding square at the outset; but this is not necessary. We can draw the curve first and produce the square from it. All that's required is to be able to sweep a line vertically at a constant velocity and another line with a constant angular velocity.

11. In 1699 Giovanni Ceva (1647–1734) introduced another curve based on Archimedes's neusis trisection: the cycloid of Ceva (Yates, 1942, pp. 29–30).

12. Pappus stated that there are four conchoid curves. By this he is counting each component as a curve. So the curves in figure T.4 represent two of them but also, by choosing different values of k, the left-hand curve could have three different forms: it could have a loop (as in figure T.4), it could have a cusp, or it could be smooth with no self-intersections (Knorr, 1993, p. 220).

13. Newton (1769, pp. 469–70)

14. The earliest source I could find was Basset (1901, p. 196). A different trisection method using the same curve can be found in Lockwood (1961, pp. 46–47); this source also shows how to use the limaçon to construct a regular pentagon.

15. If B is the origin and C is the point $(0, -1)$, then the limaçon in figure 11.8 is the polar curve $r = 1 - 2\sin\theta$.

16. Archimedes (2002, p. 165)

17. In fact, Bernoulli wanted a logarithmic spiral on his gravestone. Unfortunately, the stone cutters carved an Archimedean spiral by mistake.

18. Milici and Dawson (2012)

19. For a construction of the cissoid and an explanation of how to use it to construct two mean proportionals, see Knorr (1993, pp. 233–63), Bos (2001, pp. 44–47), or Heath (1921a, pp. 264–66).

20. Pappus (2010, pp. 150–55)

21. Heath (1931, pp. 145–46)

TANGENT: CARPENTER'S SQUARES

1. Milton (1910, p. 232)

2. In 1940 Yates (1940) showed it is possible to double the cube using a (marked) carpenter's square.

3. Scudder (1928)

4. In an expanded version of this chapter we show that the curve can be expressed algebraically as $x^2 = (y - 2)^2(y + 1)/(3 - y)$ (Richeson, 2017).

5. Brooks (2007)

CHAPTER 12. GETTING BY WITH LESS

1. Shakespeare (2016, p. 201)

2. Leybourn (1694, pp. 16–27) wrote about geometrical constructions using a fixed compass. The subtitle of Tract II, Chapter II of his entertaining book *Pleasure with Profit* is "Shewing How (without *Compasses*), having only a common *Meat-Fork* (or such like instrument, which will neither open wider nor shut closer), and a Plain Ruler, to perform many pleasant and delightful *Geometric Conclusions*."

3. See Jackson (1980) for an English translation of this section.

4. See Hallerberg (1959) for examples of Abu'l-Wafa's techniques.

5. This construction with an English translation can be found in Shelby (1977, pp. 116–17). *Geometria Deutsch* also contains an interesting approximate construction for inscribing a regular heptagon in a circle; in essence, it uses the fact that half the side length of an inscribed equilateral triangle is very close to the side length of the inscribed heptagon (Shelby, 1977, p. 118).

6. If we take the circles to have unit radius, with $A = (-1/2, 0)$ and $B = (1/2, 0)$, then, to three decimal places we have $H = (-0.815, 0.949)$, $I = (0.815, 0.949)$, and $J = (0, 1.528)$. The corresponding coordinates of a regular pentagon with side AB would be $(0.809, 0.951)$, $(0.809, 0.951)$, and $(0, 1.539)$. Put another way, the error in the height of the pentagon is less than 1%.

7. Dürer (1525), or in English, Dürer (1977). It was printed in 1525, then reprinted posthumously in 1538 with additional material. Dürer gave compass-and-straightedge constructions of regular polygons with sides numbering 3, 4, 5, 6, 7, 8, 9, 11, and 13 (pp. 143–51). The 7-, 9-, 11-, and 13-gons are not constructible, so his techniques are necessarily approximations. He mentions only that the constructions for 11 and 13 are approximate. Dürer also gave a method of trisecting an angle (pp. 150–51), which he did not say was approximate, and squaring the circle (pp. 178–79), which he did ("The *quadratura circui*, which means squaring the circle so that both square and circle have the same surface area, has not been demonstrated by scholars. But it can be done approximately for minor applications or small areas in the following manner.") He used an approximation of $\pi \approx 3\frac{1}{8}$ in the first edition and $\pi \approx 3\frac{1}{7}$ in the second. The pentagon construction also appeared in Danielle Barbaro's 1569 *La practica della perspecttiva* (Barbaro, 1980, p. 27).

8. Mackay (1886), paraphrasing Cardano.

9. The questions were published on April 21, 1547, in Tartaglia (1547).

10. Ferrari (1547)

11. The propositions that end with a circle are propositions 25 and 33 of book III and propositions 4, 5, 8, 9, 13, and 14 in book IV. See Mackay (1886) for Ferarri's and Tartaglia's solutions.

12. Tartaglia (1556), fifth part, third book.

13. Cardano (1550)

14. Benedetti (1553)

15. According to Hallerberg (1959).

16. Mohr (1673)

17. According to Hallerberg (1960) the manuscript contains 29 of the 49 construction problems in *Elements*. They are presented without proof, but all are correct. The ones that are omitted can clearly be solved using the same techniques.

18. As quoted in Hallerberg (1959).

19. Poncelet (1822)

20. Court (1958)

21. Mascheroni (1797)

22. According to Court (1958), the idea to investigate compass-only constructions may be due to Giambattista Benedetti, who wrote about the rusty compass.

23. Adler (1890)

24. Cheney (1953) gave Mascheroni's solution, which requires six swipes of a locking compass, and one with a collapsing compass.

25. As quoted in Court (1958).

26. Court (1958)

27. We accomplished this construction with only three arcs, but one was with a locking compass. Mascheroni used five arcs, but they were made with a collapsible compass. Cheney (1953) improved this to four collapsible arcs.

28. Mascheroni's construction required drawing seven circles with a locking compass. Cheney (1953) gave a collapsible compass procedure that required ten arcs.

29. Adler (1890), Adler (1906)

30. If efficiency is the aim, see Cheney (1953) who reduced it to nine circles.

31. Liouville called inversions the *transformation by reciprocal radii*.

32. For this construction and a more thorough introduction to inversions, see Courant and Robbins (1941, pp. 140–52). Hungerbühler (1994) gave a nice short proof of the Mohr–Mascheroni theorem for collapsible compasses without using inversions. Hudson (1961, ch. 8) and Gardner (1992) also contain examples of compass-only constructions. And Kostovskii (1961) is a short book about compass-only constructions, together with constructions with some other restrictions.

33. Edwards (1987)

34. Mohr (1672)

35. Seidenberg (2008)

36. "Georgius Mohr Danus in Geometria et Analysi versatissimus." (Gerhardt (1849), quoted in Court (1958))

37. As quoted in Hallerberg (1960).

38. Andersen (1980)

39. See Kac and Ulam (1968, pp. 18–19).

40. Poncelet (1822, pp. 187–90)

41. Steiner (1833) (see Steiner (1950) for an English translation).

42. Dörrie (1965, p. 165)

43. As quoted in Burckhardt (2008).

44. Burckhardt (2008)

45. See Dörrie (1965, pp. 165–70) for a simplified version of Steiner's proof.

46. Steiner (1950, p. 54)

47. The special cases, such as if $D = E$ or $F = E$, if CD is parallel to AB, or if $P = E$ or $P = F$, are handled in an elementary way.

48. Hilbert's comment is in Archibald's introduction to Steiner (1950, p. 2).

49. Cauer (1912, 1913)

50. Severi (1904)

Tangent: Origami

1. Alford (1839, p. 28)

2. For the full set of origami folds see Lang (2010).

3. Geretschlager (1995), Alperin (2000)

4. If p_1 is not on the line l_1, then they determine a parabola with focus p_1 and directrix l_1. Folding p_1 onto l_1 produces a line tangent to the parabola. Thus, if p_1 is not on l_1 and p_2 is not on l_2, then this folding move produces a line that is tangent to two parabolas. It is possible to draw two parabolas in the plane such that there are three common tangents. Thus this folding move is equivalent to solving a cubic equation, a feat that cannot be accomplished with a compass and straightedge. See Belloch (1936), Geretschlager (1995), and Hull (2011).

5. Messer (1986)

6. Fusimi (1980)

7. See Alperin (2000).

8. Both Huzita (1994) and Dureisseix (2006) give instructions for folding a heptagon and a nonagon.

9. For more information, see Madachy (1979, pp. 58–61).

Chapter 13. The Dawn of Algebra

1. Rebière (1893, p. 36)

2. Chace (1927, pp. 74–75)

3. According to Blåsjö (2016).

4. Wallis (1685, p. 3)

5. Heath (1908a, p. 379)

6. See Blåsjö (2016) for a defense of geometrical algebra.

7. As quoted in Dantzig (2007, p. 29).

8. Plofker (2007)

9. Plofker (2007)

10. We have only a Latin translation of this book, called *De numero indorum*.

11. In fact, because *al-jabr* means restoration, it is the origin of the now-obsolete Spanish word *algebrista*, which was the name given to a bone setter.

12. This was the example Cardano gave in *Ars magna*. See Cardano (1968, p. 99).

13. In fact, we can get all three roots directly from this formula. A number has three cube roots (possibly complex). For instance, 1 has three cube roots: 1 and $\frac{1}{2} \pm \frac{\sqrt{3}}{2}i$. We can obtain all of the roots of an equation by choosing the correct cube roots in the formula. Moreover, in general, if there are two complex roots, they are complex conjugates. See Turnbull (1947, ch. IX).

14. Cardano (2002, p. 96)

15. Cardano (2002, p. 49)

16. As quoted in Dunham (1990, p. 140).

17. Gliozzi (2008)

18. As quoted in Dunham (1990, p. 140).

19. To see a complete discussion of solving the cubic and quartic equations, see Dickson (1922, ch. IV).

20. Translation from Smith (1959, p. 206).

21. Cardano (1545, p. 250)

22. Cardano (2002, p. 127)
23. Cardano (1968, p. 96)

Tangent: Nicholas of Cusa

1. This handwritten limerick was found by Charles Petzold in a used copy of Hobson (1913). See Petzold (2007).
2. Albertini (2004, pp. 376–77)
3. Translation from Meschkowski (1964, p. 28).
4. Translation from Meschkowski (1964, p. 27).
5. Translation from Meschkowski (1964, p. 28).
6. As quoted in Albertini (2004, pp. 386–87).
7. Translation from Meschkowski (1964, p. 29).
8. Folkerts (1996)

Chapter 14. Viète's Analytic Art

1. Atiyah (2001)
2. As quoted in Busard (2008).
3. As quoted in Heath (1908a, pp. 138–39).
4. This problem is due to Marino Ghetaldi (1568–1626), a protégé of Viète's. He gave the construction in 1607, but he explained his analysis of the problem in 1630. See Bos (2001, pp. 81–83, 102–4).
5. Viète (1646, p. 12), or in English, Viète (1983, p. 32).
6. Viète (1983, p. 24)
7. Viète (1646, p. 91)
8. Viète (1646, p. 8)
9. Viète (1646, p. 2), or in English, Viète (1983, p. 15).
10. In fact, he gave formulas for the expansion of $(a + b)^n$ for $n = 2, 3, 4, 5, 6$ and factorization formulas for $a^n + b^n$ for $n = 3, 5$ and $a^n - b^n$ for $n = 2, 3, 4, 5$. More specifically, he gave the factorizations $a^n + b^n = (a + b)(a^{n-1} - a^{n-2}b + \cdots + b^{n-1})$ and $a^n - b^n = (a - b)(a^{n-1} + a^{n-2}b + \cdots + b^{n-1})$. These formulas appeared in Viète (1646, pp. 16–23), or in English, Viète (1983, pp. 39–50).
11. Mahoney (1994, p. 36)
12. See Viète (1646, pp. 248–49) or in English, Viète (1983, p. 403).
13. Viète's diagram contains some additional lines that appear in his proof.
14. If we substitute $x = \sqrt{p/3} \cdot y$ into the equation $x^3 = px + q$ and simplify, we obtain $y^3 - 3y = q\sqrt{27/p^3}$. So take $b = q\sqrt{27/p^3}$. Furthermore, notice that if the equation is an irreducible cubic, then $(q/2)^2 - (p/3)^3 < 0$, which is equivalent to $b = q\sqrt{27/p^3} < 2$.
15. Gleason (1988)
16. Viète (1646, pp. 90–91), or in English, Viète (1983, pp. 173–75). Because the cubic is irreducible, it has three real roots, and Viète showed how to find all three.

He did not consider negative roots, but instead of ignoring them, he found the positive roots of $z^3 = pz - q$, which are the absolute value of the negative roots of $x^3 = px + q$. To be clear, Viète did not mention sines or cosines in his work; rather he described how to use the coefficients of the cubic equation to construct certain right triangles, and from these triangles he could find the roots of the cubic. The first explicit trigonometric solution was given by Albert Girard (1595–1632) in 1629 in his *Invention nouvelle en l'algebra* (Girard, 1629).

17. This time use the substitution $x = \sqrt{4p/3} \cdot y$.

18. We will prove this trigonometric identity in chapter 18.

19. Give it a try. Use this trigonometric technique to show that the roots of $4x^3 = 3x + 0$ are 0 and $\pm\sqrt{3}/2$ and that $x^3 = 3x + 2$ has roots 2 and -1 (this second root has multiplicity 2).

20. Viète (1983, p. 416)

21. Viète (1646, p. 400). Interestingly, this expression can be found using the quadratrix by noting that $2/\pi = \lim_{n\to\infty} r_n$, where r_n is the distance of the quadratrix from the origin at an angle $\pi/2^n$. See O'Leary (2010, pp. 237–38) for more details. According to Rummler (1993), if P_n is the perimeter of the regular 2^n-gon inscribed in the unit circle, then $\frac{P_1}{P_2}\frac{P_2}{P_3}\frac{P_3}{P_4}\cdots = \frac{P_1}{P_\infty} = \frac{2}{\pi}$. Here $P_1 = 4$ is twice the length of the diameter and $P_\infty = 2\pi$ is the circumference of the circle.

22. In a 2017 blog post, John Baez and Greg Egan used the Viète approach, but they started with an inscribed pentagon instead of a square as Viète did (Baez and Egan, 2017). They obtained the following infinite product, which relates π to the golden ratio ϕ:

$$\pi = \frac{5}{\phi} \frac{2}{\sqrt{2+\sqrt{2+\phi}}} \frac{2}{\sqrt{2+\sqrt{2+\sqrt{2+\phi}}}} \cdots.$$

Tangent: Galileo's Compass

1. Balzac (1900, pp. 134–35)

2. According to Williams and Tomash (2003), the sector may have been invented by a mathematician named Thomas Hood.

3. The Museo Galileo–Istituto e Museo di Storia della Scienza has a nice interactive web page on Galileo's compass (Museo Galileo, 2008a). They also have a printable template so you can make your own compass (Museo Galileo, 2008b).

4. If the circle marking is c units from the hinge, then the marking for n is $2c\sqrt{\pi \tan(\pi/n)/n}$ units from the end.

5. In fact, there were other lines on the compass that allowed the user to square the sector of a circle. See Galilei (1606, p. 27).

6. The numbers on the polygraphic lines are not evenly spaced. If a_n is the distance of the mark for n from the hinge of the compass, then $a_n = a_6/(2\sin(\pi/n))$.

7. If the 1 is a units from the hinge, then the n is $a\sqrt[3]{n}$ units from the hinge.

8. It can also be used to find two mean proportionals (Galilei, 1606, p. 19).

9. Galilei (1606)

10. See Williams and Tomash (2003) for details.

CHAPTER 15. DESCARTES'S COMPASS-AND-STRAIGHTEDGE ARITHMETIC

1. Chasles (1875, p. 94)

2. See Richeson (2008) for Descartes's work related to Euler's polyhedron formula, a century before Euler's discovery.

3. Grabiner (1995)

4. Boyer (1947)

5. Serfati (2008)

6. Euclid's and Descartes's work was formalized by Hilbert in 1899. Hilbert (1903) gave a complete set of axioms (20 of them) for Euclidean geometry (Hilbert (1959) in English). He also developed the theory of arithmetic on line segments.

7. Descartes (1954, pp. 2–5)

8. Descartes (1954, pp. 5–7)

9. We would write this as $\sqrt[3]{a^3 - b^3 + ab^2}$.

10. Negative and fractional exponents were introduced by John Wallis in his 1656 *Arithmetica infinitorum*.

11. In his *Whetstone of Witte*, Robert Recorde introduced an equal sign a little longer than ours, justifying it by saying, "I will sette as I doe often in woorke use, a paire of paralleles, or Gemowe lines of one lengthe, thus: =, bicause noe. 2. thynges can be moare equalle."

12. Descartes (1954, pp. 12–13). Descartes does not include the details presented here—he sets up the diagram and leaves the details to the reader.

13. The arithmetic operations must also satisfy the commutative property, the associative property, and the distributive property.

14. Descartes (1954, p. 13) stated, "And if it can be solved by ordinary geometry, that is, by the use of straight lines and circles traced on a plane surface, when the last equation shall have been entirely solved there will remain at most only the square of an unknown quantity, equal to the product of its root by some known quantity, increased or diminished by some other quantity also known."

However, Lützen (2010) writes, "Descartes started by formulating the theorem that if the geometric problem is solvable by ruler and compass then the end equation is a quadratic equation (the impossibility result formulated in contraposed form), but then he went on to prove the converse, namely that if the final equation is quadratic then the problem can be solved by ruler and compass.... One may wonder if Descartes was genuinely so weak in logic that he confused a statement and its converse."

15. Clearly, $\left(\sqrt[3]{7 + 5\sqrt{2}}\right)^3 = 7 + 5\sqrt{2}$, and, although it is a little more work to show, $\left(1 + \sqrt{2}\right)^3 = 7 + 5\sqrt{2}$, as well.

16. See Bos (2001, p. 278) for an example of such a drawing approach.

17. The letter is dated March 26, 1619. As quoted in Mancosu (2008, p. 104).

18. Descartes (1954, pp. 40, 43)

19. Descartes (1954, p. 43)

20. He also wrote about curves that could be described pointwise—that is, curves for which every point can be constructed separately—and the use of strings to draw curves. See Bos (1981).

21. Kempe (1876). In fact, Kempe's proof had a flaw. The error was discovered and corrected in the twenty-first century (Demaine and O'Rourke, 2007, pp. 31–40).

22. The details of this compass can be found in what have now become known as his *Cogitationes privitae*. See Serfati (1993) or Bos (2001, pp. 237–39).

23. This drawing is given twice in *Geometry* (Descartes, 1954, pp. 46, 154).

24. See Descartes (1954, pp. 44–47, 152–66) for his discussion of his mesolabe. See also Mancosu (2008, pp. 104–5), Sasaki (2003, pp. 112–21), O'Leary (2010, pp. 271–73), and Bos (2001, pp. 240–45).

25. Bos (1988)

26. As quoted in Bos (1988).

27. See Blåsjö (2015). Leibniz's contemporaries agreed that geometry should be larger than Descartes's geometry, but they still believed Leibniz took things too far. L'Hôpital wrote that Leibniz's "machine is so very complicated and so cumbersome that it cannot be of any use in practice, and what is more, this sheds no new light whatever on the [problem of] inverse of tangents [that is, integration]" (Bos, 1988).

TANGENT: LEGISLATING π

1. Lines 215–217 of Pope (1711).

2. The full bill can be found in Hallerberg (1977).

3. This appeared in Goodwin's 1892 monograph *Universal Inequality Is the Law of All Creation*. This quote is taken from Hallerberg (1977).

4. Goodwin (1894). Note that the *American Mathematical Monthly* is now the flagship journal of the Mathematical Association of America (MAA), but at this point the *Monthly* was privately run. The MAA was not founded until 1915.

5. From the *Indiana Sentinel* on January 20, 1897 (Singmaster, 1985).

6. From the *Indianapolis Journal* on February 6, 1897 (Singmaster, 1985).

7. Waldo (1916)

8. February 13, 1897, as quoted in Edington (1935).

9. Goodwin (1895)

CHAPTER 16. DESCARTES AND THE PROBLEMS OF ANTIQUITY

1. Alighieri (1867, pp. 222–23, Canto XXXIII)

2. Descartes (1954, p. 17)

3. Descartes (1954, pp. 206–8)

4. Descartes uses the letter z.

5. See Bos (2001, pp. 256–57) for details.

6. Bos (2001, p. 255)

7. If we had x- and y-axes with the origin at A, then the equations of the parabola and the circle would be $y = -x^2$ and $(y+2)^2 + (x - q/2)^2 = 4 + q^2/4$, respectively. Eliminating y, we obtain $x(x^3 - 3x + q) = 0$. Thus, the x-coordinates of the points of intersection are the three roots of the cubic equation and $x = 0$.

8. Descartes (1954, pp. 204–7)

9. We can show that the parabola has an equation $y = -x^2/a$ and the circle has equation $(x + q/2)^2 + (y + a/2)^2 = q^2/4 + a^2/4$. When we solve for the x-coordinate of the point of intersection (ignoring $x = 0$), we obtain $x^3 = a^2q$.

10. Descartes (1954, pp. 216–19)

11. Lützen (2010) wrote, "The concept of reducibility was still somewhat fluid in Descartes' program."

12. Descartes (1954, p. 219)

13. Lützen (2010)

14. Lützen (2010)

15. In 1754, more than a century after Descartes's death, Jean Étienne Montucla (1725–1799) attempted to finish Descartes's algebraic proof of impossibility. But this proof too had problems. See Lützen (2010).

16. In a letter to Mersenne, March 31, 1638, as quoted in Mancosu (2008).

17. Descartes (1954, p. 91)

18. Ross (1936, p. 426)

19. Clagett (1964, p. 69)

20. Clagett (1964, p. 171)

21. Descartes (1908, pp. 304–5). Hobson (1913, p. 32) gave the procedure and showed it is equivalent to the equality $\frac{4}{\pi} = \tan\left(\frac{\pi}{4}\right) + \frac{1}{2}\tan\left(\frac{\pi}{8}\right) + \frac{1}{4}\tan\left(\frac{\pi}{16}\right) + \cdots$.

22. See Boyer (1964) for a history of early rectifications of curves.

23. Martin (2010) has a history of the cycloid.

24. Traub (1984, p. 76)

25. The length of this curve from $(0, 0)$ to (a, a) is $a(13\sqrt{13} - 8)/27$, which can be constructed with a compass and straightedge.

26. Bos (1981). Mancosu (1999) disagreed; he wrote, "Although the algebraic rectification of algebraic curves was essential in destroying the Aristotelian dogma, it did not really undermine the foundation of Descartes' *Geometry*, nor, to my knowledge, did anybody at that time claim this to be the case."

TANGENT: HOBBES, WALLIS, AND THE NEW ALGEBRA

1. From Kepler's 1619 *Harmonices mundi*, as quoted in Bos (2001, p. 191).

2. This tale, which is likely apocryphal, appears in Aubrey (1898, p. 332).

3. Wallis (1656). English translation in Struik (1969, pp. 244–53).

4. For more about his life and his mathematical contributions see Scriba (2008).

5. These quotes come from "Six lessons to the professors of the mathematics of geometry, the other of astronomy" (Hobbes, 1845, pp. 248, 316, 330).

6. "A *point* is that with no part," and "a *line* is a breadthless length" (Heath, 1908a, p. 153).

7. Hobbes (1839, p. 111)

8. As quoted in Jesseph (1999, p. 360).

9. As quoted in Jesseph (1999, pp. 3–4). Omitted from this quote are his references to his publications and two items that are not relevant to our study.

10. Pycior (2006, p. 144)

11. Newton (1972, p. 429). The notes were edited by Newton's successor William Whiston and were published against Newton's wishes in 1707. The original publication was in Latin; an English translation by Joseph Raphson (ca. 1648–ca. 1715) appeared in 1720, after Raphson's death. Newton's name first appeared on this work in 1761, many years after his death. This quote is from a paragraph in the appendix, which Newton marked for deletion.

12. Bos (2001, p. 134)

Chapter 17. Seventeenth-Century Quadratures of the Circle

1. Kline (1972, p. 392). In a letter to David Gregory, nephew of James Gregory.

2. In modern terminology, the arc length of the parabola $y = x^2$ from $x = a$ to $x = b$ is $\int_a^b \sqrt{1 + 4x^2}\, dx$, which is the area under the hyperbola $y^2 - 4x^2 = 1$ on that same interval.

3. If we slice the paraboloid perpendicular to its axis of symmetry, we obtain a bowl-shaped surface with depth d and radius (across the top of the bowl) r. Then, using today's notation, its surface area is $\pi r(\sqrt{(r^2 + 4d^2)^3} - r^3)/(6d^2)$. The result was discovered at almost the same time, and most likely independently, by the Dutch mathematician Hendrik van Heuraet. See Yoder (1988, pp. 119–26) for discussion of the priority dispute between Huygens and van Heuraet.

4. Yoder (1988, p. 138)

5. Gregory Saint-Vincent (1647)

6. Meskens (1994): "In contrast with classical Greek mathematics, Gregory thus accepts, for the first time in the history of mathematics, the existence of a limit."

7. Gregory Saint-Vincent (1647, pp. 602–3) had a geometrical statement and proof, but we will express everything using coordinates. We want to find the two mean proportionals between a and b; that is, we wish to find x_0 and y_0 so that $a/x_0 = x_0/y_0 = y_0/b$. Construct the circle that circumscribes an $a \times b$ rectangle. If we use Cartesian coordinates with opposite corners of the rectangle going through the origin and (b, a), then the equation of the circle is $x^2 - bx + y^2 - ay = 0$. Now construct the hyperbola with equation $xy = ab$. It passes through (b, a) with the x- and y-axes as asymptotes. Then the second point of intersection of the circle and the hyperbola is (x_0, y_0).

8. As quoted in Dhombres (1993).

9. Archimedes's method yields similar-looking formulas, but they were for perimeters, not areas. If the perimeters of the inscribed and circumscribed regular n-gons are i_n and c_n, respectively, then $i_{2n} = \sqrt{c_{2n} i_n}$ and $c_{2n} = 2/(1/c_n + 1/i_n)$. For a short proof of Gregory's theorem, see Edgar and Richeson (2020).

10. See Scriba (1983) for information on Gregory's work and Huygens's criticisms.

11. Dehn and Hellinger (1943)

12. Scriba (1983)

13. Thoren (2008)

14. The most accurate value at the time was due to van Ceulen. See page 282.

15. See van Maanen (1986).

16. As quoted in Yoder (1988, p. 138).

17. Our evidence of this discovery was a fifteenth-century book by Nilakantha Somayaji (1444–1544) and a commentary on his book. See Roy (1990).

18. As quoted in Horváth (1983), which also has a description of Leibniz's proof.

19. In an October 24, 1676, letter from Newton to Henry Oldenburg, as quoted in Roy (1990).

20. As quoted in Rickey (1987).

21. Rickey (1987)

22. Guicciardini (2009, p. 7)

23. Newton's argument can be found in *Methodus fluxionum et serierum infinitarum*, or in the 1736 English translation (Newton, 1736, pp. 94–95).

24. When n is a positive integer, the coefficients for the expansion come from the $(n + 1)$st row of Pascal's triangle.

25. The expression for the circle can be written as

$$y = \sqrt{x - x^2}$$
$$= x^{1/2}(1 - x)^{1/2}$$
$$= x^{1/2}\left(1 - \frac{1}{2}x - \frac{1}{8}x^2 - \frac{1}{16}x^3 - \frac{5}{128}x^4 - \frac{7}{256}x^5 - \cdots\right)$$
$$= x^{1/2} - \frac{1}{2}x^{3/2} - \frac{1}{8}x^{5/2} - \frac{1}{16}x^{7/2} - \frac{5}{128}x^{9/2} - \frac{7}{256}x^{11/2} - \cdots.$$

26. Beckmann (1971, p. 142)

TANGENT: DIGIT HUNTERS

1. James (1909, p. 203)

2. De Morgan (1915b, p. 65)

3. For more on normalcy see Wagon (1985) or Bailey and Borwein (2014).

4. Bailey and Borwein (2014)

5. For more information on the digit hunters see, for instance, Arndt and Haenel (2000), Bailey et al. (1997), Bailey and Borwein (2014), Castellanos (1988a), Castellanos (1988b), Hobson (1913), and Wrench (1960). The website Wikipedia (2018) maintains a thorough and current list of all π record holders.

6. As quoted in Plofker (2007, pp. 221–22).

7. Hayashi et al. (1990)

8. The key to his method was the observation that $3\sin\theta/(2 + \cos\theta) < \theta < 2\sin(\theta/3) + \tan(\theta/3)$.

9. Huygens (1654)

10. Euler (1744b). Castellanos (1988a) shows how to obtain second arctangent series from this formula (which he incorrectly attributes to Charles Dodgson, better known as Lewis Carroll). He also gives many more arctangent formulas for π.

11. For more about the pi room, see Huylebrouck (1996).

12. George Reitwiesner, Clyde V. Hoff, Homé S. McAllister, and W. Barkley Fritz. See Reitwiesner (1950).

13. For more about this algorithm, see Arndt and Haenel (2000, ch. 7).

14. Ramanujan discovered many formulas for π (and approximations for π). See Arndt and Haenel (2000, pp. 57–58, 226–27) and references therein.

15. Rabinowitz and Wagon (1995)

CHAPTER 18. COMPLEX NUMBERS

1. Klein (1908, p. 138), with English translation from Klein (2009, pp. 55–56).

2. Painlevé (1900)

3. The first translation is by Vera Sanford (Smith, 1959, p. 202). The second is T. Richard Witmer's translation (Cardano, 1968, pp. 219–20).

4. All of these quotes are from Cardano (1545, ch. 37).

5. In 1777 Euler wrote, "In the following I shall denote the expression $\sqrt{-1}$ by the letter i so that $ii = -1$." It was published posthumously in 1794 (Euler, 1794). Translation from Struik (1969, p. 248).

6. Bombelli (1572)

7. Both quotes are found in Jayawardene (2008a).

8. To justify these equalities, simply cube both sides: $\left(\sqrt[3]{2+11i}\right)^3 = 2 + 11i$ and $(2+i)^3 = (4 + 2i + 2i - 1)(2 + i) = (3 + 4i)(2 + i) = 6 + 8i + 3i - 4 = 2 + 11i$. Verifying the other equality is similar.

9. Cauchy (1821, p. 180)

10. Given the root $x = 4$ we can easily find the other two real roots. Polynomial long division shows that $x^3 - 15x - 4 = (x - 4)(x^2 + 4x + 1)$. Then the quadratic formula gives the other two roots to be $-2 \pm \sqrt{3}$.

11. Crossley (1987, p. 93)

12. Wantzel (1843)

13. Descartes (1954, p. 175)

14. Leibniz (1850)

15. Leibniz (1863), with English translation from Remmert (1990a).

16. Leibniz (1858). English translation based on that in Alexander (2011) and conversations with Christopher Francese and Travis Ramsey. Note that the phrase "double life" comes from Leibniz's word *Amphibio* which others have translated as amphibian or hermaphrodite.

17. Euler (1770) from the English translation Euler (1882, p. 43).

18. Grabiner (1974)

19. Gauss (1831)

20. Gauss (1831), with the English translation from Ferraro (2008, p. 328).

21. Wallis (1685) hinted at this geometrical description. The Norwegian surveyor Casper Wessel (1745–1818) gave the first modern interpretation in a presentation to the Royal Academy of Denmark in 1797 (Wessel, 1799). However, Wessel's article was largely unknown until it was translated into French in 1897.

22. Euler (1988, p. 106)

23. See Wells (1990) for instance.

24. Remmert (1990b) used this relationship to define π, and from it he derived all the familiar properties of π.

25. Napier (1614)

26. In fact, we could call π the ellipse constant. The area of an ellipse $x^2/a^2 + y^2/b^2 = 1$ is πab. A circle of radius r is the special case $a = b = r$.

27. Bernoulli (1690) shows that $e = \lim_{n\to\infty}(1 + 1/n)^n$.

28. His use of e first appeared in an article Euler wrote in 1727 or 1728, but it wasn't published until 1862 (Euler, 1862). As far as we know, Euler's next use of e was in a letter to Christian Goldbach (1690–1764) on November 25, 1731 (Fuss, 1843, pp. 56–59). It first appeared in print in Euler's physics book *Mechanica* in 1736 (Euler, 1736).

29. Euler (1748) has an English translation Euler (1988).

30. Boyer (1951)

31. Euler (1748, p. 90, ch. VII, sect. 122)

32. Euler did not discover the series for the exponential function. In 1665–66 the young Newton found the first several terms of the series for the inverse of the function $\ln(x + 1)$, which we now know is the function $e^x - 1$ (Newton, 2008, p. 235). Later, Leibniz wrote about the full series in letters to other mathematicians.

33. For an English translation of Euler's proof see Euler (1988, p. 112). While Euler was the first to write down this formula, another British mathematician in Newton's inner circle, Roger Cotes (1682–1716), came up with an equivalent formulation (Cotes, 1714). In 1714 he gave the formula $ix = \ln(\cos x + i \sin x)$. If we apply the exponential function to both sides, it yields Euler's formula.

34. When we plug $i\theta$ in for z in the series $e^z = 1 + z/1! + z^2/2! + z^3/3! + \cdots$, we obtain

$$e^{i\theta} = \frac{1}{0!} + \frac{i\theta}{1!} + \frac{(i\theta)^2}{2!} + \frac{(i\theta)^3}{3!} + \frac{(i\theta)^4}{4!} + \frac{(i\theta)^5}{5!} + \cdots$$

$$= \left(\frac{1}{0!} - \frac{\theta^2}{2!} + \frac{\theta^4}{4!} - \frac{\theta^6}{6!} + \cdots\right) + i\left(\frac{\theta}{1!} - \frac{\theta^3}{3!} + \frac{\theta^5}{5!} + \cdots\right)$$

$$= \cos\theta + i\sin\theta.$$

35. Actually, as Euler pointed out, i^i does not have a single value; rather, it takes on infinitely many real values. The angle i makes with the real axis can be expressed as $2\pi k + \pi/2$ for any integer k. Thus, using the reasoning above, $i^i = e^{-2\pi k - \pi/2}$.

36. Peirce (1859)

37. Peirce (1881). Note that $i^{-i} = (e^{\frac{\pi}{2}i})^{-i} = e^{\frac{\pi}{2}} = \sqrt{e^\pi}$.

TANGENT: THE τ REVOLUTION

1. From Stifel (1544). Translation from Beckmann (1971, p. 166).

2. Palais (2001)

3. Hartl (2010)

4. In 2011 Michael Cavers wrote "The pi manifesto," giving this and other reasons why we should stick with π (Cavers, 2011).

CHAPTER 19. GAUSS'S 17-GON

1. "Die Mathematik ist die Königin der Wissenschaften und die Zahlentheorie ist die Königin der Mathematik" (von Waltershausen, 1856, p. 79).

2. See Hayes (2006) for a history of this anecdote.

3. Translation from Fauvel and Gray (1987, p. 487). Interesting fact: According to Dunnington (2004, p. 30), Gauss gave, as a souvenir, the slate that he used to prove the constructibility of the 17-gon to his friend and classmate at Göttingen, Farkas (Wolfgang) Bolyai, father of János Bolyai, co-discoverer of non-Euclidean geometry.

4. Gauss (1796) as translated by and quoted in Fauvel and Gray (1987, p. 492).

5. Neumann (2005, pp. 304, 314)

6. Gauss (1965, p. 458)

7. Goldenring (1915)

8. Archibald (1916)

9. Richmond presented this construction in Richmond (1893) and elaborated on it in Richmond (1909). It can be found in Goldenring (1915, p. 28).

10. Fermat primes should not be confused with Mersenne primes, which have the form $2^m - 1$.

11. Gauss (1965, pp. 407–60)

12. Richelot (1832). See DeTemple (1991) who used Carlyle circles.

13. This is sequence number A003401 in the *On-line Encyclopedia of Integer Sequences*; the set of nonconstructible n-gons is sequence number A004169.

14. Gauss (1796) as translated by and quoted in Fauvel and Gray (1987, p. 493).

15. Wantzel (1837)

16. Fermat (1894, vol. 2, pp. 205–6), translated by A. Bergeron and D. Zhao.

17. Fuss (1843, p. 10), translation from Sandifer (2007b).

18. Euler (1738), translation by Jordan Bell.

19. From an 1813 prospectus, as quoted in Zerah Colburn's memoir (Colburn, 1833, p. 38). See Mitchell (1907) for a discussion of this incident.

20. Euler (1750). Dunham (1990, pp. 223–45) has a nice account of Euler's proof.

21. Euler's first of four proofs of Fermat's little theorem and its generalization appears in Euler (1741). We now know that Leibniz proved Fermat's little theorem some time before 1683, but never published it.

22. Weil (1984, p. 58)

23. Euler (1751), Vandermonde (1774)

24. This excerpt comes from a letter that Gauss wrote to his former student Christian Ludwig Gerling on January 6, 1819 (Frei, 2007).

25. If z is on the unit circle, then $z = e^{i\theta}$ for some θ. Then $1/z = 1/e^{i\theta} = e^{i(-\theta)} = \overline{z}$.

26. In particular, after the vertex z_k we list the vertex z_{3k}, where $3k$ is reduced modulo 17. So for instance, after z_9 comes $z_{27} = z_{10}$, and after this comes $z_{30} = z_{13}$.

27. To see where this equality comes from, notice that because s is odd we have $2^{rs} + 1 = (2^r)^s - (-1)^s$. Then we use the identity $a^s - b^s = (a - b)(a^{s-1} + a^{s-2}b + \cdots + ab^{s-2} + b^{s-1})$ with $a = 2^r$ and $b = -1$.

TANGENT: MIRRORS

1. See Dayoub and Lott (1977) for a variety of geometry problems for the Mira.

2. If this Tangent reads a lot like the Tangent on origami (page 192), it should. There is a clear connection between folding a piece of origami paper across a line and reflecting an object across a mirror. Thus, Mira constructions are origami constructions, and vice versa.

3. See Dayoub and Lott (1977, pp. 54–60) and Emert et al. (1994).

4. In fact, Hochstein (1963) even took refraction into account—the bending of the light as it passes through the glass. We ignore that factor.

CHAPTER 20. PIERRE WANTZEL

1. This appeared in Heinrich Weber's obituary for Kronecker, "Die ganzen Zahlen hat der liebe Gott gemacht, alles andere ist Menschenwerk" (Weber, 1893).

2. Hayes (2007)

3. de Lapparent (1895)

4. As quoted in Cajori (1918).

5. Saint-Venant (1848)

6. Saint-Venant (1848)

7. Saint-Venant (1848)

8. Saint-Venant (1848)

9. de Lapparent (1895)

10. Smale (1981) pointed out that Gauss's first and fourth proofs had a topological flaw that is an "immense gap" and yet "a subtle point even today." Alexander Ostrowski (1893–1986) fixed the flaw in 1920.

11. For example, *Eisenstein's criterion* is a very well-known technique. It was proved independently by Theodor Schönemann in 1846 and Eisenstein in 1850. Suppose $f(x) = a_n x^n + \cdots + a_1 x + a_0$ has integer coefficients, and there is a prime p that divides $a_{n-1}, \ldots, a_0$ but not a_n and p^2 does not divide a_0. Then f is irreducible over the rational numbers. For example, 7 divides 21 and 14, but not 2, and $7^2 = 49$ does not divide 14. Thus $f(x) = 2x^2 + 21x + 14$ is irreducible.

12. This is article 42 in Gauss (1965, p. 25).

13. Notice that when we factored out $x - \frac{3}{2}$, the other factor also had rational coefficients. This will always happen. If a polynomial with rational coefficients and degree of at least 2 has a rational root, then it is reducible.

14. See Wantzel (1837). Hayes (2007) contains a link to an English translation. Hayes also gives a plausible explanation for why Wantzel's name is listed as "M. L. Wantzel" instead of "P. L. Wantzel" in the byline for the article. The "M." is an honorific for Monsieur, which was somewhat common in this publication at

this time. Hayes wonders if Wantzel was familiarly known by his middle name Laurent, rather than his first name Pierre.

15. Hartshorne (1998) pointed out in an online forum that he had discovered an error in Wantzel's proof. This prompted an interesting discussion about whether we should still credit Wantzel with the first proof of the impossibility result. The prominent mathematician John Conway weighed in asserting that yes, it is an error, and if presented to Wantzel at the time, he would have recognized that it was an error. Conway speculates that Wantzel would have been able to quickly find a fix. Thus, in his view, Wantzel still deserves credit (Conway, 1998).

16. Suzuki (2008) has an extended explanation of Wantzel's proof.

17. This example is from Hartshorne (2000a, p. 246).

18. Thank you to Bill Dunham for pointing this out to me.

19. Observe that Wantzel wrote the polynomial as $x^3 âĿŠ \frac{3}{4}x + \frac{1}{4}a$.

20. Here's the sketch due to Eisenstein using his irreducibility criterion (see note 11 on page 324). Let $f(z) = z^{p-1} + z^{p-2} + \cdots + z + 1$ and let $g(z) = f(z+1)$. One can show that f is irreducible if and only if g is irreducible. Eisenstein's criterion does not help us with f, but it does with g. If we multiply out $g(z) = (z+1)^{p-1} + (z+1)^{p-2} + \cdots + (z+1) + 1$, we obtain $g(z) = z^{p-1} + a_{p-2}z^{p-2} + \cdots + a_1 z + p$, where p divides all of the coefficients (except the leading coefficient). Because p^2 does not divide the constant term, then g is irreducible, and hence so is f.

21. Recall that on page 314 we showed that if a prime has the form $2^l + 1$, then it has the form $2^{2^j} + 1$.

22. All of the nonzero coefficients in the examples shown in the chapter are 1 or -1, but this is not always the case. However, the smallest cyclotomic polynomial to break this pattern is $\Phi_{105}(z)$: the coefficients of z^7 and z^{41} are both -2.

23. Weintraub (2013) contains several classical proofs of the irreducibility of the nth cyclotomic polynomial, including Gauss's proof when n is prime.

24. Francis (1978) extended this observation to show that an angle of rational-degree measure is constructible if and only if it has the form $3n/r$ where n and r are relatively prime, and $120r$ is a product of the type presented in Gauss's theorem.

25. This is a modified version of an example in Buckley and MacHale (1985), which contains other examples as well. They also give criteria ensuring that an angle is not constructible and cannot be trisected.

26. This example can be found on the Alexander Bogomolny's Cut-the-Knot website (Bogomolny, 2017a). He credits Andrew Schultz with the example, but in a follow-up post (Bogomolny, 2017b) Ed Fisher argues that the proof was invalid, and proceeds to give a rigorous one.

27. Buckley and MacHale (1985)

28. Graves (1889, pp. 433–35)

29. According to Lützen (2009), see Petersen (1877, pp. 161–77).

30. Klein (1897, p. 2)

31. Klein (1897, p. 2)

32. Hobson (1913, p. 50)

33. Archibald (1914a)

34. Archibald (1914b)

35. Pierpont (1895)

36. Cajori (1918)

37. Bell (1986, p. 67)

38. In 1945 Bell correctly acknowledged that Wantzel proved it is impossible to trisect an arbitrary angle and to double the cube. See Francis (1986).

39. Francis (1986) stated that Coxeter's 1961 *Introduction to Geometry*, Boyer's 1968 *History of Mathematics*, and Burton's 1976 *Elementary Number Theory* claim that Gauss proved the converse. This misstatement also appears in Eves's 1990 *An Introduction to the History of Mathematics* (Eves, 1990, p. 152).

40. Francis (1986)

41. Lützen (2009)

42. Lützen (2010)

43. Lützen (2009)

44. Wantzel (1837)

TANGENT: WHAT CAN WE CONSTRUCT WITH OTHER TOOLS?

1. Alford (1839, pp. 555–56)

2. Wernick (1971) gives many such examples.

3. Poncelet (1822). See also Martin (1998, pp. 99–101) and Wernick (1971).

4. See Bos (2001, p. 169, footnote 6).

5. This sequence is A005109 in the *On-line Encyclopedia of Integer Sequences*.

6. Gleason (1988) gives constructions of the heptagon and the 13-gon. Conway and Guy (1996, pp. 199–200) gives these and the nonagon.

7. This prime was discovered in 2009 by Andy Brady's computer as part of Paul Underwood's 321 Prime Search program, which crowdsources the search for primes of the form $3 \cdot 2^k + 1$. See PrimeGrid (2018).

8. Gleason (1988)

9. Videla (1997) and Bainville and Genevés (2000) define conic constructible.

10. Knorr (1993, p. 128)

11. Videla (1997)

12. Pierpont (1895) or Videla (1997). Eric Bainville and Bernard Genevés gave explicit constructions of n-gons for $n = 5, 7, 9, 13, 17, 19, 37, 73$, and 97 (Bainville and Genevés, 2000). Plus, they give a general procedure to follow for constructing other polygons (it won't necessarily work for all n).

13. Emert et al. (1994)

14. Alperin (2000)

15. Martin (1998, ch. 9)

16. See Benjamin and Snyder (2014, theorem 2) for details.

17. Baragar (2002)

18. Benjamin and Snyder (2014)

CHAPTER 21. IRRATIONAL AND TRANSCENDENTAL NUMBERS

1. Joyce (2002, p. 484). Notice that the year mentioned (1882) in this excerpt is the year that Lindemann solved the problem of squaring the circle.

2. One-third of *Arithmetica integra* was essentially an algebraic presentation of book X of Euclid's *Elements* (Stifel, 1544). English translation in Kline (1972, p. 251).

3. As quoted in the Zukav (1984, footnote p. 208).

4. Bos (2001, p. 138)

5. Stevin (1958, p. 532)

6. Newton (1972, p. 493)

7. Struik (1969, p. 633)

8. Grabiner (1983)

9. Euler's proof was presented to the St. Petersburg Academy on March 7, 1737, and it appeared in print in 1744 (Euler, 1744a). There is an English translation by M. F. Wyman and B. F. Wyman (Euler, 1985). See Sandifer (2007c) for details.

10. de Stainville (1815, pp. 339–41)

11. Titchmarsh (1948, p. 159)

12. As quoted in Knorr (1993, p. 363).

13. Berggren et al. (2004, p. 753)

14. This English translation of Euler (1748) is found in Struik (1969, p. 347).

15. Lambert's 1761 presentation was published in 1768 (Lambert, 1768). The results also appeared in 1766 (Lambert, 1770). See Struik (1969, pp. 369–74) for an English translation. Wallisser (2000) looks at the rigor of Lambert's proof and concludes that it is rigorous. Laczkovich (1997) gives a streamlined proof.

16. Translation from Struik (1969, p. 374).

17. Legendre (1794, p. 304)

18. Niven (1947), Niven (1956). See Jones (2010) for a variation.

19. For another short proof, see Breusch (1954).

20. Hermite (1873)

21. See Niven (1956, p. 27) for further references.

22. See Erdős and Dudley (1983), for instance.

23. Petrie (2012) proposes that "Euler considered constant quantities to be transcendental if the function describing their relationship to the unity was transcendental."

24. Liouville's theorem states that if α is an irrational algebraic number of degree $d > 1$ (that is, α is the root of an irreducible polynomial of degree d), then there exists $c > 0$ such that for every rational number p/q, we have $c/q^d \leq |\alpha - p/q|$.

25. He wrote two short notes on this topic (Liouville, 1844a,b).

26. Liouville (1851)

27. See Yandell (2002, pp. 172–74) for a more detailed intuitive explanation. To illustrate the argument, he uses the transcendental

$$\frac{1}{10} + \frac{1}{10^{10}} + \frac{1}{10^{10^{10}}} + \frac{1}{10^{10^{10^{10}}}} + \cdots,$$

which has zeros even farther apart.

28. See Petrie (2012).

29. Presented to the St. Petersburg Academy in 1775 (Euler, 1785).

30. Legendre (1794, pp. 303–4)

31. In one sense there are many Liouville numbers: they are uncountably infinite. But in another sense the set is small: it has measure zero.

32. As quoted in Freudenthal (2008).

33. Baker (1990, p. 3)

34. Hermite (1873)

35. Freudenthal (2008)

36. As quoted in Bell (1986, p. 464).

37. The emphasis was in Lindemann's original (Lindemann, 1882). This translation is found in Remmert (1990b).

38. Recall that $e^a = b$ is equivalent to $\ln(b) = a$.

39. Suppose π is algebraic. Then because the algebraic numbers are a field (see the proof in Niven (1956, p. 84)), πi is algebraic, which is a contradiction. Thus π is transcendental.

40. Weierstrass (1885), as quoted in Shidlovskii (1989, p. 4).

41. Weierstrass (1885). Using the language of linear algebra, the Lindemann–Weierstrass theorem says that $e^{a_1}, \ldots, e^{a_m}$ are linearly independent over the algebraic numbers.

42. See Mary Winston Newson's English translation (Hilbert, 1902).

43. Euler (1748, sect. 105)

44. Gelfond (1929)

45. See Yandell (2002, p. 194).

46. Kuzmin (1930)

47. We can draw the weaker conclusion that an irrational number raised to an irrational power can be rational without invoking this powerful theorem. We know $\sqrt{2}$ is irrational, so either $\sqrt{2}^{\sqrt{2}}$ is rational, in which case we have found our example, or it is not, in which case we can use the argument in the text.

48. According to Hille (1942, p. 198), Gelfond posted an outline of his proof on April 1, 1934. Schneider learned of Gelfond's proof on May 28, the day that he submitted his paper (Gelfond, 1934; Schneider, 1935).

49. Baker proved that if $a_1, \ldots, a_k$ are algebraic numbers not equal to 0 or 1, and $b_1, \ldots, b_k$ are irrational numbers such that $1, b_1, \ldots, b_k$ are linearly independent over the rational numbers, then $a_1^{b_1} \cdots a_k^{b_k}$ is transcendental. See Baker (1990) and Baker (1984), for instance.

50. Cantor (1874)

51. Cantor proved this in his 1874 article and then gave another proof—the famous "diagonal slash" proof—in 1891 (Cantor, 1891).

Tangent: Top 10 Transcendental Numbers

1. Written by W. S. Gilbert and music composed by Arthur Sullivan.

2. Siegel (2014)

3. Mahler (1937)

Epilogue: Sirens or Muses?

1. Hobson (1913, p. 12)

2. Yates (1942, p. 6)

References

Abbott, E. A. (1884). *Flatland: A Romance of Many Dimensions*. London: Seely.

Académie des Sciences France (1778). *Histoire de l'Académie Royale, année 1775*. Paris: De L'Imprimerie Royale.

Acerbi, F. (2008). Archimedes. In C. C. Gillispie (ed.), *Complete Dictionary of Scientific Biography*, Volume 19, pp. 85–91. Detroit: Charles Scribner's Sons.

Adler, A. (1890). Zur Theorie der Mascheronischen Konstruktionen. *Situngsberichte der Wiener Akademie 99*, 910–16.

Adler, A. (1906). *Theorie der geometrischen Konstruktionen*. Leipzig: G. J. Göschensche.

Albertini, T. (2004). Mathematics and astronomy. In C. M. Bellitto, T. M. Izbicki, G. Christianson (eds.), *Introducing Nicholas of Cusa: A Guide to a Renaissance Man*, pp. 373–406. New York: Paulist Press.

Aleklett, K., D. J. Morrissey, W. Loveland, P. L. McGaughey, and G. T. Seaborg (1981). Energy dependence of ^{209}Bi fragmentation in relativistic nuclear collisions. *Phys. Rev. C 23*(3), 1044–46.

Alexander, J. C. (2011). Blending in mathematics. *Semiotica 187*, 1–48.

Alford, H. (1839). *The Works of John Donne*, Volume I. London: John W. Parker.

Alighieri, D. (1867). *The Divine Comedy of Dante Alighieri. Volume III: Paradiso (Translated by Henry Wadsworth Longfellow)*. Boston: Ticknor & Fields.

Alperin, R. C. (2000). A mathematical theory of origami constructions and numbers. *New York J. Math. 6*, 119–33.

Andersen, K. (1980). An impression of mathematics in Denmark in the period 1600–1800. *Centaurus 24*, 316–34. Special issue dedicated to Olaf Pedersen on his sixtieth birthday.

Andersen, K. and H. Meyer (1985). Georg Mohr's three books and the *Gegenübung auf Compendium Euclidis curiosi*. *Centaurus 28*(2), 139–44.

Archibald, R. C. (1914a). Book review: "Squaring the circle," a history of the problem. *Bull. Amer. Math. Soc. (N.S.) 21*(2), 82–93.

Archibald, R. C. (1914b). Remarks on Klein's "Famous problems of elementary geometry." *Amer. Math. Monthly 21*(8), 247–59.

Archibald, R. C. (1916). Book review: Die elementargeometrischen Konstruktionen des regelmässigen Siebzehnecks: Eine historisch-kritische Darstellung. *Bull. Amer. Math. Soc. 22*(5), 239–46.

Archimedes (2002). *The Works of Archimedes*. Mineola, NY: Dover Publications. Reprint of the 1897 edition and the 1912 supplement, edited by T. L. Heath.

Aristophanes (2000). *Aristophanes: Birds. Lysistrata. Women at the Thesmophoria. Edited and Translated by Jeffrey Henderson*. Cambridge, MA: Harvard University Press.

Aristotle (1869). *The Nicomachean Ethics of Aristotle: Newly Translated into English by Robert Williams*. Longmans, Green.

Arndt, J. and C. Haenel (2000). *Pi—Unleashed*. Berlin: Springer.

Artmann, B. (1999). *Euclid—The Creation of Mathematics*. New York: Springer.

Atiyah, M. (2001, August–September). Mathematics in the 20th century. *Amer. Math. Monthly 108*(7), 654–66.

Aubrey, J. (1898). *'Brief Lives,' Chiefly of Contemporaries, Set Down by John Aubrey, between the Years 1669 & 1696*, Volume I (A–H). Oxford: Clarendon Press.

Austin, J. D. and K. A. Austin (1979, April). Constructing and trisecting angles with integer angle measures. *Math. Teacher 72*, 290–93.

Badger, L. (1994). Lazzarini's lucky approximation of π. *Math. Mag. 67*(2), 83–91.

Baez, J. (1998). The crackpot index. http://math.ucr.edu/home/baez/crackpot .html.

Baez, J. and G. Egan (2017, March 7). Azimuth: Pi and the golden ratio. https: //johncarlosbaez.wordpress.com/2017/03/07/pi-and-the-golden-ratio/.

Bailey, D. H. and J. Borwein (2014). Pi Day is upon us again and we still do not know if pi is normal. *Amer. Math. Monthly 121*(3), 191–206.

Bailey, D. H., J. M. Borwein, P. B. Borwein, and S. Plouffe (1997). The quest for pi. *Math. Intelligencer 19*(1), 50–57.

Bainville, E. and B. Genevés (2000). Constructions using conics. *Math. Intelligencer 22*(3), 60–72.

Baker, A. (1966). Linear forms in the logarithms of algebraic numbers IV. *Mathematika 15*, 204–16.

Baker, A. (1984). *A Concise Introduction to the Theory of Numbers*. Cambridge: Cambridge University Press.

Baker, A. (1990). *Transcendental Number Theory*. Cambridge: Cambridge University Press.

Ball, W.W.R. (1914). *Mathematical Recreations and Essays*. London: Macmillan.

Balzac, H. de (1900). *Honoré de Balzac in Twenty-Five Volumes: The First Complete Translation into English*, Volume 25. New York: P. F. Collier & Sons.

Baragar, A. (2002). Constructions using a compass and twice-notched straightedge. *Amer. Math. Monthly 109*(2), 151–64.

Barbaro, D. (1569 (Arnaldo Forni reprint, 1980)). *La practica della perspecttiva*, Volume 8. Biblioteca di architettura e urbanistica. Teoria e storia. Collana diretta da Roberto Fregna.

Barner, K. (2001). How old did Fermat become? *NTM: International Journal for History and Ethics of Natural Sciences, Technology and Medicine (New series) 8*(4), 209–28.

Bashmakova, I. and G. Smirnova (2000). *The Beginnings and Evolution of Algebra*. Washington, DC: Mathematical Association of America.

Basset, A. B. (1901). *An Elementary Treatise on Cubic and Quartic Curves*. Cambridge: Deighton Bell.

Bassetto, C. D. (1937). *London Music in 1888–89: As Heard by Corno Di Bassetto (Later Known as Bernard Shaw) with Some Further Autobiographical Particulars*. London: Constable.

Beckmann, P. (1971). *A History of Pi*. New York: St. Martin's Press.

Bell, E. T. (1986). *Men of Mathematics*. New York: Simon & Schuster.

Belloch, M. P. (1936). Sul metodo del ripiegamento della carta per la risoluzione dei problemi geometrici. *Periodico di Mathematische Ser. 4 16*(2), 104–8.

Benedetti, G. (1553). *Resolutio omnium Euclidis problematum, aliorumque ad hoc necessario inventorum, una tantummodo circini data apertura*. Venice: Venetiis.

Benjamin, E. and C. Snyder (2014). On the construction of the regular hendecagon by marked ruler and compass. *Math. Proc. Cambridge Philos. Soc. 156*(3), 409–24.

Berger, E. J. (1951). A simple trisection device. *Math. Teacher 44*, 319–20.

Berggren, L., J. Borwein, and P. Borwein (2004). *Pi: A Source Book* (3rd ed.). New York: Springer.

Bernoulli, D. (1724). *Exercitationes quaedam mathematicae*, Chapter "Problema aliquod geometricum." Venice: Venetiis.

Bernoulli, J. (1690, May). Quæstiones nonnullæ de usuris, cum solutione problematis de sorte alearum, propositi in ephem. gall. a. 1685. *Acta Eruditorum*, 219–23.

Bhojane, K. A. (1987). 71.40 an apparatus for trisecting an angle. *Math. Gaz. 71*(458), 299–300.

Blåsjö, V. (2015). The myth of Leibniz's proof of the fundamental theorem of calculus. *Nieuw Archief voor Wiskunde 16*(1), 46–50.

Blåsjö, V. (2016). In defence of geometrical algebra. *Arch. Hist. Exact Sci. 70*(3), 325–59.

Blonder, G. (2015, September 4). Trisecting the angle with a straightedge. https://plus.maths.org/content/trisecting-angle-ruler.

Bogomolny, A. (2017a). There are trisectable angles that are not constructible. http://www.cut-the-knot.org/do_you_know/trisect.shtml.

Bogomolny, A. (2017b). There are trisectable angles that are not constructible: Some corrections. http://www.cut-the-knot.org/do_you_know/trisect1 .shtml.

Bombelli, R. (1572). *L'algebra parte maggiore dell'arimetica*. Bologna: Giovanni Rossi.

Bos, H.J.M. (1981). On the representation of curves in Descartes' *Géométrie*. *Arch. Hist. Exact Sci. 24*(4), 295–338.

Bos, H.J.M. (1988). Tractional motion and the legitimation of transcendental curves. *Centaurus 31*(1), 9–62.

Bos, H.J.M. (2001). *Redefining Geometrical Exactness: Descartes' Transformation of the Early Modern Concept of Construction*. Sources and Studies in the History of Mathematics and Physical Sciences. New York: Springer.

Bos, H.J.M. (2008). Huygens, Christiaan (also Hughens, Christian). In C. C. Gillispie (ed.), *Complete Dictionary of Scientific Biography*, Volume 6, pp. 597–613. Detroit: Charles Scribner's Sons.

Boyer, C. B. (1947). Cartesian geometry from Fermat to Lacroix. *Scripta Math. 13*, 133–53.

Boyer, C. B. (1951). The foremost textbook of modern times. *Amer. Math. Monthly 58*(4), 223–26.

Boyer, C. B. (1964). *L'aventure de l'esprit: Mélanges Alexandre Koyré*, Volume I, Chapter "Early rectification of curves," pp. 30–39. Hermann.

Boyer, C. B. and U. C. Merzbach (2011). *A History of Mathematics* (3rd ed.). Hoboken, NJ: John Wiley & Sons.

Bretschneider, C. A. (1870). *Die Geometrie und die Geometer vor Euklides*. Leipzig: B. G. Teubner.

Breusch, R. (1954). A proof of the irrationality of π. *Amer. Math. Monthly 61*, 631–32.

Brooks, D. A. (2007). A new method of trisection. *College Math. J. 38*(2), 78–81.

Buckley, S. and D. MacHale (1985). Dividing an angle into equal parts. *Math. Gaz. 69*(447), 9–12.

Bulmer-Thomas, I. (1976). *Selections Illustrating the History of Greek Mathematics with an English Translation by Ivor Thomas, Vol. 2 of Loeb Classical Library.* Cambridge, MA: Harvard University Press.

Bulmer-Thomas, I. (2008a). Euclid. In C. C. Gillispie (ed.), *Complete Dictionary of Scientific Biography,* Volume 4, pp. 414–37. Detroit: Charles Scribner's Sons.

Bulmer-Thomas, I. (2008b). Hippias of Elis. In C. C. Gillispie (ed.), *Complete Dictionary of Scientific Biography,* Volume 6, pp. 405–10. Detroit: Charles Scribner's Sons.

Bulmer-Thomas, I. (2008c). Hippocrates of Chios. In C. C. Gillispie (ed.), *Complete Dictionary of Scientific Biography,* Volume 6, pp. 410–18. Detroit: Charles Scribner's Sons.

Bulmer-Thomas, I. (2008d). Menaechmus. In C. C. Gillispie (ed.), *Complete Dictionary of Scientific Biography,* Volume 9, pp. 268–77. Detroit: Charles Scribner's Sons.

Bulmer-Thomas, I. (2008e). Oenopides of Chios. In C. C. Gillispie (ed.), *Complete Dictionary of Scientific Biography,* Volume 10, pp. 179–82. Detroit: Charles Scribner's Sons.

Bulmer-Thomas, I. (2008f). Theaetetus. In C. C. Gillispie (ed.), *Complete Dictionary of Scientific Biography,* Volume 13, pp. 301–7. Detroit: Charles Scribner's Sons.

Bulmer-Thomas, I. (2008g). Theodorus of Cyrene. In C. C. Gillispie (ed.), *Complete Dictionary of Scientific Biography,* Volume 15, p. 503. Detroit: Charles Scribner's Sons.

Bulmer-Thomas, I. (2008h). Theodorus of Cyrene. In C. C. Gillispie (ed.), *Complete Dictionary of Scientific Biography,* Volume 14, pp. 314–19. Detroit: Charles Scribner's Sons.

Burckhardt, J. J. (2008). Steiner, Jakob. In C. C. Gillispie (ed.), *Complete Dictionary of Scientific Biography,* Volume 13, pp. 12–22. Detroit: Charles Scribner's Sons.

Burkert, W. (1972). *Lore and Science in Ancient Pythagoreanism.* Cambridge, MA: Harvard University Press.

Burton, D. M. (2007). *The History of Mathematics: An Introduction* (6th ed.). Boston: McGraw-Hill.

Busard, H.L.L. (2008). Viète, François. In C. C. Gillispie (ed.), *Complete Dictionary of Scientific Biography,* Volume 14, pp. 18–25. Detroit: Charles Scribner's Sons.

Cajori, F. (1918). Pierre Laurent Wantzel. *Bull. Amer. Math. Soc.* 24(7), 339–47.

Cajori, F. (1929). A forerunner of Mascheroni. *Amer. Math. Monthly* 36(7), 364–65.

Cajori, F. (1991). *A History of Mathematics* (5th ed.). New York: Chelsea.

Cajori, F. (2007a). *A History of Mathematical Notations,* Volume I. New York: Cosimo.

Cajori, F. (2007b). *A History of Mathematical Notations,* Volume II. New York: Cosimo.

Caldwell, C. (2017). The PrimeNumbers' Crackpot index. https://primes.utm.edu/notes/crackpot.html.

Cantor, G. (1874). Über eine Eigenschaft des Inbegriffs aller reellen algebraischen Zahlen. *J. reine angew. Math.* 77, 258–62.

Cantor, G. (1891). Ueber eine elementare Frage der Mannigfaltigkeitslehre. *Jahresber. Dtsch. Math.-Ver. 1*, 75–78.

Cappon, L. J. (2012). *The Adams-Jefferson Letters: The Complete Correspondence between Thomas Jefferson and Abigail and John Adams*. Chapel Hill: University of North Carolina Press.

Cardano, G. (1545). *Ars magna*. Nüremberg: Johann Petreius.

Cardano, G. (1550). *De subtilitate libri XXI*. Nuremberg: Johannes Petreius.

Cardano, G. (1968). *The Great Art or Rules of Algebra, Translated by T. R. Witmer*. Cambridge, MA: MIT Press.

Cardano, G. (2002). *The Book of My Life (De vita propria liber)*. New York: New York Review Books.

Carpenter, F. B. (1872). *The Inner Life of Abraham Lincoln: Six Months at the White House*. Cambridge: Hurd and Houghton.

Carroll, L. (1917). *Through the Looking-Glass: And What Alice Found There*. Chicago: Rand McNally.

Castellanos, D. (1988a). The ubiquitous π. *Math. Mag. 61*(2), 67–98.

Castellanos, D. (1988b). The ubiquitous π. *Math. Mag. 61*(3), 148–63.

Cauchy, A.-L. (1821). *Cours d'analyse de l'École Royale Polytechnique*. Paris: Imprimerie Royale.

Cauer, D. (1912). Über die Konstruktion des Mittelpunktes eines Kreises mit dem Lineal allein. *Math. Ann. 73*(1), 90–94.

Cauer, D. (1913). Über die Konstruktion des Mittelpunktes eines Kreises mit dem Lineal allein. *Math. Ann. 74*(3), 462–64.

Cavers, M. (2011). The pi manifesto. http://www.thepimanifesto.com/.

Cawthorne, S. and J. Green (2009). Harold and the purple heptagon. *Math Horiz. 17*(1), 5–9.

Chace, A. B. (1927). *The Mathematical Rhind Papyrus*, Volume I. Oberlin, Ohio: Mathematical Association of America.

Chasles, M. (1875). *Aperçu historique sur l'origine et le développement des méthodes en géométrie*. Paris: Gauthier-Villars.

Cheney, Jr., W. F. (1953, March). Can we outdo Mascheroni? *Math. Teacher 46*, 152–56.

Cicero, M. T. (1886). *Tusculan Disputations: Translated with an Introduction and Notes by Andrew P. Peabody*. Boston: Little, Brown.

Clagett, M. (1964). *Archimedes in the Middle Ages. Vol. I: The Arabo-Latin Tradition*. Madison, WI: University of Wisconsin Press.

Clagett, M. (2008). Archimedes. In C. C. Gillispie (ed.), *Complete Dictionary of Scientific Biography*, Volume 1, pp. 213–31. Detroit: Charles Scribner's Sons.

Clary, D. A. (2004). *Rocket Man: Robert H. Goddard and the Birth of the Space Age*. Hyperion.

Clausen, T. (1840, Jan). Vier neue mondförmige Flächen, deren Inhalt quadrirbar ist. *J. reine angew. Math. 1840*(21), 375–76.

Colburn, Z. (1833). *A Memoir of Zerah Colburn: Written by Himself*. Springfield, MA: G. and C. Merriam.

Cole, J. H. (1925). *Survey of Egypt Paper No. 39: The Determination of the Exact Size and Orientation of the Great Pyramid of Giza*. Cairo: Government Press.

Conway, J. H. (1998, November 23). Re: [HM] Wantzel. http://mathforum.org/kb/message.jspa?messageID=1175545.

Conway, J. H. and R. Guy (1996). *The Book of Numbers*. New York: Springer.

Coolidge, J. L. (1949). *The Mathematics of Great Amateurs*. Oxford: Clarendon Press.

Corry, L. (2015). *A Brief History of Numbers*. Oxford: Oxford University Press.

Cotes, R. (1714). Logometria. *Phil. Trans. 29*, 5–45.

Courant, R. and H. Robbins (1941). *What Is Mathematics?* New York: Oxford University Press.

Court, N. A. (1958, May). Mascheroni constructions. *Math. Teacher 51*, 370–72.

Cox, D. A. (2012). *Galois Theory* (2nd ed.). Hoboken, NJ: John Wiley & Sons.

Crombie, A. C. (2008). Descartes, René du Perron. In C. C. Gillispie (ed.), *Complete Dictionary of Scientific Biography*, Volume 4, pp. 51–55. Detroit: Charles Scribner's Sons.

Crossley, J. N. (1987). *Emergence of Number* (2nd ed.). Singapore: World Scientific.

Crowe, D. W. (1992). Albrecht Dürer and the regular pentagon. In I. Hargittai (ed.), *Fivefold Symmetry*, pp. 465–87. Singapore: World Scientific.

da Vinci, L. (1939). *The Literary Works of Leonardo da Vinci* (enl. 2nd ed.), Volume 1. Oxford University Press.

Dantzig, T. (1955). *The Bequest of the Greeks*. George Allen & Unwin.

Dantzig, T. (2007). *Number: The Language of Science*. New York: Plume.

Dawson, T. R. (1939). "Match-stick" geometry. *Math. Gaz. 23*(254), 161–68.

Dayoub, I. M. and J. W. Lott (1977). *Geometry: Constructions and Transformations*. Dale Seymour Publications.

de Lapparent, A. (1895). Wantzel (1814–1848). In École Polytechnique (ed.), *Livre du centenaire, 1794–1894*, Volume I: L'école et la science., pp. 133–35. Paris: Gauthier-Villars et Fils.

de Lavoisier, A. L. (1777). Rapport fait a l'Académie Royale des Sciences, par MM. Fougeroux, Cadet, & Lavoisier d'une observation communiquée par M. l'Abbé Bachelay sur une pierre qu'on prétend être tombée du ciel pendant un orage. *J. Physique 2*, 251–55.

De Morgan, A. (1915a). *Budget of Paradoxes*, Volume I. Chicago: Open Court.

De Morgan, A. (1915b). *Budget of Paradoxes*, Volume II. Chicago: Open Court.

de Stainville, J. (1815). *Mélanges d'analyse algébrique et de géométrie*. Paris: Veuve Courcier.

Dedekind, R. (1872). *Stetigkeit und irrationale Zahlen*. Brunswick: F. Vieweg und sohn.

Dehn, M. and E. D. Hellinger (1943). Certain mathematical achievements of James Gregory. *Amer. Math. Monthly 50*, 149–63.

Demaine, E. D. and J. O'Rourke (2007). *Geometric Folding Algorithms: Linkages, Origami, Polyhedra*. Cambridge: Cambridge University Press.

Descartes, R. (1908). *Oeuvres de Descartes: Publiées par Charles Adam & Paul Tannery*, Volume 10. Paris: Léopold Cerf.

Descartes, R. (1954). *The Geometry of René Descartes: Translated from French and Latin by David Eugene Smith and Marcia L. Latham*. New York: Dover.

DeTemple, D. W. (1991). Carlyle circles and the Lemoine simplicity of polygon constructions. *Amer. Math. Monthly 98*(2), 97–108.

Dhombres, J. (1993). Is one proof enough? Travels with a mathematician of the Baroque period. *Educational Studies in Mathematics* 24(4), 401–19.

Dickson, L. E. (1911). Constructions with ruler and compasses; regular polygons. In J.W.A. Young (ed.), *Monographs on Topics of Modern Mathematics Relevant to the Elementary Field*, pp. 351–86. New York: Longsmans, Green.

Dickson, L. E. (1922). *First Course in the Theory of Equations*. New York: John Wiley & Sons.

Dijksterhuis, E. J. (1987). *Archimedes*. Princeton, NJ: Princeton University Press. Translated from the Dutch by C. Dikshoorn, reprint of the 1956 edition, with a contribution by Wilbur R. Knorr.

Dorodnov, A. V. (1947). On circular lunes quadrable with the use of ruler and compass. *Doklady Akad. Nauk SSSR (N. S.) 58*, 965–68.

Dörrie, H. (1965). *100 Great Problems of Elementary Mathematics: Their History and Solution*. New York: Dover. Translated from the German by David Antin.

Dudley, U. (1962, May). π_t: 1832–1879. *Math. Mag. 35*(3), 153–54.

Dudley, U. (1983). What to do when the trisector comes. *Math. Intelligencer 5*(1), 20–26.

Dudley, U. (1992). *Mathematical Cranks*. MAA Spectrum. Washington, DC: Mathematical Association of America.

Dudley, U. (1994). *The Trisectors* (rev. ed.). MAA Spectrum. Washington, DC: Mathematical Association of America.

Dudley, U. (1999, February). Legislating pi. *Math Horiz. 6*(3), 10–13.

Dudley, U., P. J. Campbell, P. D. Straffin, and D. P. Chavey (2008, Spring). How to call a crank a crank (and win if you get sued). *UMAPJ. 29*(1), 59–74.

Dun, L. (1996). A comparison of Archimedes' and Liu Hui's studies of circles. In F. Dainian, R. S. Cohen (eds.), *Chinese Studies in the History and Philosophy of Science and Technology*, pp. 279–87. Dordrecht: Springer Netherlands.

Dunham, W. (1990). *Journey through Genius: The Great Theorems of Mathematics*. New York: John Wiley & Sons.

Dunnington, G. W. (2004). *Carl Friedrich Gauss: Titan of Science. A Study of His Life and Work*. Washington, DC: Mathematical Association of America.

Dureisseix, D. (2006). Folding optimal polygons from squares. *Math. Mag. 79*(4), 272–80.

Dürer, A. (1525). *Underweysung der Messung mit dem Zirckel und Richtscheyt*. Nuremberg: Hieronymus Andreas Formschneider.

Dürer, A. (1977). *The Painter's Manual: A Manual of Measurement of Lines, Areas, and Solids by Means of Compass and Ruler (Translated and with a Commentary by Walter L. Strauss)*. New York: Abaris Books.

Duvernoy, S. (2008). Leonardo and theoretical mathematics. *Nexus Network J. 10*(1), 39–49.

Edgar, T. and D. Richeson (2020). A visual proof of Gregory's theorem. To appear in *Math. Mag.*

Edington, W. E. (1935). House bill no. 246, Indiana State Legislature, 1897. *Proc. Indiana Acad. Sci. 45*, 206–10.

Edwards, H. (1987, March). The return of the put-together man. *Spin 2*(12), 54–60.

Emerson, R. W. (1893). *Natural History of Intellect and Other Papers*. Boston: Houghton Mifflin.

Emert, J. W., K. I. Meeks, and R. B. Nelson (1994). Reflections on a Mira. *Amer. Math. Monthly 101*(6), 544–49.

Erdős, P. and U. Dudley (1983). Some remarks and problems in number theory related to the work of Euler. *Math. Mag. 56*(5), 292–98.

Euler, L. (1729). Tentamen explicationis phaenomenorum aeris. *Comment. Acad. Sci. Petropol. 2*, 347–68.

Euler, L. (1736). *Mechanica sive motus scientia analytice exposita, Tomus I*. St. Petersburg: Aoademiae Scientiarum.

Euler, L. (1738). Observationes de theoremate quodam Fermatiano aliisque ad numeros primos spectantibus. *Comment. Acad. Sci. Petropol. 6*, 103–7.

Euler, L. (1741). Theorematum quorundam ad numeros primos spectantium demonstratio. *Comment. Acad. Sci. Petropol. 8*, 141–46.

Euler, L. (1744a). De fractionibus continuis dissertatio. *Comment. Acad. Sci. Petropol. 9*, 98–137.

Euler, L. (1744b). De variis modis circuli quadraturam numeris proxime exprimendi. *Comment. Acad. Sci. Petropol. 9*, 222–36.

Euler, L. (1744c). Solutio problematis geometrici circa lunulas a circulis formatas. *Comm. Acad. Sci. Imp. Petropol. 9*, 207–21.

Euler, L. (1748). *Introductio in analysin infinitorum, 2 vols*. Lausanne: M.-M. Bousquet.

Euler, L. (1750). Theoremata circa divisores numerorum. *Novi Comment. Acad. Sci. Imp. Petropol. 1*, 20–48.

Euler, L. (1751). De extractione radicum ex quantitatibus irrationalibus. *Comment. Acad. Sci. Petropol. 16*, 16–60.

Euler, L. (1770). *Vollständige Anleitung zur Algebra*. St. Petersburg: Kays. Acad. der Wissenschaften.

Euler, L. (1772). Considerationes cyclometricae. *Novi Comment. Acad. Sci. Imp. Petropol. 16*, 160–70.

Euler, L. (1785). De relatione inter ternas pluresve quantitates instituenda. *Opuscula Analytica 2*, 91–101.

Euler, L. (1794). De formulis differentialibus angularibus maxime irrationalibus, quas tamen per logarithmos et arcus circulares integrare licet. *Institutiones calculi integralis 4*, 183–94.

Euler, L. (1862). Meditatio in experimenta explosione tormentorum nuper instituta. In P. H. Fuss, N. Fuss (eds.), *Opera Postuma mathematica et physica*, Volume 2, pp. 800–804. St. Petersburg: St. Petersburg Academy of Science.

Euler, L. (1882). *Elements of Algebra (Translation from the French by Rev. John Hewlett)* (3rd ed.). London: Longman, Hurst, Rees, Orme.

Euler, L. (1985). An essay on continued fractions (translated by Myra F. Wyman and Bostwick F. Wyman). *Math Systems Theory 18*, 295–328.

Euler, L. (1988). *Introduction to Analysis of the Infinite: Book I. Translated by John D. Blanton*. New York: Springer.

Eves, H. (1990). *An Introduction to the History of Mathematics* (6th ed.). Thompson Brooks/Cole.

Farrington, O. C. (1900). The worship and folk-lore of meteorites. *J. American Folklore 13*(50), 199–208.

Fauvel, J. and J. Gray (1987). *The History of Mathematics: A Reader*. Palgrave Macmillan.

Fermat, P. de (1894). *Oeuvres de Fermat*. Paris: Gauthier-Villars.

Ferrari, L. (1547, October). *Quinto Cartello di Lodovico Ferraro contra Messer Nicolo Tartaglia*. Milan.

Ferraro, G. (2008). *The Rise and Development of the Theory of Series up to the Early 1820s*. New York: Springer.

Finlay, A. H. (1959). Zig-zag paths. *Math. Gaz. 43*(345), 199.

Folkerts, M. (1996). Regiomontanus' role in the transmission and transformation of Greek mathematics. In F. J. Ragep, S. P. Ragep, with S. Livesey (eds.), *Tradition, Transmission, Transformation: Proceedings of Two Conferences on Pre-Modern Science Held at the University of Oklahoma*, pp. 89–114. Leiden: E. J. Brill.

Fowler, D. H. (1979). Ratio in early Greek mathematics. *Bull. Amer. Math. Soc. (N.S.) 1*(6), 807–46.

Francis, R. L. (1978). A note on angle construction. *College Math. J. 9*(2), 75–80.

Francis, R. L. (1986, April). Did Gauss discover that too? *Math. Teacher 59*, 288–93.

Frei, G. (2007). The unpublished section eight: On the way to function fields over a finite field. In *The Shaping of Arithmetic after C. F. Gauss's "Disquisitiones arithmeticae,"* pp. 159–98. Berlin: Springer.

Freudenthal, H. (2008). Hermite, Charles. In C. C. Gillispie (ed.), *Complete Dictionary of Scientific Biography*, Volume 6, pp. 306–9. Detroit: Charles Scribner's Sons.

Fusimi, K. (1980, October). Trisection of angle by Abe. *Saiensu (supplement)*, 8.

Fuss, P. H. (1843). *Correspondance mathématique et physique de quelques célèbres géomètres du XVIIIème siècle*. L'Académie Impériale des Sciences.

Galilei, G. (1606). *Operations of the Geometric and Military Compass (Le operazioni del compasso geometrico et militare): 1977 Translation by Stillman Drake*. Padua.

Gardner, M. (1985). *The Magic Numbers of Dr. Matrix*, Chapter "The King James Bible," pp. 173–89. Buffalo, NY: Mathematical Association of America.

Gardner, M. (1992). *Mathematical Circus*, Chapter "Mascheroni Constructions," pp. 216–31. MAA Spectrum. Washington, DC: Mathematical Association of America.

Gauss, C. F. (1796, June 1). Neue Entdeckungen. *Intelligenzblatt der allgemeinen Literaturzeitung 66*, 554.

Gauss, C. F. (1831). Theoria residuorum biquadraticorum: Commentatio secunda. *Göttingische gelehrte Anzeigen, Stück 64*, 625–38.

Gauss, C. F. (1965). *Disquisitiones arithmeticae. Translated by Arthur A. Clarke*. New Haven: Yale University Press.

Gelfond, A. (1929). Sur les nombres transcendants. *Rendus Acad. Sci. Paris 189*, 189.

Gelfond, A. O. (1934). Sur le septième problème de D. Hilbert. In *Dokl. Akad. Nauk. SSSR*, Volume 2, pp. 1–6.

Geretschlager, R. (1995). Euclidean constructions and the geometry of origami. *Math. Mag. 68*(5), 357–71.

Gerhardt, C. I. (1849). *Leibnizens Mathematische Schriften*, Volume I. A. Asher & Comp.

Girard, A. (1629). *L'invention nouvelle en l'algébre*. Amsterdam: G. J. Blaeuw.

Girstmair, K. (2003). Hippocrates' lunes and transcendence. *Expo. Math. 21*(2), 179–83.

Gleason, A. M. (1988). Angle trisection, the heptagon, and the triskaidecagon. *Amer. Math. Monthly 95*(3), 185–94.

Gliozzi, M. (2008). Cardano, Girolamo. In C. C. Gillispie (ed.), *Complete Dictionary of Scientific Biography*, Volume 3, pp. 64–67. Detroit: Charles Scribner's Sons.

Goldenring, R. (1915). *Die elementargeometrischen Konstruktionen des regelmässigen Siebzehnecks*. Leipzig: B. G. Teubner.

Goodwin, E. J. (1894, July). Quadrature of the circle. *Amer. Math. Monthly 1*(7), 246–47.

Goodwin, E. J. (1895, November). Queries and information. Letter. *Amer. Math. Monthly 2*(11), 337.

Grabiner, J. (1974, Apr.). Is mathematical truth time-dependent. *Amer. Math. Monthly 81*(4), 354–65.

Grabiner, J. (1983, September). The changing concept of change: The derivative from Fermat to Weierstrass. *Math. Mag. 56*(4), 195–206.

Grabiner, J. (1995). Descartes and problem-solving. *Math. Mag. 68*(2), 83–97.

Graesser, R. F. (1956). Archytas' duplication of the cube. *Math. Teacher 49*(5), 393–95.

Grattan-Guinness, I. (1996). Numbers, magnitudes, ratios, and proportions in Euclid's *Elements*: How did he handle them? *Historia Math. 23*(4), 355–75.

Graves, R. P. (1889). *Life of Sir William Rowan Hamilton*, Volume III. Dublin: Hodges, Figgis.

Gregory of Saint-Vincent (1647). *Opus geometricum quadraturae circuli*. Antwerp.

Guicciardini, N. (2009). *Isaac Newton on Mathematical Certainty and Method*. Transformations: Studies in the History of Science and Technology. Cambridge, MA: MIT Press.

Gurjar, L. V. (1942). The problem of squaring the circle as solved in the Śulvasutras. *J. Univ. Bombay (N.S.) 10*(part 5), 11–16.

Hallerberg, A. E. (1959, April). The geometry of the fixed-compass. *52*, 230–44.

Hallerberg, A. E. (1960, February). Greg Mohr and *Euclidis curiosi*. *Math. Teacher 53*, 127–32.

Hallerberg, A. E. (1977). Indiana's squared circle. *Math. Mag. 50*(3), 136–40.

Hardy, G. H. and E. M. Wright (1975). *An Introduction to the Theory of Numbers* (4th ed.). Oxford: Clarendon Press.

Hartl, M. (2010). The tau manifesto (June 28, 2010; updated March 14, 2011). http://tauday.com/tau-manifesto.

Hartshorne, R. (1998, November 22). [HM] Wantzel. http://mathforum.org/kb/message.jspa?messageID=1175542.

Hartshorne, R. (2000a). *Geometry: Euclid and Beyond*. Undergraduate Texts in Mathematics. New York: Springer.

Hartshorne, R. (2000b). Teaching geometry according to Euclid. *Notices Amer. Math. Soc. 47*(4), 460–65.

Hayashi, T., T. Kusuba, and M. Yano (1990). The correction of the Mādhava series for the circumference of a circle. *Centaurus 33*(2-3), 149–74 (1991).

Hayes, B. (2006, May-June). Gauss's day of reckoning. *American Scientist 94*(3), 200.

Hayes, B. (2007). Foolproof: Mathematical proof is foolproof, it seems, only in the absence of fools. *American Scientist 95*(1), 10–15.

Heath, T. L. (1908a). *The Thirteen Books of Euclid's Elements. Volume I. Books I–II.* Cambridge: Cambridge University Press.

Heath, T. L. (1908b). *The Thirteen Books of Euclid's Elements. Volume II. Books III–IX.* Cambridge: Cambridge University Press.

Heath, T. L. (1908c). *The Thirteen Books of Euclid's Elements. Volume III. Books X–XIII and appendix.* Cambridge: Cambridge University Press.

Heath, T. L. (1921a). *A History of Greek Mathematics. Vol. I: From Thales to Euclid.* Oxford: Clarendon Press.

Heath, T. L. (1921b). *A History of Greek Mathematics. Vol. II: From Aristarchus to Diophantus.* Oxford: Clarendon Press.

Heath, T. L. (1931). *A Manual of Greek Mathematics.* Oxford: Clarendon Press.

Heath, T. L. (1980). *Mathematics in Aristotle.* New York: Garland.

Henry, C. (1886). Lettres inédites d'euler à d'Alembert. *Bullett. Bibliogr. Sc. Matem. 19*, 136–48.

Hermite, C. (1873). Sur la fonction exponentielle. *C. R. Math. Acad. Sci. Paris 77*, 18–24, 74–79, 285–93, 285–93.

Herz-Fischler, R. (2000). *The Shape of the Great Pyramid.* Waterloo, Ontario: Wilfrid Laurier University Press.

Hilbert, D. (1902, July). Mathematical problems. *Bull. Amer. Math. Soc 8*, 437–79.

Hilbert, D. (1903). *Grundlagen der Geometrie.* Leipzig: B. G. Teubner.

Hilbert, D. (1959). *The Foundations of Geometry. Authorized Translation by E. J. Townsend.* La Salle, IL: Open Court.

Hille, E. (1942). Gelfond's solution of Hilbert's seventh problem. *Amer. Math. Monthly 49*(10), 654–61.

Hlavaty, J. H. (1957, November). Mascheroni constructions. *Math. Teacher 50*, 482–87.

Hobbes, T. (1839). *The English Works of Thomas Hobbes,* Volume I. London: John Bohn.

Hobbes, T. (1845). *The English Works of Thomas Hobbes,* Volume VII. London: Longsman, Brown, Green and Longmans.

Hobson, E. W. (1913). *"Squaring the Circle": A History of the Problem.* London: Cambridge University Press.

Hochstein, A. E. (1963, November). Trisection of an angle by optical means. *Math. Teacher 56*(7), 522–24.

Hofmann, J. E. (2008). Saint Vincent, Gregorius. In C. C. Gillispie (ed.), *Complete Dictionary of Scientific Biography,* Volume 12, pp. 74–76. Detroit: Charles Scribner's Sons.

Hogendijk, J. P. (1984). Greek and Arabic constructions of the regular heptagon. *Arch. Hist. Exact Sci. 30*(3-4), 197–330.

Horadam, A. F. (1960). 2926. Constructions possible by ruler and compasses. *Math. Gaz. 44*(350), 270–76.

Horváth, M. (1983). On the Leibnizian quadrature of the circle. *Ann. Univ. Sci. Budapest. Sect. Comput. 4*, 75–83 (1984).

Hudson, H. P. (1961). *Ruler & Compasses.* Longsman Modern Mathematical Series. New York: Longsmans, Green.

Hughes, B. B. (1989). Hippocrates and Archytas double the cube: A heuristic interpretation. *College Math. J. 20*(1), 42–48.

Hull, T. C. (2011). Solving cubics with creases: The work of Beloch and Lill. *Amer. Math. Monthly 118*(4), 307–15.

Hungerbühler, N. (1994). A short elementary proof of the Mohr-Mascheroni theorem. *Amer. Math. Monthly 101*(8), 784–87.

Huxley, G. L. (2008). Eudoxus of Cnidus. In C. C. Gillispie (ed.), *Complete Dictionary of Scientific Biography*, Volume 4, pp. 465–67. Detroit: Charles Scribner's Sons.

Huygens, C. (1654). *De circuli magnitudine inventa*. Johannes and Daniel Elzevier.

Huylebrouck, D. (1996). The π-room in Paris. *Math. Intelligencer 18*(2), 51–53.

Huzita, H. (1994). Drawing the regular heptagon and the regular nonagon by origami (paper folding). *Symmetry: Culture and Science 5*(1), 69–83.

Isaacson, W. (2017). *Leonardo da Vinci*. New York: Simon & Schuster.

Jackson, A. (2005). Uncovering new views on Archimedes. *Notices Amer. Math. Soc. 52*(5), 522.

Jackson, D. E. P. (1980). Towards a resolution of the problem of τὰ ἐνὶ διαστήματι γραφόμενα in Pappus' Collection Book VIII. *Classical Quarterly 30*(2), 523–33.

Jacob, M. (2005). Interdire la quadrature du cercle à L'Académie: Une décision autoritaire des lumières. *Revue d'histoire des mathématiques 11*, 89–139.

James, W. (1909). *The Meaning of Truth: A Sequel to "Pragmatism"*. New York: Longmans, Green.

Jayawardene, S. A. (2008a). Bombelli, Rafael. In C. C. Gillispie (ed.), *Complete Dictionary of Scientific Biography*, Volume 2, pp. 279–81. Detroit: Charles Scribner's Sons.

Jayawardene, S. A. (2008b). Ferrari, Ludovico. In C. C. Gillispie (ed.), *Complete Dictionary of Scientific Biography*, Volume 4, pp. 586–88. Detroit: Charles Scribner's Sons.

Jefferson, T. (2008). *The Papers of Thomas Jefferson. Retirement Series, Vol. 5, 1 May 1812 to 10 March 1813*. Princeton, NJ: Princeton University Press.

Jesseph, D. M. (1999). *Squaring the Circle: The War between Hobbes and Wallis*. Chicago, IL: University of Chicago Press.

Johnson, C. (1975). A construction for a regular heptagon. *Math. Gaz. 59*(407), 17–21.

Johnson, W. W. and W. E. Story (1879). Notes on the "15" puzzle. *American J. Mathematics 2*(4), 397–404.

Jones, P. S. (1954a). Complex numbers: An example of recurring themes in the development of mathematics—I. *Math. Teacher 47*(2), 106–14.

Jones, P. S. (1954b). Complex numbers: An example of recurring themes in the development of mathematics—II. *Math. Teacher 47*(4), 257–64.

Jones, P. S. (1954c). Complex numbers: An example of recurring themes in the development of mathematics—III. *Math. Teacher 47*(5), 340–45.

Jones, P. S. (1956a). Irrationals or incommensurables I: Their discovery, and a "logical scandal." *Math. Teacher 49*(2), 123–27.

Jones, P. S. (1956b). Irrationals or incommensurables III: The Greek solution. *Math. Teacher 49*(4), 282–85.

Jones, P. S. (1956c). Irrationals or incommensurables IV: The transitional period. *Math. Teacher 49*(6), 469–71.

Jones, P. S. (1956d). Irrationals or incommensurables V: Their admission to the realm of numbers. *Math. Teacher 49*(7), 541–543.

Jones, T. W. (2010). Discovering and proving that π is irrational. *Amer. Math. Monthly 117*(6), 553–57.

Jones, W. (1706). *Synopsis palmariorum mathesios*. London.

Joyce, J. (2002). *Ulysses*. Mineola, NY: Dover.

Kac, M. and S. M. Ulam (1968). *Mathematics and Logic*. New York: Dover.

Kazarinoff, N. D. (1968). On who first proved the impossibility of constructing certain regular polygons with ruler and compass alone. *Amer. Math. Monthly 75*(6), 647.

Kazarinoff, N. D. (1970). *Ruler and the Round: Or, Angle Trisection and Circle Division*. Prindle, Weber & Schmidt.

Keats, J. (1820). *The Eve of St. Agnes*. London: Taylor and Hessey.

Kemp, M. (1981). *Leonardo da Vinci: The Marvellous Works of Nature and Man*. Cambridge, MA: Harvard University Press.

Kempe, A. B. (1876). On a general method of describing plane curves of the nth degree by linkwork. *Proc. Lond. Math. Soc. 7*, 213–16.

Klein, F. (1897). *Famous Problems of Elementary Geometry: The Duplication of the Cube, the Trisection of an Angle, the Quadrature of the Circle*. Boston: Ginn. Translated by Wooster Woodruff Beman and David Eugene Smith.

Klein, F. (1908). *Elementarmathematik vom höheren Standpunkte aus: I: Arithmetik, Algebra, Analysis*. Leipzig: B. G. Teubner.

Klein, F. (2009). *Elementary Mathematics from an Advanced Standpoint: Arithmetic, Algebra, Analysis*. New York: Cosimo Classics.

Kline, M. (1972). *Mathematical Thought from Ancient to Modern Times*. Oxford: Oxford University Press.

Knorr, W. R. (1975). *The Evolution of the Euclidean Elements: A Study of the Theory of Incommensurable Magnitudes and Its Significance for Early Greek Geometry*. Dordrecht: D. Reidel.

Knorr, W. R. (1975/76). Archimedes and the measurement of the circle: A new interpretation. *Arch. Hist. Exact Sci. 15*(2), 115–40.

Knorr, W. R. (1986). Archimedes' *Dimension of the Circle*: A view of the genesis of the extant text. *Arch. Hist. Exact Sci. 35*(4), 281–324.

Knorr, W. R. (1989). On Archimedes' construction of the regular heptagon. *Centaurus 32*(4), 257–71.

Knorr, W. R. (1993). *The Ancient Tradition of Geometric Problems*. New York: Dover. Corrected reprint of the 1986 original.

Knorr, W. R. (2004). On the early history of axiomatics: The interaction of mathematics and philosophy in Greek antiquity. In J. Christianidis (ed.), *Classics in the History of Greek Mathematics*, pp. 81–109. Dordrecht: Springer Science & Business Media.

Kostovskii, A. (1961). *Geometrical Constructions Using Compasses Only*. New York: Blaisdell.

Kuzmin, R. A. (1930). Sur un nouvelle classe de nombres transcendants. *Bulletin Acad. Sci. Leningrad 3*(7), 585–97.

La Nave, F. and B. Mazur (2002). Reading Bombelli. *Math. Intelligencer* 24(1), 12–21.

Laczkovich, M. (1997). On Lambert's proof of the irrationality of π. *Amer. Math. Monthly* 104(5), 439–43.

Lam, L.-Y. and T.-S. Ang (1986). Circle measurements in ancient China. *Historia Math.* 13(4), 325–40.

Lambert, J.-H. (1768). Memoire sur quelques propriétés remarquables des quantités transcendantes circulaires et logarithmiques. *Mémoires de l'Acad. de Berlin, année 1761* 17, 265–322.

Lambert, J.-H. (1770). *Beyträge zum Gebrauche der Mathematik und deren Anwendung II*, Chapter "Vorläufige Kenntnisse für die, so die Quadratur und Rectification des Circuls suchen," pp. 140–69. Berlin: Buchhandlung der Realschule.

Landau, E. (1903). Über quadrierbare kreisbogenzweiecke. *Sitzungsberichte der Berliner Mathematischen Gesellschaft* 2, 1–6.

Lang, R. J. (2010). Origami and geometric constructions. https://langorigami.com/article/huzita-justin-axioms/.

Lange, L. J. (1999). An elegant continued fraction for π. *Amer. Math. Monthly* 106(5), 456–58.

Langton, S. G. (2007). The quadrature of lunes, from Hippocrates to Euler. In R. E. Bradley, L. A. D'Antonio, C. E. Sandifer (eds.), *Euler at 300: An Appreciation*, pp. 53–62. Washington, DC: Mathematical Association of America.

Lawson, D. (2016, April 7). A trig-free proof of Crockett Johnson's theorem. https://divisbyzero.com/2016/04/07/a-trig-free-proof-of-crockett-johnsons-theorem/-johnsons-theorem/.

Lazzarini, M. (1901). Un' applicazione del calcolo della probabilità alla ricerca sperimentale di un valure approssimato di π. *Periodico di Matematica* 4, 140–43.

Leclerc, Georges-Louis, Comte de Buffon (1829). *Oeuvres complètes de Buffon*, Volume XV, Chapter "Essai d'arithmétique morale," pp. 338–447. Paris: Verdière et Ladrange.

Legendre, A. M. (1794). *Éléments de géométrie* (1st ed.). Paris: Firmin Didot.

Leibniz, G. W. (1850). I. Leibniz an Hugens. In C. I. Gerhardt (ed.), *Leibnizens mathematische Schriften, Volume 1, II*, pp. 11–14. Berlin: A. Asher & Comp.

Leibniz, G. W. (1858). Specimen novum analyseos pro scientia infiniti circa summa et quadraturas. In C. I. Gerhardt (ed.), *Leibnizens mathematische Schriften, Volume 5, I*, pp. 350–61. Halle: Druck und Verlag von H. W. Schmidt.

Leibniz, G. W. (1863). Mathesis universalis. In C. I. Gerhardt (ed.), *Leibnizens mathematische Schriften, Volume 7, III*, pp. 49–76. Halle: Druck und Verlag von H. W. Schmidt.

Leibniz, G. W. (2005). *The Early Mathematical Manuscripts of Leibniz*. Mineola, NY: Dover.

Leybourn, W. (1694). *Pleasure with Profit: Consisting of Recreations of Divers Kinds, viz. Numerical, Geometrical, Mechanical, Statical, Astronomical, Horometrical, Cryptographical, Magnetical, Automatical, Chymical, and Historical: Published to Recreate Ingenious Spirits; and to induce them to make farther scrutiny into these (and the like) Sublime Sciences. And To divert them from following such Vices, to which Youth (in this Age) are so much Inclin'd*. London: Printed for R. Baldwin and J. Dunton.

Lindemann, F. (1882). Über die Zahl π. *Math. Ann.* 20(2), 213–25.

Liouville, J. (1844a). Sur des classes très étendues de quantités dont la valeur n'est ni algébrique, ni même réductible à des irrationnelles algébriques. *C. R. Acad. Sci. Paris 18*, 883–85.

Liouville, J. (1844b). Nouvelle démonstration d'un théorème sur les irrationnelles algébriques inséré dans le compte-rendu de la dernière séance. *C. R. Acad. Sci. Paris 18*, 910–11.

Liouville, J. (1851). Sur des classes très étendues de quantités dont la valeur n'est ni algébrique, ni même réductible à des irrationnelles algébriques. *J. Math. Pures et Appl. 1*, 133–42.

Livy (1972). *The War with Hannibal: Books XXI–XXX of the History of Rome from Its Foundation*. Harmondsworth: Penguin Books.

Lockwood, E. H. (1961). *A Book of Curves*. New York: Cambridge University Press.

Longrigg, J. (2008). Anaxagoras. In C. C. Gillispie (ed.), *Complete Dictionary of Scientific Biography*, Volume 1, pp. 149–50. Detroit: Charles Scribner's Sons.

Lützen, J. (2009). Why was Wantzel overlooked for a century? The changing importance of an impossibility result. *Historia Math. 36*(4), 374–94.

Lützen, J. (2010). The algebra of geometric impossibility: Descartes and Montucla on the impossibility of the duplication of the cube and the trisection of the angle. *Centaurus 52*(1), 4–37.

Mackay, J. S. (1886). Solutions of Euclid's problems, with a rule and one fixed aperture of the compasses, by the Italian geometers of the sixteenth century. *Proc. Edinb. Math. Soc. 5*, 2–22.

Maclean, I. (2008). Cardano, Girolamo. In C. C. Gillispie (ed.), *Complete Dictionary of Scientific Biography*, Volume 20, pp. 36–38. Detroit: Charles Scribner's Sons.

Madachy, J. S. (1979). *Madachy's Mathematical Recreations*. New York: Dover.

Mahler, K. (1937). Arithmetische Eigenschaften einer Klasse von Dezimalbrüchen. *Proc. Konin. Neder. Akad. Wet. Ser. A 40*(5), 421–28.

Mahoney, M. S. (1994). *The Mathematical Career of Pierre de Fermat, 1601–1665* (2nd ed.). Princeton, NJ: Princeton University Press.

Mancosu, P. (1999). *Philosophy of Mathematics and Mathematical Practice in the Seventeenth Century*. Oxford: Oxford University Press.

Mancosu, P. (2008). Descartes and mathematics. In J. Broughton, J. Carriero (eds.), *A Companion to Descartes*, pp. 103–23. Malden, MA: Blackwell.

Marinoni, A. (1974). The writer. In L. Reti (ed.), *The Unknown Leonardo*. New York: McGraw-Hill.

Markowsky, G. (1992). Misconceptions about the golden ratio. *College Math. J. 23*(1), 2–19.

Martin, G. E. (1998). *Geometric Constructions*. Undergraduate Texts in Mathematics. New York: Springer.

Martin, J. (2010). The Helen of geometry. *College Math. J. 41*(1), 17–28.

Marvin, U. B. (1996). Ernst Florens Friedrich Chladni (1756–1827) and the origins of modern meteorite research. *Meteoritics and Planetary Science 31*, 545–88.

Mascheroni, L. (1797). *Geometria del compasso*. Pavia.

Masià, R. (2016). A new reading of Archytas' doubling of the cube and its implications. *Arch. Hist. Exact Sci. 70*(2), 175–204.

Masotti, A. (2008a). Ferro (or Ferreo, dal Ferro, del Ferro), Scipione. In C. C. Gillispie (ed.), *Complete Dictionary of Scientific Biography*, Volume 4, pp. 595–97. Detroit: Charles Scribner's Sons.

Masotti, A. (2008b). Tartaglia (also Tartalea or Tartaia), Niccolò. In C. C. Gillispie (ed.), *Complete Dictionary of Scientific Biography*, Volume 13, pp. 258–62. Detroit: Charles Scribner's Sons.

Mather, W. (1710). *The Young Mans Companion: Or, Arithmetick Made Easie. The Eighth Edition: With Many Additions.* London: Printed for S. Clarke.

Merriam-Webster.com (2017, May 3). "Impossible." https://www.merriam-webster.com/dictionary/impossible.

Meschkowski, H. (1964). *Ways of Thought of Great Mathematicians: An Approach to the History of Mathematics.* San Francisco: Holden-Day.

Meskens, A. (1994). Gregory of Saint Vincent: A pioneer of the calculus. *Math. Gaz.* 78(483), 315–19.

Messer, P. (1986, December). Problem 1054. *Crux Math.* 12(10), 284–85.

Milici, P. and R. Dawson (2012). The equiangular compass. *Math. Intelligencer* 34(4), 63–67.

Millay, E. S. V. (1923). *The Harp Weaver and Other Poems.* New York: Harper and Brothers.

Milton, J. (1910). *The Poetical Works of John Milton: Paradise Lost.* London: Macmillan.

Mintz, S. I. (2008). Hobbes, Thomas. In C. C. Gillispie (ed.), *Complete Dictionary of Scientific Biography*, Volume 6, pp. 444–51. Detroit: Charles Scribner's Sons.

Mitchell, F. D. (1907, April). Mathematical prodigies. *Amer. J. Psych.* 18(1), 61–143.

Mitchell, U. G. and M. Strain (1936, Jan). The number *e*. *Osiris* 1, 476–96.

Mohr, G. (1672). *Euclides Danicus.* Amsterdam.

Mohr, G. (1673). *Compendium Euclidis curiosi.* Amsterdam.

Moser, L. (1947). The watch as angle trisector. *Scripta Math.* 13, 57.

Murdoch, J. (2008). Euclid: Transmission of the *Elements*. In C. C. Gillispie (ed.), *Complete Dictionary of Scientific Biography*, Volume 4, pp. 437–59. Detroit: Charles Scribner's Sons.

Museo Galileo (2008a). Galileo's compass: Museo Galileo—Istituto e Museo di Storia della Scienza. http://brunelleschi.imss.fi.it/esplora/compasso/index.html.

Museo Galileo (2008b). How to make Galileo's compass: Museo Galileo—Istituto e Museo di Storia della Scienza. http://brunelleschi.imss.fi.it/esplora/compasso/dswmedia/risorse/ecostruire_compasso.pdf.

Napier, J. (1614). *Mirifici logarithmorum canonis descriptio.* Edinburgh: Andrew Hart.

National Security Agency (2012, January 27). NSA press release: National Cryptologic Museum opens new exhibit on Dr. John Nash.

Netz, R. (2004). *The Works of Archimedes: The Two Books "On the Sphere and the Cylinder,"* Volume I. Cambridge: Cambridge University Press.

Netz, R. and W. Noel (2007). *The Archimedes Codex: How a Medieval Prayer Book Is Revealing the True Genius of Antiquity's Greatest Scientist.* Philadelphia, PA: Da Capo Press.

Neumann, O. (2005). Carl Friedrich Gauss, *Disquisitiones arithmeticae* (1801). In I. Grattan-Guinness (ed.), *Landmark Writings in Western Mathematics 1640–1940*, pp. 303–15. Amsterdam: Elsevier.

Neumann, O. (2007a). Cyclotomy: From Euler through Vandermonde to Gauss. In R. E. Bradley, C. E. Sandifer (eds.), *Leonhard Euler: Life, Work and Legacy*, Volume 5, pp. 323–62. Elsevier.

Neumann, O. (2007b). The *Disquisitiones arithmeticae* and the theory of equations. In *The Shaping of Arithmetic after C. F. Gauss's "Disquisitiones arithmeticae,"* pp. 107–27. Berlin: Springer.

Newcomb, S. (1903, October 22,). The outlook for the flying machine. *Independent: A Weekly Magazine*, 2508–12.

Newton, I. (1736). *The Method of Fluxions and Infinite Series: With Its Application to the Geometry of Curve-Lines*. London: Henry Woodfall.

Newton, I. (1769). *The Universal Arithmetick: Or, a Treatise of Arithmetical Composition and Resolution. Written in Latin by Sir Isaac Newton. Translated by The Late Mr. Ralphson; and Revised and Corrected by Mr. Cunn. To which is Added, a Treatise Upon the Measures of Ratios, by James Maguire, A.M. The Whole Illustrated and Explained, in a Series of Notes, by the Rev. Theaker Wilder, D.D. Senior Fellow of Trinity College, Dublin*. London: W. Johnston.

Newton, I. (1972). *The Mathematical Papers of Isaac Newton, Volume V 1683–1684: Edited by D. T. Whiteside*, Volume V. Cambridge: Cambridge University Press.

Newton, I. (2008). The 'De analysi per aequationes infinitas'. In D. T. Whiteside (ed.), *The Mathematical Papers of Isaac Newton, Volume II 1667–1670*, pp. 206–76. New York: Cambridge University Press.

Niven, I. (1947). A simple proof that π is irrational. *Bull. Amer. Math. Soc. 53*, 509.

Niven, I. (1956). *Irrational Numbers*. Carus Mathematical Monographs, No. 11. Mathematical Association of America. Distributed by John Wiley and Sons, New York, NY.

O'Grady, P. F. (2008). *The Sophists: An Introduction*. London: Duckworth.

O'Leary, M. (2010). *Revolutions of Geometry*. Hoboken, NJ: Wiley.

O'Shaughnessy, L. (1983). Putting God back in math. *Math. Intelligencer 5*(4), 76–78.

Ovid (2004). *Metamorphoses: A New Translation by Charles Martin*. New York: W. W. Norton.

P., J. (1906). Review of *The Seven Follies of Science: A Popular Account of the Most Famous Scientific Impossibilities and the Attempts Which Have Been Made to Solve Them. Nature 75*(1932), 25.

Painlevé, P. (1900). *Analyse des travaux scientifiques*. Paris: Gauthier-Villars.

Palais, B. (2001). π is wrong. *Math. Intelligencer 23*(3), 7–8.

Pappus of Alexandria (2010). *Pappus of Alexandria: Book 4 of the "Collection" (Edited with Translation and Commentary by Heike Sefrin-Weis)*. London: Springer.

Peirce, B. (1859). Note on two new symbols. *Math. Monthly 1*(5), 167–68.

Peirce, B. (1881). Linear associative algebra. *Amer. J. Math. 4*(1), 97–229.

Peng-Yoke, H. (2008). Liu Hui. In C. C. Gillispie (ed.), *Complete Dictionary of Scientific Biography*, Volume 8, pp. 418–25. Detroit: Charles Scribner's Sons.

Petersen, J. (1877). *De algebraiske ligningers theori*. Copenhagen: Andr. Fred. Host & Son.

Petrie, B. J. (2012). Leonhard Euler's use and understanding of mathematical transcendence. *Historia Math. 39*, 280–91.

Petzold, C. (2007, December 13). E. W. Hobson's "Squaring the Circle." http://www.charlespetzold.com/blog/2007/11/221214.html.

Pierpont, J. (1895). On an undemonstrated theorem of the *Disquisitiones arithmeticæ*. *Bull. Amer. Math. Soc. 2*(3), 77–83.

Pittsburgh Press (1931). Insists angle was trisected.

Plato (1901). *The Republic of Plato: An Ideal Commonwealth, Translated by Benjamin Jowett*. New York: Willey.

Plato (1992). *Theaetetus. Edited, with Introduction by Bernard Williams. Translated by M. J. Levett. Revised by Myles Burnyeat*. Indianapolis: Hackett.

Plofker, K. (2007). Mathematics in India. In V. J. Katz (ed.), *The Mathematics of Egypt, Mesopotamia, China, India, and Islam: A Sourcebook*, pp. 384–514. Princeton, NJ: Princeton University Press.

Plutarch (1917). *Plutarch's Lives: With an English Translation by Bernadotte Perrin*, Volume V. London: William Heinemann.

Poincaré, H. (1895). Analysis situs. *J. Ec. Polytech. Math. Ser. 2 1*(2), 1–123.

Poncelet, J. V. (1822). *Traité des propriétés projectives des figures*. Paris: Bachelier.

Pope, A. (1711). *Essay on Criticism*. London: W. Lewis.

Popper, K. R. (1963). *The Open Society and Its Enemies: The Spell of Plato*, Volume 1. Princeton, NJ: Princeton University Press.

PrimeGrid (2018, July 13). PrimeGrid's 321 prime search. http://www.primegrid.com/download/321-5082306.pdf.

Proclus (1992). *A Commentary on the First Book of Euclid's "Elements": Translated with Introduction and Notes by Glenn R. Morrow*. Princeton, NJ: Princeton University Press.

Pycior, H. M. (1987, Apr.–Jun.). Mathematics and philosophy: Wallis, Hobbes, Barrow, and Berkeley. *J. Hist. Ideas 48*(2), 265–86.

Pycior, H. M. (2006). *Symbols, Impossible Numbers, and Geometric Entanglements: British Algebra through the Commentaries on Newton's Universal Arithmetick*. Cambridge: Cambridge University Press.

Rabinowitz, S. and S. Wagon (1995). A spigot algorithm for the digits of π. *Amer. Math. Monthly 102*(3), 195–203.

Rebière, A. (1893). *Mathématiques et mathématiciens: Pensées et curiosités. Deuxième édition*. Paris: Librairie Nony.

Reitwiesner, G. W. (1950). An ENIAC determination of π and e to more than 2000 decimal places. *MTAC 4*, 11–15.

Remmert, R. (1990a). Complex numbers. In J. H. Ewing (ed.), *Numbers (With an Introduction by K. Lamotke. Translated by H.L.S. Orde.)*, pp. 55–96. New York: Springer.

Remmert, R. (1990b). What is π? In J. H. Ewing (ed.), *Numbers (With an introduction by K. Lamotke. Translated by H.L.S. Orde.)*, pp. 123–53. New York: Springer.

Richelot, F. J. (1832). De resolutione algebraica aequationis $x^{257} = 1$, sive de divisione circuli per bisectionam anguli septies repetitam in partes 257 inter se aequales commentatio coronata. *J. reine angew. Math. 9*, 1–26, 146–61, 209–30, 337–56.

Richeson, D. S. (2008). *Euler's Gem: The Polyhedron Formula and the Birth of Topology*. Princeton, NJ: Princeton University Press.

Richeson, D. S. (2015). Circular reasoning: Who first proved that C divided by d is a constant? *College Math. J.* 46(3), 162–71.

Richeson, D. S. (2017, February). A trisectrix from a carpenter's square. *Math. Mag.* 90(1), 8–11.

Richmond, H. W. (1893). A construction for a regular polygon of seventeen sides. *Quarterly J. Pure Applied Mathematics* 26, 206–7.

Richmond, H. W. (1909). To construct a regular polygon of 17 sides. *Math. Ann.* 67(4), 459–61.

Rickey, V. F. (1987). Isaac Newton: Man, myth, and mathematics. *College Math. J.* 18(5), 362–89.

Ross, W. D. (1936). *Aristotle's Physics: A Revised Text with Introduction and Commentary by W. D. Ross*. Oxford: Clarendon Press.

Roy, R. (1990). The discovery of the series formula for π by Leibniz, Gregory and Nilakantha. *Math. Mag.* 63(5), 291–306.

Rummler, H. (1993). Squaring the circle with holes. *Amer. Math. Monthly* 100(9), 858–60.

Russell, B. (1967). *The Autobiography of Bertrand Russell*, Volume 1. Boston: Little, Brown.

Saint-Venant, J. C. (1848). Biographie: Wantzel. *Nouvelles Annales de Mathématiques Série 1 7*, 321–31.

Sanders, S. T. (1931). The angle-trisection chimera once more. *Mathematics News Letter 6*(3), 1–6.

Sandifer, C. E. (2007a). *The Early Mathematics of Leonhard Euler*. MAA Spectrum. Washington, DC: Mathematical Association of America.

Sandifer, C. E. (2007b). *How Euler Did It*, Chapter "Fermat's little theorem" (November 2003), pp. 45–48. MAA Spectrum. Washington, DC: Mathematical Association of America.

Sandifer, C. E. (2007c). *How Euler Did It*, Chapter "Who proved e is irrational?" (February 2006), pp. 185–90. MAA Spectrum. Washington, DC: Mathematical Association of America.

Sasaki, C. (2003). *Descartes's Mathematical Thought*. Dordrecht, Netherlands: Kluwer Academic.

Schepler, H. C. (1950a). The chronology of pi. *Math. Mag.* 23(4), 216–28.

Schepler, H. C. (1950b). The chronology of pi. *Math. Mag.* 23(3), 165–70.

Schepler, H. C. (1950c). The chronology of pi. *Math. Mag.* 23(5), 279–83.

Schneider, T. (1935). Transzendenzuntersuchungen periodischer Funktionen I. Transzendenz von Potenzen. *J. reine angew. Math. 172*, 65–69.

Scott, J. F. (1976). *The Scientific Work of René Descartes (1596–1650)*. London: Taylor and Francis. With a foreword by H. W. Turnbull, reprinted.

Scriba, C. J. (1983). Gregory's converging double sequence: A new look at the controversy between Huygens and Gregory over the "analytical" quadrature of the circle. *Historia Math. 10*(3), 274–85.

Scriba, C. J. (2008). Wallis, John. In C. C. Gillispie (ed.), *Complete Dictionary of Scientific Biography*, Volume 14, pp. 146–55. Detroit: Charles Scribner's Sons.

Scudder, H. T. (1928). Discussions: How to trisect an angle with a carpenter's square. *Amer. Math. Monthly* 35(5), 250–51.

Seidenberg, A. (1961). The ritual origin of geometry. *Arch. Hist. Exact Sci.* 1(5), 488–527.

Seidenberg, A. (1972). On the area of a semi-circle. *Arch. Hist. Exact Sci.* 9(3), 171–211.

Seidenberg, A. (1981). The ritual origin of the circle and square. *Arch. Hist. Exact Sci.* 25(4), 269–327.

Seidenberg, A. (2008). Mohr, Georg. In C. C. Gillispie (ed.), *Complete Dictionary of Scientific Biography*, Volume 9, pp. 446–47. Detroit: Charles Scribner's Sons.

Serfati, M. (1993). Les compas cartésiens. *Archives de Philosophie* 56(2), 197–230.

Serfati, M. (2008). A note on the *Geometry* and Descartes's mathematical work. *Notices Amer. Math. Soc.* 55(1), 50–53.

Sesiano, J. (2008). Diophantus of Alexandria. In C. C. Gillispie (ed.), *Complete Dictionary of Scientific Biography*, Volume 15, pp. 118–22. Detroit: Charles Scribner's Sons.

Sesiano, J. (2009). *An Introduction to the History of Algebra: Solving Equations from Mesopotamian Times to the Renaissance. Translated from the 1999 French Original by Anna Pierrehumbert.* Providence, RI: American Mathematical Society.

Severi, F. (1904). Sui problemi determinati risolubili colla riga e col compasso. *Rend. Circ. Mat. Palermo* 18, 256–59.

Shakespeare, W. (1966). *Shakespeare: Julius Caesar. Edited by A. W. Verity.* Cambridge: Cambridge University Press Archive.

Shakespeare, W. (2016). *King Henry IV, Part 2.* London: Bloomsbury Arden Shakespeare.

Shanks, D. (1992). Improving an approximation for pi. *Amer. Math. Monthly* 99(3), 263.

Shelby, L. R. (1977). *Gothic Design Techniques: The Fifteenth-Century Design Booklets of Mathes Roriczer and Hanns Schmuttermayer.* Carbondale, IL: Southern Illinois University.

Shidlovskii, A. B. (1989). *Transcendental Numbers*, Volume 12, de Gruyter Studies in Mathematics. Berlin: Walter de Gruyter. Translated from the Russian by Neal Koblitz, with a foreword by W. Dale Brownawell.

Siegel, C. L. (2014). Über einige Anwendungen diophantischer Approximationen. In U. Zannier (ed.), *On Some Applications of Diophantine Approximations*, pp. 81–138. Pisa: Scuola Normale Superiore.

Simoson, A. J. (2009). Solomon's sea and π. *College Math. J.* 40(1), 22–32.

Simpson, J. and E. Weiner (eds.) (1989). *Oxford English Dictionary* (2nd ed.). Oxford: Clarendon Press.

Singmaster, D. (1985). The legal values of pi. *Math. Intelligencer* 7(2), 69–72.

Slocum, J. and D. Sonneveld (2006). *The 15 Puzzle: How it Drove the World Crazy. The Puzzle that Started the Craze of 1880. How Amercia's Greatest Puzzle Designer, Sam Loyd, Fooled Everyone for 115 Years.* Beverly Hills, CA: Slocum Puzzle Foundation.

Smale, S. (1981). The fundamental theorem of algebra and complexity theory. *Bull. Amer. Math. Soc. (N.S.)* 4(1), 1–36.

Smeur, A.J.E.M. (1970). On the value equivalent to π in ancient mathematical texts. A new interpretation. *Arch. Hist. Exact Sci.* 6(4), 249–70.

Smith, D. E. (1958). *History of Mathematics. Vol. II. Special Topics of Elementary Mathematics.* New York: Dover.

Smith, D. E. (1959). *A Source Book in Mathematics.* Mineola, NY: Dover.

Smithsonian: The National Museum of American History (2018). Mathematical paintings of Crockett Johnson. http://americanhistory.si.edu/collections/object-groups/mathematical-paintings-of-crockett-johnson.

Stark, M. E. (1948). Mathematical miscellany. *Math. Mag.* 22(1), 52–55.

Steiner, J. (1833). *Die geometrischen Constructionen ausgeführt mittelst der geraden Linie und eines festen Kreises.* Berlin: Ferdinand Dümmler.

Steiner, J. (1950). *Geometrical Constructions with a Ruler, Given a Fixed Circle with Its Center.* New York: Scripta Mathematica. Translated from the first German edition (1833) by Marion Elizabeth Stark, edited with an introduction and notes by Raymond Clare Archibald.

Stern, M. D. (1985). A remarkable approximation to π. *Math. Gaz.* 69, 218–19.

Stevin, S. (1958). *The Principal Works of Simon Stevin, Mathematics, Edition Vol. IIB: Edited by D. J. Struik.* Amsterdam: C. V. Swets & Zeitlinger.

Stifel, M. (1544). *Arithmetica integra.* Nüremberg: Iohan Petreium.

Stillwell, J. (2010). *Mathematics and Its History* (3rd ed.). New York: Springer.

Stroud, J. B. (2008). Crockett Johnson's geometric paintings. *J. Math. Arts* 2(2), 77–99.

Struik, D. J. (1969). *A Source Book in Mathematics, 1200–1800.* Cambridge, MA: Harvard University Press.

Suzuki, J. (2007). Euler and number theory: A study in mathematical invention. In R. E. Bradley, C. E. Sandifer (eds.), *Leonhard Euler: Life, Work and Legacy*, pp. 363–83. Amsterdam: Elsevier.

Suzuki, J. (2008). A brief history of impossibility. *Math. Mag.* 81(1), 27–38.

Tartaglia, N. F. (1547). *Seconda riposta data da Nicolo Tartalea Brisciano a Messer Lodovico Ferraro.* Venice.

Tartaglia, N. F. (1556). *General trattato di numeri et misure.* Venice.

Taton, R. (2008). Poncelet, Jen Victor. In C. C. Gillispie (ed.), *Complete Dictionary of Scientific Biography*, Volume 11, pp. 76–82. Detroit: Charles Scribner's Sons.

Thoreau, H. (1985). *A Week on the Concord and Merrimack Rivers / Walden; Or, Life in the Woods / The Maine Woods / Cape Cod.* New York: Literary Classics of the United States.

Thoren, V. E. (2008). Severin, Christian. In C. C. Gillispie (ed.), *Complete Dictionary of Scientific Biography*, Volume 12, pp. 332. Detroit: Charles Scribner's Sons.

Time (1931). Angle trisected? *Time* 18(8), 34.

Titchmarsh, E. C. (1948). *Mathematics for the General Reader.* London: Hutchinson's University Library.

Tolstoy, L. (2002). *War and Peace.* New York: Modern Library.

Toomer, G. J. (2008). Nicomedes. In C. C. Gillispie (ed.), *Complete Dictionary of Scientific Biography*, Volume 10, pp. 114–16. Detroit: Charles Scribner's Sons.

Traub, G. (1984). *The Development of the Mathematical Analysis of Curve Length from Archimedes to Lebesgue.* PhD thesis, New York University.

Tsaban, B. and D. Garber (1998). On the rabbinical approximation of π. *Historia Math.* 25(1), 75–84.

Tschakaloff (Chakalov), L. (1929). Beitrag zum Problem der quadrierbaren Kreis-bogenzweiecke. *Math. Z. 30*(1), 552–59.

Tschebotaröw (Chebotarëv), N. (1935). Über quadrierbare Kreisbogenzweiecke I. *Math. Z. 39*(1), 161–75.

Turnbull, H. W. (1947). *Theory of Equations* (4th ed.). Edinburgh: Oliver and Boyd.

Twain, M. (Samuel L. Clemens)(1917). *Life on the Mississippi.* New York: P. F. Collier & Sons.

van der Waerden, B. L. (1954). *Science Awakening.* Groningen, Holland: P. Noord-hoff.

van der Waerden, B. L. (1983). *Geometry and Algebra in Ancient Civilizations.* Berlin: Springer.

van der Waerden, B. L. (1985). *A History of Algebra: From al-Khwārizmī to Emmy Noether.* Berlin: Springer.

van Maanen, J. A. (1986). The refutation of Longomontanus' quadrature by John Pell. *Annals of Science 43*(4), 315–52.

Vandermonde, A.-T. (1774). Mémoire sur la résolution des équations. *Histoire de l'Académie Royale des Sciences, année 1771*, 365–416.

Vasari, G. (1998). *The Lives of the Artists: Translated with an Introduction and Notes by Julia Conaway Bondanella and Peter Bondanella.* New York: Oxford University Press.

Verheyen, H. F. (1992). The icosahedral design of the Great Pyramid. In I. Hargittai (ed.), *Fivefold Symmetry*, pp. 333–59. Singapore: World Scientific.

Videla, C. R. (1997). On points constructible from conics. *Math. Intelligencer 19*(2), 53–57.

Viète, F. (1646). *Francisci Vietae opera mathematica in unum volumen congesta ac recog-nita, operâ atque studio Francisci à Schooten.* Leiden: Bonaventurae & Abrahami Elzeviriorum.

Viète, F. (1983). *The Analytic Art (Translated by T. Richard Witmer).* Kent, Ohio: Kent State University Press.

Vogel, K. (1958). *Vorgriechische Mathematik. I. Vorgeschichte und Ägypten*, Volume 1 of *Mathematische Studienhefte.* Hannover: Hermann Schroedel.

Vogel, K. (2008). Diophantus of Alexandria. In C. C. Gillispie (ed.), *Complete Dic-tionary of Scientific Biography*, Volume 4, pp. 110–19. Detroit: Charles Scribner's Sons.

von Fritz, K. (1945). The discovery of incommensurability by Hippasus of Metapontum. *Ann. of Math. 46*(2), 242–64.

von Fritz, K. (2008). Pythagoras of Samos. In C. C. Gillispie (ed.), *Complete Dictio-nary of Scientific Biography*, Volume 11, pp. 219–25. Detroit: Charles Scribner's Sons.

von Waltershausen, W. S. (1856). *Gauss zum Gedächtniss.* Leipzig: S. Hirzel.

Wagon, S. (1985). Is π normal? *Math. Intelligencer 7*(3), 65–67.

Waldo, C. A. (1916). What might have been. *Proc. Indiana Acad. Sci. 26*, 445–46.

Wallis, J. (1656). *Arithmetica infinitorum.* Oxford: Leon Lichfield.

Wallis, J. (1685). *A Treatise of Algebra Both Historical and Practical.* London: John Playford.

Wallisser, R. (2000). On Lambert's proof of the irrationality of π. In *Algebraic Num-ber Theory and Diophantine Analysis (Graz, 1998)*, pp. 521–30. Berlin: de Gruyter.

Wantzel, P. L. (1837). Recherches sur les moyens de reconnaître si un problème de géométrie peut se résoudre avec la règle et le compas. *J. Math. Pures Appl.* 2(1), 366–72.

Wantzel, P. L. (1843). Classification des nombres incommensurables d'origine algébrique. *Nouv. Ann. Math.* 2, 117–27.

Wasserstein, A. (1959). Some early Greek attempts to square the circle. *Phronesis* 4(2), 92–100.

Weber, H. (1893). Leopold Kronecker. In *Jahresbericht der Deutschen Mathematiker-Vereinigung*, pp. 5–31. Berlin: Druck und Verlag von George Reimer.

Weierstrass, K. (1885). Zu Lindemann's Abhandlung "Über die Ludolph'sche Zahl." *Sitzungber. Königl. Preuss. Akad. Wissensch. zu Berlin* 2, 1067–86.

Weik, J. W. (1922). *The Real Lincoln: A Portrait.* Cambridge, MA: Houghton Mifflin.

Weil, A. (1984). *Number Theory: An Approach through History from Hammurapi to Legendre.* Boston: Birkhäuser.

Weintraub, S. H. (2013, June–July). Several proofs of the irreducibility of the cyclotomic polynomials. *Amer. Math. Monthly* 120(6), 537–45.

Wells, D. (1990). Are these the most beautiful? *Math. Intelligencer* 12(3), 37–41.

Wernick, W. (1971, December). The double straightedge. *Math. Teacher* 64, 697–704.

Wessel, C. (1799). Om directionens analytiske Betegning, et Forsøg, anvendt fornemmelig til plane og sphaeriske Polygoners Opløsning. *Nye Samling af det Kongelige Danske Videnskabernes Selskabs Skrifter* 5, 469–518.

Westlake, D. E. (1990). *Drowned Hopes.* New York: Mysterious Press.

Whiteside, D. T. (2008). Gregory (more correctly Gregorie), James. In C. C. Gillispie (ed.), *Complete Dictionary of Scientific Biography*, Volume 5, pp. 524–30. Detroit: Charles Scribner's Sons.

Wijnquist, D. (1766). *Lunulas quasdam circulares quadrabiles.* PhD thesis, Royal Academy of Abo.

Wikipedia contributors (2018). Chronology of computation of π. Accessed August 16, 2018. https://en.wikipedia.org/w/index.php?title=Chronology_of_computation_of_%CF%80&oldid=830591209.

Williams, M. R. and E. Tomash (2003, Jan.–March). The sector: Its history, scales, and uses. *IEEE Annals of the History of Computing* 25(1), 34–47.

Wrench, J.W.J. (1960, December). The evolution of extended decimal approximations to π. *Math. Teacher* 53, 644–50.

Wussing, H. (2008). Lindemann, Carl Louis Ferdinand. In C. C. Gillispie (ed.), *Complete Dictionary of Scientific Biography*, Volume 8, pp. 367–68. Detroit: Charles Scribner's Sons.

Yandell, B. H. (2002). *The Honors Class: Hilbert's Problems and Their Solvers.* A. K. Peters.

Yates, R. C. (1940). The angle ruler, the marked ruler and the carpenter's square. *National Math. Mag.* 15(2), 61–73.

Yates, R. C. (1942). *The Trisection Problem.* Baton Rouge, LA: Franklin Press.

Yoder, J. G. (1988). *Unrolling Time: Christiaan Huygens and the Mathematization of Nature.* Cambridge: Cambridge University Press.

Young, G. D. (1984). The Arabic textual traditions of Euclid's *Elements. Historia Math.* 11(2), 147–60.

Youschkevitch, A. P. (2008). Abū'l-Wafāᵓ Al-Būzjānī, Muḥammad Ibn Muḥammad Ibn Yaḥyā Ibn Ismāᶜīl Ibn Al-ᶜAbbās. In C. C. Gillispie (ed.), *Complete Dictionary of Scientific Biography*, Volume 1, pp. 39–43. Detroit: Charles Scribner's Sons.

Zubov, V. P. (1968). *Leonardo da Vinci. Translated from the Russian by David H. Kraus*. Cambridge, MA: Harvard University Press.

Zukav, G. (1984). *The Dancing Wu Li Masters: An Overview of the New Physics*. New York: Bantam Books.

Index